Recent Advances in the Chemistry of Anti-infective Agents

Edited by

P. H. Bentley
Smith and Nephew Research Ltd.

R. Ponsford
SmithKline Beecham Pharmaceuticals

ROYAL
SOCIETY OF
CHEMISTRY

The Proceedings of the First International Symposium, arranged by the Fine Chemicals and Medicinals Group of the Industrial Division of the Royal Society of Chemistry, Cambridge, England, 5–8th July 1992.

Special Publication No. 119

ISBN 0-85186-245-4

A catalogue record for this book is available from the British Library

Published by The Royal Society of Chemistry,
Thomas Graham House, Science Park, Cambridge CB4 4WF

Printed in Great Britain by Bookcraft (Bath) Ltd.

Preface

The first International Symposium on Recent Advances in the Chemistry of Anti-Infective Agents followed the highly successful symposia devoted to the chemistry of the β-lactam antibiotics held in 1976, 1980, 1984 and 1988. Since developments in the latter area have slowed in recent years, the wider coverage of anti-infective agents was considered a progressive step. This culminated in a symposium, which succeeded in blending the key areas of chemotherapy resulting in a highly stimulating and rewarding meeting.

The first day, devoted to antibacterials, began with Professor Baldwin's meticulous account of the further progression in elucidating β-lactam biosynthesis and paved the way for presentations describing the most recent discoveries in the carbapenem area (Drs. Menard, Perboni and Wollmann). This β-lactam section was appropriately concluded by highly visual insights into the interaction of these agents with penicillin binding proteins (Professor Knox). Presentations in the quinolone series, an increasingly important commercial class of antibacterials, was prefaced by Dr. Keith's account of the dual-action cephalosporins and was followed by an excellent review of the most recent advances in the SAR of the quinolones by Dr. Chu. The chemistry of the other classes of antibiotics was represented by presentations on macrolides, (Dr. Kirst) and t-RNA synthetase inhibitors (Dr. Walker). Professor Evans, unable to present his scheduled paper on the vancomycin antibiotics, treated his audience to a memorable lecture based on the potent tumour promoter (+) calyculin A.

The transition from antibacterials to antifungals, on the second day, was appropriately bridged by Professor Miller's detailed account of iron-transport mediated drug delivery processes. It provided further appreciation of the complex processes required to transport a chemotherapeutic agent to its site of action.

Professor Hay highlighted the principal clinical opportunities for new antifungal agents, with relevance to both western medicine and the needs of the under-developed world. Two lectures by Drs. Richardson and Girijavallabhan on azole inhibitors of cytochrome P450 provided a review and update on the status of the search for drugs which will offer advances over those currently available, such as fluconazole and itraconazole. Alternative approaches towards novel agents were described in presentations of studies seeking inhibitors of chitin synthetase or 2,3-oxidosqualene-lanosterol cyclase (Dr. Jolidon), or modification of amphotericin B to improve its selectivity (Dr. MacPherson). In the final lecture, Professor Barrett presented his elegant synthetic work on papulacandin D, thus illustrating the ability of the organic chemist to meet the continued synthetic challenges posed by chemotherapeutic agents.

Presentations in the antiviral section on the third day covered a variety of different topics and whilst HIV figured prominently in the programme, the audience were reminded of the importance of other viral targets.

Dr. Storer described the preparation and anti-herpetic activity of carbocyclic nucleoside analogues in development, while Dr. Colman reviewed the 3D-structure and function of influenza virus neuraminidase and discussed novel inhibitors of the enzyme, which show promise as potential drug candidates.

Work described in the anti-HIV area addressed a variety of molecular targets and in each case highlighted a compound undergoing clinical evaluation. Drs. Kukla and Adams each described the discovery and optimisation of series of non-nucleoside inhibitors of reverse transcriptase. Novel oxathiolane nucleoside analogues targetting the same enzyme were the subject of a presentation by Dr. Mansour, whilst Dr. Kempf described the design of specific inhibitors of HIV protease and presented a structural basis for the observed SAR. Dr. Tam described the discovery and pre-clinical evaluation of a novel TAT inhibitor. Overall, the programme gave an excellent indication of the breadth of efforts in medicinal chemistry directed to the discovery of effective anti-HIV agents. These presentations were complemented by that of Dr. Hudson who described work with atovaquone, a novel drug for the treatment of malaria, toxoplasmosis and PCP, the latter being a major opportunistic agent in AIDS patients.

We thank all the speakers for their excellent presentations, the participants in the poster programme, our session chairpersons, and the staff of Churchill College for contributing to an enjoyable symposium. Finally, we acknowledge with gratitude the sustained help of our fellow organisers, Colin Greengrass, Dick Storer and Elaine Wellingham.

P.H. Bentley   Smith and Nephew Research Ltd.

R.J. Ponsford SmithKline Beecham Pharmaceuticals

October 1992

Editor's Note

Professor Baldwin's presentation is not reported in these proceedings.

Contents

Chemistry of Antibacterials/Antibiotics

Chemistry of Antifungals

Chemistry of Antivirals and Antiparasitics

Chemistry of Antibacterials/Antibiotics

1

1-β-Aminoalkyl *versus* 6-Aminoalkyl Carbapenems: Synthesis and SAR

M. Ménard,* J. Banville, A. Martel, J. Desiderio,
J. Fung-Tomc, and R. A. Partyka
BRISTOL-MYERS SQUIBB CO., PHARMACEUTICAL RESEARCH INSTITUTE, 100
INDUSTRIAL BLVD., CANDIAC, QUEBEC J5R 1J1, CANADA

1 INTRODUCTION

The Anti-infective Chemistry Group of Bristol-Myers Squibb in Canada, like so many other groups since the disclosure of thienamycin,[1] has been involved in an extensive carbapenem modification effort. The original purpose of one of our exploration programs consisted of the development of synthetic methodology that would allow the preparation of new carbapenems like **2,** with the only proviso being that the group at position 6 be different from the hydroxyethyl group of imipenem **1**. The rationale behind this goal was that the likelihood of discovering a patentable compound that was microbiologically distinct from imipenem would be better with a 6-modified rather than a 2-modified carbapenem. The amines **3** were first obtained as intermediates in this program; we found that the

aminoethyl group at position 6 conferred a number of highly desirable properties to the carbapenem nucleus the most noteworthy being oral absorbability and high anti-pseudomonal activity. A serious drawback was the poor chemical stability, in solution at physiological pH, of the most active members in this series. A potential solution to this inherent instability was sought in additional modifications at position

1; this ultimately resulted in a program shift to 1-modified carbapenems. Compounds of type **4** were prepared and were also found to possess unique antibacterial properties. What follows is a brief account of the most important phases of this program studying carbapenems bearing an amine residue in the side chains.

2   6-AMINOALKYL CARBAPENEMS

The starting material for the preparation of 6-aminoethyl carbapenems of type **3** was the (1'R)-hydroxyethyl diazoazetidinone **5**.[2] It is readily converted into its (1'S)-isomer **8** via a Mitsunobu[3] reaction as shown in Scheme 1. Both isomers can be elaborated into isomeric 6-hydroxyethyl-2-substituted carbapenems, **7** and **10**, following described methodology[4].

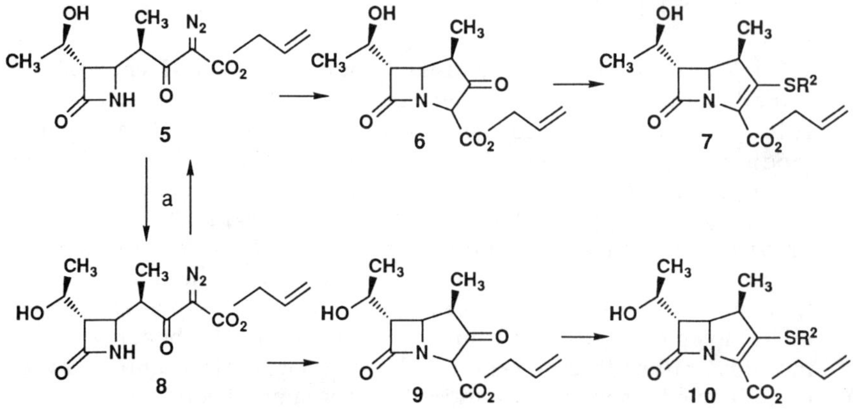

a) PPh$_3$, DEAD, HCO$_2$H then H$_3$O$^+$

<u>Scheme 1</u>:   Key intermediates for elaborations at position 1'

The isomeric 6-(1'-aminoethyl) carbapenems were obtained from **5, 7, 8** or **10** following either Schemes 2 or 3 with equal success. The reduction of the azido group in **11** could be achieved by catalytic hydrogenation after removal of the allyl protective group but it was more conveniently done via a Staudinger reaction followed by formation of the Schiff base[5] **12**; removal of the allyl group followed by an aqueous work-up then gave the appropriate 6-(1'-aminoethyl) carbapenem in good yield.

Several compounds with a variety of substituents at position 2 were prepared and their *in vitro* antibacterial activity was evaluated. The 2-SMe series of compounds is presented as a typical model (Table 1). In general the 6-hydroxyethyl carbapenems displayed superior *in vitro* potency against gram-positive organisms as compared with the

Table 1   Amines *vs* Alcohols in the 6-side chain

Organism	R1' *	OH R	OH S	NH₂ R	NH₂ S	Imipenem
		MIC (microdilution, µg/ml, Nutrient Broth, 5×10^5 CFU/ml)				
Str. pneumoniae A9585		.06	1	16	4	.002
Str. Pyogenes A9604		.06	1	16	4	.002
E. faecalis A20688		4	125	125	125	.5
S. aureus A9537		.25	1	16	8	.004
S. aureus P+ A15090		.13	4	16	16	.016
E. coli A15119		.03	2	.13	.13	.03
K. pneumoniae A9664		.06	4	.5	.5	.06
Ent. cloacae A9659		.13	4	.5	.5	1
P. mirabilis A9900		.03	2	.06	.06	1
P. vulgaris A21559		.06	4	.13	.06	1
P. rettgeri A22424		.5	8	8	4	1
Ps. aeruginosa A9843		16	125	1	.5	2
t1/2, pH 7.4, 37°, 10^{-4}M, (h.)		175	160	6.8	4.5	15.0
t1/2, pH 2, 37°, 10^{-4}M, (min.)		23	28	212	36	2
DHP-1 (imipenem = 1)		2.9	17.6	0.19	19.2	1

corresponding 6-aminoethyl derivatives. In contrast, the latter compounds were as active against gram-negative organisms and substantially more effective against *Pseudomonas aeruginosa*. Contrary to the 6-hydroxyethyl compounds where the (1'R)-isomer is more active than its (1'S)-isomer only little difference was observed

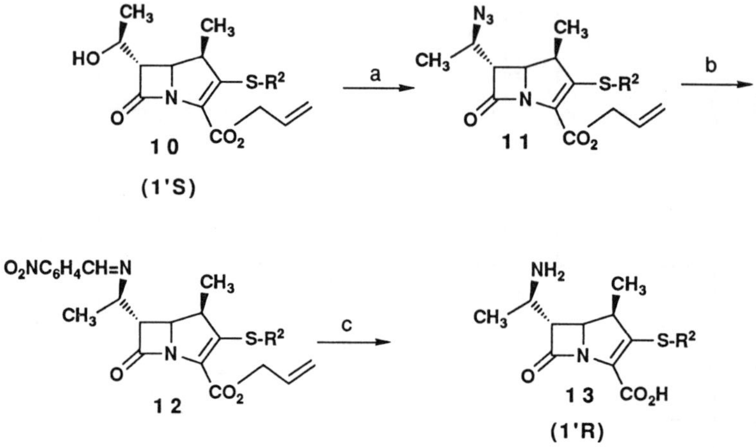

a) PPh$_3$, DEAD, HN$_3$. b) PPh$_3$ then p-O$_2$NC$_6$H$_4$CHO. c) Pd(PPh$_3$)$_4$, K 2-ethylhexanoate then pH 6.

Scheme 2:   Preparation of 6-(1'-aminoethyl) carbapenems

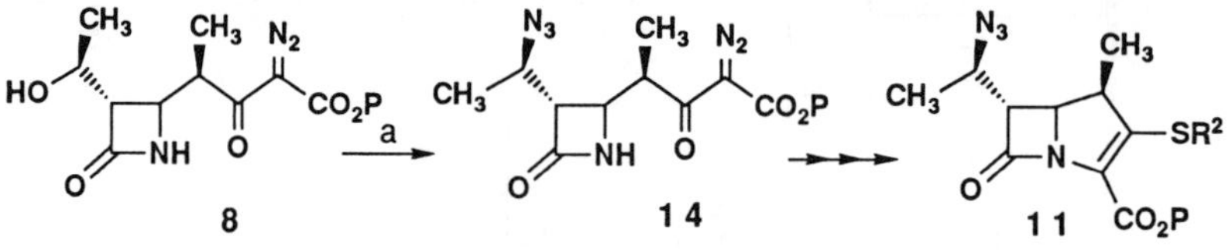

a) PPh$_3$, DEAD, HN$_3$ P = Allyl

Scheme 3:   Alternate preparation of 6-(1'-aminoethyl) carbapenems

between the two 6-aminoethyl isomers. Replacement of the alcohol function by the more nucleophilic amine residue was found to cause a drastic reduction in the chemical stability of the compounds in solution at physiological pH. However the 6-(1'R)-aminoethyl isomer showed an unexpectedly large increase in solution stability at acidic pH and in resistance to hydrolysis by the DHP-1 enzyme. In addition several 6-(1'R)-aminoethyl-1-methyl carbapenems with an appropriate substitutent at position 2 were found to be well absorbed following oral administration to mice.

Methylation (Me$_2$SO$_4$) of the (1'S)-amine gave a separable mixture of the (1'S)-substituted amines listed in Table 2. Mono and dimethylation had almost no influence on the stability of the molecule whereas permethylation to the quaternary ammonium compounds resulted in a marked increase in stability. It is worth noting that quaternization of the amine residue at position 6, contrary to what is normally observed at position 2,[6] resulted in a significant reduction in antibacterial activity.

Because of the similar microbiological properties of the two amine epimers at 1' the corresponding 6-aminomethyl compound

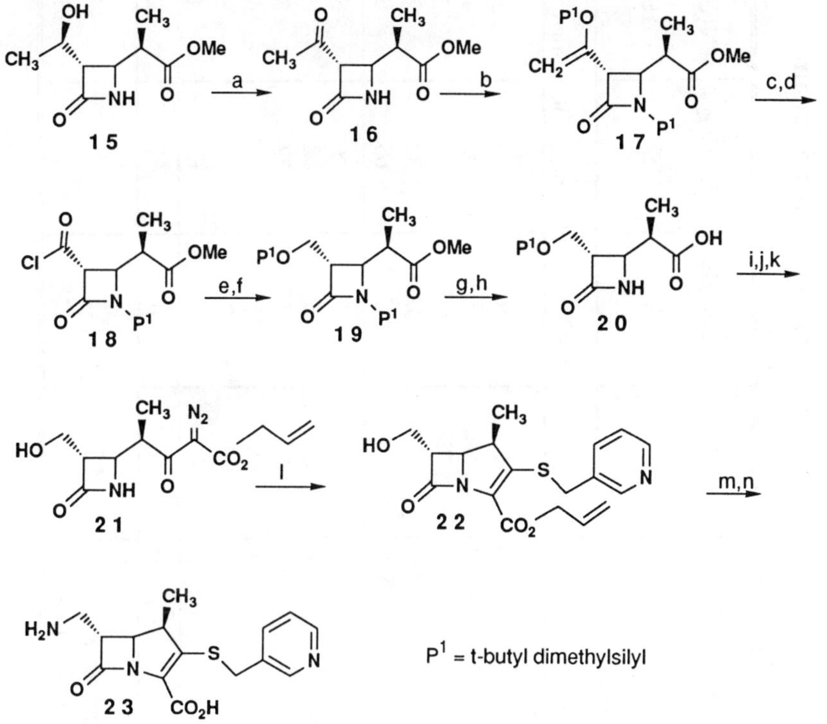

a) Jones oxidation, (83%). b) TBDMSOTf (2 equiv.), Et$_3$N (2 equiv.), (84%). c) O$_3$ then Me$_2$S, (85%). d) (COCl)$_2$-DMF, (90%). e) n-Bu$_4$NBH$_4$, CH$_2$Cl$_2$, -78°, (57%). f) TBDMSOTf, Et$_3$N, (95%). g) TBAF, AcOH, THF, (98%). h)1N NaOH, MeOH, (84%). i) (imidazole)$_2$CO then Mg(allyl malonate)$_2$ (93%). j) p-dodecylbenzenesulfonyl azide, Et$_3$N (100%). k) 1N HCl, MeOH (82%). l) Rh(octanoate)$_2$ then (PhO)$_2$POCl, i-Pr$_2$EtN then 3-picolylthiol (47%). m) PPh$_3$, DEAD, HN$_3$ (68%). n) PPh$_3$, C$_6$H$_6$, then p-O$_2$NC$_6$H$_4$CHO, then K 2-ethylhexanoate, Pd(PPh$_3$)$_4$ then H$_3$O$^+$ (44%).

<u>Scheme 4:</u>　　　Preparation of 6-aminomethyl carbapenems

Table 2   Substituted amines in the 6-side chain

Organism	R1' *	NH2 S	NHCH3 S	N(CH3)2 S	+N(CH3)3 S	+N(CH3)3 R	Imipenem
		MIC (microdilution, µg/ml, Nutrient Broth, 5x10^5 CFU/ml)					
Str. pneumoniae	A9585	4	8	16	.5	8	.002
Str. Pyogenes	A9604	4	16	32	1	16	.002
E. faecalis	A20688	125	125	16	125	125	.5
S. aureus	A9537	8	32	32	2	8	.004
S. aureus P+	A15090	16	63	63	8	125	.016
E. coli	A15119	.13	2	8	8	63	.03
K. pneumoniae	A9664	.5	8	63	16	125	.06
Ent. cloacae	A9659	.5	4	32	8	63	1
P. mirabilis	A9900	.06	1	8	4	32	1
P. vulgaris	A21559	.06	2	8	8	63	1
P. rettgeri	A22424	4	8	63	16	125	1
Ps. aeruginosa	A9843	.5	8	32	16	125	2
t1/2, pH 7.4, 37°, 10^{-4}M, (h.)		4.5	4	2	20	140	15
t1/2, pH 2, 37°, 10^{-4}M, (min.)		36	42	48	288	1680	2

was prepared to better understand the role of stereochemistry in these basic analogs of the natural side chain. It was synthesized,[7] as described in Scheme 4, from the corresponding hydroxymethyl derivative **22** which could be obtained by an oxidative degradation of the usual hydroxyethyl starting material; ozonolysis of the silyl enol ether **17** was the key step in this sequence. Reduction of the resulting carboxylic acid, through the intermediacy of the acid chloride **18**, to the protected hydroxymethyl derivative **19** gave the standard key intermediate **20** for elaboration to the final carbapenem **22**.

In Table 3 a 6-aminomethyl carbapenem is compared to its two chiral aminoethyl homologs. As might be expected the antibacterial activity of the three compounds did not differ markedly. It was the chemical stability that was affected by the absence of chirality in the side chain at position 6; the aminomethyl compound resembled the 6-(1'S)-aminoethyl isomer rather than the R-epimer.

Table 3   Aminomethyl *vs* Aminoethyl at position 6

R6	NH₂ / CH₃	NH₂	CH₃ / NH₂
Organism	MIC (microdilution, μg/ml, Nutrient Broth)		
Str. pneumoniae A9585	.5	.5	2
Str. Pyogenes A9604	.5	.5	.25
E. faecalis A20688	63	125	2
S. aureus A9537	1	.5	.5
S. aureus P+ A15090	2	.5	2
E. coli A15119	.03	.008	.03
K. pneumoniae A9664	.13	.03	.06
Ent. cloacae A9659	.13	.016	.06
P. mirabilis A9900	.008	.004	.016
P. vulgaris A21559	.016	.016	.03
P. rettgeri A22424	2	.25	.13
Ps. aeruginosa A9843	1	1	.5
t1/2, pH 7.4 , 37°, 10⁻⁴M, (h.)	3	2	2
t1/2, pH 2, 37°, 10⁻⁴M, (min.)	144	22	31

Slight modulation of activity and stability could be induced with changes in the group at position 2, but in general the 6-(1'-R)-aminoethyl-1-methyl-carbapenems were found to be highly active antibacterial agents that were well absorbed orally. However their unacceptable lack of solution stability prevented their further development as such.

A solution to the instability problem was foreseen in additional modification at position 1 which has been known to have an important influence on the stability of carbapenems.[4] The next section will describe our search for an approriate group for position 1 that would increase the stability of the nucleus while conferring antimicrobial activity comparable to that of the 1-β-methyl compounds.

3   1-AMINOALKYL CARBAPENEMS

The methodology[8] we had developed for the stereoselective preparation of 1-β-alkyl carbapenems was tested and found to extrapolate to a variety of functional groups for position 1. This methodology, as described in Scheme 5, was found to be compatible with olefins, protected alcohols and amines, azides, ethers and thioethers. The size or the length of the chain in R^1 did not have much influence on the high stereoselectivity of the reaction which consistently gave azetidinone 26 with a β to α ratio greater than 85/15 as has been described for the 1-methyl case.[8] The only limitation was found to be a competing elimination in the formation of the enol silyl ether 25 when a propionyl thioester (24) was used that had a leaving group at position 3 (for example $R^1 = -CH_2N_3$).

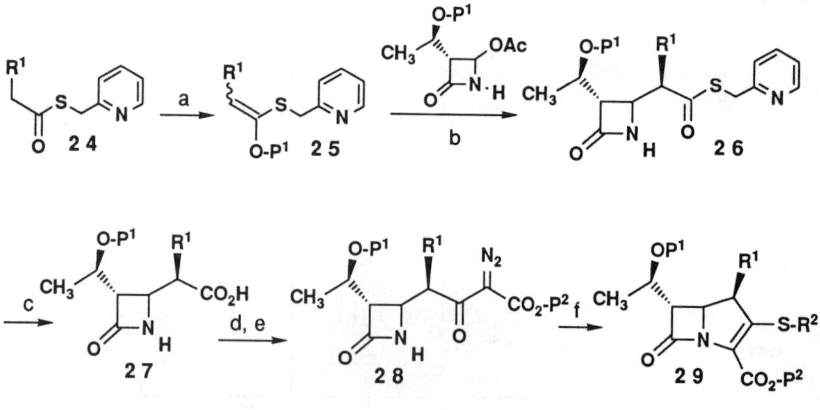

$R^1 = (CH2)_n-X$; n = 2 to 5; X = N_3, OP^1, OR, SR, $NHCO_2P^2$, $CH_2CH=CH_2$
P^1 = TBDMS; P^2 = allyl

a) Et_3N, TBDMSOTf, CH_2Cl_2. b) $ZnCl_2$, CH_2Cl_2. c) NaOH, H_2O_2, THF, H_2O.
d) $(imidazole)_2CO/CH_3CN$, Mg(allyl malonate)$_2$/C_6H_6. e) TsN_3,Et_3N/CH_3CN.
f) $Rh(OAc)_2$/C_6H_6, $(PhO)_2POCl$/i-Pr_2EtN/CH_3CN, R^2SH/i-Pr_2EtN/CH_3CN

Scheme 5:   Stereoselective preparation of 1-substituted carbapenems.

For aminoalkyl derivatives at position 1 the 1-azidoalkyl intermediates **30** (n= 2 to 5) were readily deprotected to the 1-azidoalkyl carbapenems **31** which were reduced to the corresponding amines **32** under a variety of conditions (Scheme 6). Most zwitterionic compounds could be purified by crystallization from aqueous alcohol; otherwise purifications were achieved by reversed-phase chromatography.

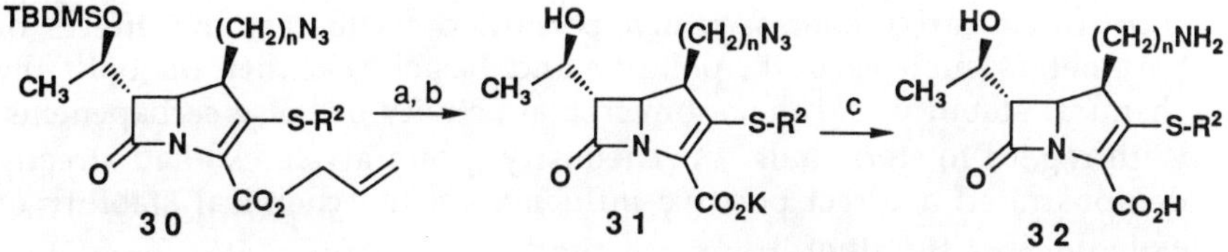

a) TBAF/AcOH/THF, 120h. b) Pd(PPh₃)₄, K 2-ethylhexanoate/EtOAc.
c) Pd/Celite, H₂O, pH 6 or H₂, Pd/Al, H₂O

<u>Scheme 6</u>:   Preparation of 1-aminoalkyl carbapenems

The azido-intermediates **30** could also be obtained from the corresponding alcohols via their mesylates or triflates. Actually, alcohols like **33** (n = 2 to 5) were found to be important intermediates

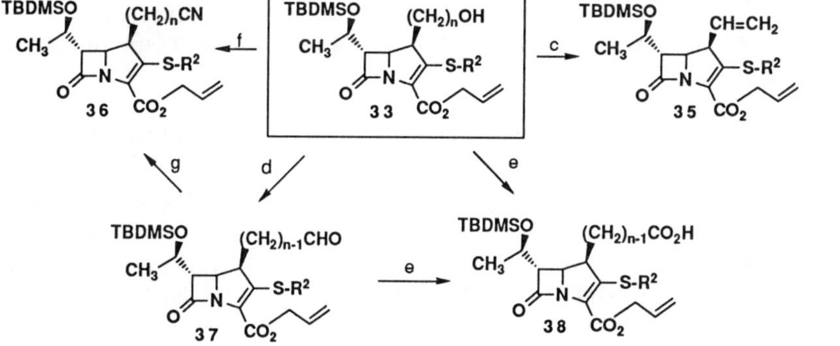

a) PPh₃, DEAD, HN₃/THF. b)Tf₂O, i-Pr₂EtN, HNR₂/CH₂Cl₂. c) 2-NPSC, PBu₃/THF, H₂O₂/THF.
d) Swern oxidation. e) PDC/DMF. f) Tf₂O, i-Pr₂EtN/CH₂Cl₂, LiCN/DMF. g) TMSONH₂, Tf₂O,
i-Pr₂EtN/CH₂Cl₂.

<u>Scheme 7</u>:   Transformations at position 1

for the preparation of a variety of 1-β-substituted carbapenems, as illustrated in Scheme 7. S_N2 type displacement afforded, in addition to the azido derivatives 30, the open chain or cyclic tertiary amines 34, ethers and thioethers, and the nitriles 36. Elimination gave the vinyl derivatives 35 as well as allyl type compounds. Selective oxidation gave either the aldehydes 37 or the acids 38, which could in turn be converted to other derivatives.

Using this general methodology we prepared a number of 1-β-substituted carbapenems, which permitted us to test the effects of parameters such as bulk, polarity and basicity/acidity on both the chemical stability and the antimicrobial activity of these carbapenems. With regard to steric bulk, the chemistry group at Merck[4] had already demonstrated a direct positive influence on the chemical stability of carbapenems by alkyl groups in the 1–β orientation. We found that this effect plateaued at the equivalent of two methylene units since the 1-propyl compound showed a stability profile similar to that of its lower homolog (Table 4). A 1-β-vinyl group had an effect that was more like that of a methyl, whereas a 1-β-allyl compound was more like the corresponding ethyl homolog. Unfortunately this increase in stability was accompanied by a general decrease in antibacterial activity with increasing chain length.

In an attempt to counteract this loss of activity we prepared several carbapenems with functional groups of varied polarity attached to the 1-β-alkyl moiety. We found, as indicated in part in table 4, that groups such as -OH, -OR, -N$_3$, -CN, -CH=O and derivatives and CO_2H and derivatives restored the lost activity only to a small extent but had an overall negative influence on the chemical stability of the molecule. The presence of an amine residue at position 1 gave exceptional results: not only did it restore the gram-positive and gram-negative activities to the level of the methyl analog but it also increased anti-pseudomonal activity several folds. Unfortunately this beneficial effect was also accompanied by a redundant drastic reduction in stability.

This reduced solution stability is in fact a general property of all aminoalkyl carbapenems (Table 5). The 6- and 1-aminoalkyl analogs are particularly unstable because of the possibility of intramolecular attack of the amine residue, on the β-lactam carbonyl; compounds 40 (n = 1) resulting from intramolecular transacylation have been isolated and characterized in the case of 1-aminoethyl carbapenems 39. On the other hand the 6- and 1-aminoalkyl derivatives show prolonged solution half-lives under acidic conditions. The chemical stability of 1-aminoalkyl carbapenems is a function of both the pH of the solution (Figure 1) and the length of the spacer separating the amine moiety from the nucleus (Table 6). Several attempts were made to obtain the 1-aminomethyl homolog 39 (n = 0); it could be

Table 4   Substitution at position 1

Organism	R¹	CH₃	CH₂CH₃	CH=CH₂	(CH₂)₂CH₃	(CH₂)₂OH	CH₂CH=O	CH₂CO₂H	(CH₂)₂NH₂
		MIC (microdilution, µg/ml, Mueller-Hinton Broth, 5×10^5 CFU/ml)							
Str. pneumoniae	A9585	.008	.13	.015	.25	.25	.03	.13	.13
Str. Pyogenes	A9604	.008	.13	.015	.25	.25	.03	.13	.13
E. faecalis	A20688	2	16	4	32	16	4	64	8
S. aureus	A9537	.03	1	.06	1	1	.5	8	.5
S. aureus P+	A15090	.06	2	.25	2	1	1	16	1
E. coli	A15119	.008	.06	.015	1	.03	.06	.25	.008
K. pneumoniae	A9664	.016	.13	.03	1	.06	.13	1	.016
Ent. cloacae	A9659	.125	1	.5	8	2	2	2	.06
P. mirabilis	A9900	.06	1	.125	4	2	2	2	.5
P. vulgaris	A21559	.125	.25	.125	.5	1	.25	.25	.5
P. rettgeri	A22424	.125	1	.25	4	.5	1	1	1
Ps. aeruginosa	A9843	8	>64	32	>64	16	125	64	.25
t1/2, pH 7.4, 37°, 10⁻⁴M, (h.)		64	244	71	210	65	35	20	3.8
t1/2, pH 2, 37°, 10⁻⁴M, (min.)		29	126		117	22	15	10	188

Table 5

Amines in the 1, 2 or 6-side chains

Organism		(OH / CH₂CH₂NH₂ / SCH₃ / CO₂H) MIC (microdilution, µg/ml, 5×10⁵ CFU/ml) Mueller-Hinton Broth	(OH / CH₃ / SCH₂CH₂NH₂ / CO₂H) MIC (microdilution, µg/ml, 5×10⁵ CFU/ml) Mueller-Hinton Broth	(NH₂ / CH₃ / SCH₃ / CO₂H) MIC (microdilution, µg/ml, 5×10⁵ CFU/ml) Nutrient Broth	(OH / CH₃ / SCH₃ / CO₂H) MIC (microdilution, µg/ml, 5×10⁵ CFU/ml) Nutrient Broth
Str. pneumoniae	A9585	2	.007	16	.06
Str. Pyogenes	A9604	2	.003	16	.06
E. faecalis	A20688	32	1	125	4
S. aureus	A9537	4	.007	16	.25
S. aureus P+	A15090	32	.015	16	.13
E. coli	A15119	.03	.03	.13	.03
K. pneumoniae	A9664	.13	.06	.5	.06
Ent. cloacae	A9659	.5	.125	.5	.13
P. mirabilis	A9900	8	1	.06	.03
P. vulgaris	A21559	4	.5	.13	.06
P. rettgeri	A22424	8	.5	8	.5
Ps. aeruginosa	A9843	.5	1	1	16
t1/2 pH7.4, 37°, 10⁻⁴M, (h.)		1.6	32	6.8	175
t1/2 pH2, 37°, 10⁻⁴M, (min.)		129	25	212	23

obtained in a protected form (t-BOC, allyloxycarbonyl) but only **40** (n = 0) could be isolated upon deprotection. In comparison the stability of 1-β-aminopropyl carbapenems is roughly equivalent to that of the 1-β-methyl analogs.

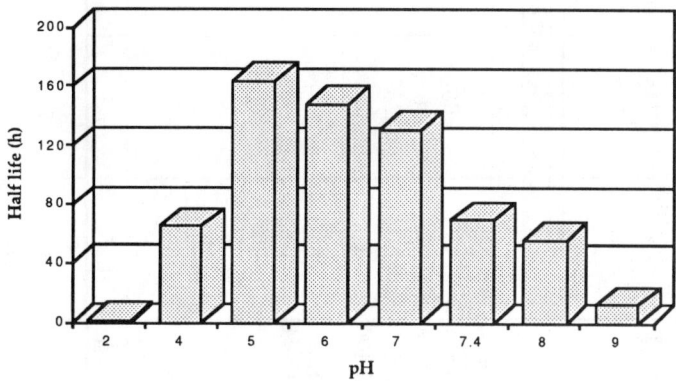

An amine residue, in either of the side chains at 6, 1 or 2, increased anti-pseudomonal activity several times over that of the corresponding non-basic carbapenem (Table 5), which indicates that the effect might be related to improved penetration through the outer cell membrane rather than to an increased intrinsic activity. On the other hand the 1-α-(aminoalkyl)-epimers are much less active and less stable than the corresponding β-isomers (Table 7).

Figure 1: Effect of pH on the stability of 1-aminopropyl carbapenems **4**, A = $(CH_2)_3$, R^2 = SCH_2CH_2CN

Measured in buffered solutions at 37°C and 10^{-4}M.

The antimicrobial activity of 1-β-aminoalkyl carbapenems was found to be less dependent on the chain length at position 1 than was the solution stability (Table 6); the overall gram-positive and gram-negative activities of these compounds remained remarkably constant over a range of 2 to 5 methylene units. Only a slight loss of anti-pseudomonal activity was observed with the longer chains. In addition going from a primary to secondary or tertiary amine, and finally to a quaternary ammonium derivative, had only a modest effect on both chemical stability and antibacterial activity (Table 8).

Table 6   Length of chain at position 1

Structure at position 1:

$$\text{OH} \quad (CH_2)_n\text{-}R^1$$

with substituents SCH₂CH₂CN → SCH_2CH_2CN, CO_2H, CH_3, and the bicyclic β-lactam (carbapenem) core.

Organism	R¹	NH₂	NH₂	NH₂	NH₂	S(CH₂)₂NH₂	NHCOCH₂NH₂
	n	2	3	4	5	2	4
		MIC (microdilution, µg/ml, Mueller-Hinton Broth, 5×10^5 CFU/ml)					
Str. pneumoniae	A9585	.13	.25	.06	.13	.25	1
Str. Pyogenes	A9604	.13	.06	.06	.13	.25	1
E. faecalis	A20688	16	16	16	8	16	32
S. aureus	A9537	.5	.25	.5	.25	1	2
S. aureus P+	A15090	1	.2	2	.25	1	4
E. coli	A15119	.008	.008	.008	.004	.13	1
K. pneumoniae	A9664	.03	.03	.008	.008	.13	1
Ent. cloacae	A9659	.06	.03	.03	.5	1	.06
P. mirabilis	A9900	.5	2	1	.5	1	4
P. vulgaris	A21559	.5	2	1	.5	1	2
P. rettgeri	A22424	1	2	1	1	2	8
Ser.marcescens	A20019	.25	1	.5	.25	.5	4
Ps. aeruginosa	A9843	.25	.5	1	1	2	16
t1/2, pH 7.4, 37°, 10^{-4} M, (h.)		3.8	70	107	136	46	150
t1/2, pH 2, 37°, 10^{-4} M, (min.)		188	137	144	130	108	131

Table 7   β *vs* α Configuration of the 1-side chain

Organism	R^1 *	OH β	OH α	NH$_2$ β	NH$_2$ α	Imipenem
		MIC (microdilution, μg/ml, Nutrient Broth, 5x10^5 CFU/ml)				
Str. pneumoniae A9585		.13	2	.25	2	.004
Str. Pyogenes A9604		.13	1.5	.03	.5	.001
E. faecalis A20688		32	125	4	125	.25
S. aureus A9537		.25	4	.13	4	.004
S. aureus P+ A15090		.5	4	.25	4	.008
E. coli A15119		.25	2	.008	1	.016
K. pneumoniae A9664		.25	4	.016	4	.03
Ent. cloacae A9659		.5	8	.03	4	.13
P. mirabilis A9900			1	.03	2	.06
P. vulgaris A21559		.5	4	.03	4	.06
P. rettgeri A22424		2	32	.13	16	.25
Ps. aeruginosa A9843		125	125	1	32	1
t1/2, pH 7.4, 37°, 10^{-4}M, (h.)		180	4.5	92	3.5	15
t1/2, pH 2, 37°, 10^{-4}M, (min.)		145		123	10	2
DHP-1 (imipenem = 1)		0.8		0.14		1

Table 8   Substitution of the amine at position 1

Organism	R¹	NH₂	NHMe	NMe₂	NMe₃⁺	(pyrrolidine)	(N-methylpyrrolidinium)
		MIC (microdilution, µg/ml, Mueller-Hinton Broth, 5×10^5 CFU/ml)					
Str. pneumoniae	A9585	.13	.03	.03	.03	.008	.5
Str. Pyogenes	A9604	.13	.03	.03	.016	.008	.5
E. faecalis	A20688	16	8	1	.4	2	63
S. aureus	A9537	.5	.5	.25	.5	.13	8
S. aureus P+	A15090	1	1	.25	1	.5	8
E. coli	A15119	.008	.007	.03	.13	.016	1
K. pneumoniae	A9664	.03	.06	.03	.03	.016	2
Ent. cloacae	A9659	.06	.25	.06	.25	.03	16
P. mirabilis	A9900	.5	.5	.5	1	.25	16
P. vulgaris	A21559	.5	.5	.5	.13	.25	8
P. rettgeri	A22424	1	1	.5	1	.5	63
Ps. aeruginosa	A9843	.25	.5	1	1	1	63
t1/2, pH 7.4, 37°, 10^{-4}M, (h.)		3.8	9.4	11	17	18	16
t1/2, pH 2, 37°, 10^{-4}M, (min.)		188	86	66	88	66

Table 9   Basic residues at positions 1 and 2

Organism	R² n	NH₂ 2	NHCH=NH 2	NHC(=NH)NH₂ 2	NMe₂ 2	(N-Me pyrrolidine) 0	Imipenem
			MIC (microdilution, μg/ml, Mueller-Hinton Broth, 5x10⁵ CFU/ml)				
Str. pneumoniae	A9585	.13	.13	.03	.25	.015	.002
Str. Pyogenes	A9604	.13	.13	.03	.25	.007	.002
E. faecalis	A20688	8	16	16	16	2	.5
S. aureus	A9537	.25	.25	.25	.25	.06	.004
S. aureus P+	A15090	.5	.5	.5	.5	.125	.016
E. coli	A15119	.13	.13	.13	.25	.015	.03
K. pneumoniae	A9664	.25	.25	.13	.5	.03	.06
Ent. cloacae	A9659	.5	.25	.5	1	.03	1
P. mirabilis	A9900	2	16	4	1	2	1
P. vulgaris	A21559	2	8	8	.5	2	1
P. rettgeri	A22424	4	16	8	2	4	1
Ps. aeruginosa	A9843	.25	.25	.5	.25	.06	2
t1/2, pH 7.4, 37°, 10⁻⁴M, (h.)		33	44	63	25	41	15
t1/2, pH 2, 37°, 10⁻⁴M, (min.)		133	172		132		2

The combination of basic side chains at position 1 and 2 (Table 9) gave carbapenems with outstanding anti-pseudomonal acitivity and good overall gram-positive and gram-negative potency, with a slight weakness against certain species of *Proteus.*; the N-formimidoyl or the corresponding guanidino derivatives did not show any improvements over the parent cysteamine substitutent at position 2. The broadest spectrum of activity was observed with cyclic amines directly attached to the sulfur atom at position 2.

4   CONCLUSION

We have shown that the carbapenem nucleus can accommodate substituents in the 1-β-position, other than the usual methyl group, without a loss in antibacterial activity. Extensive modification could be made while retaining the high level of activity. In particular the 1-β-aminoalkyl carbapenems have been found to constitute a new class of broad spectrum antibacterial agents.

ACKNOWLEDGEMENTS

The authors acknowledge the contribution of Dr. H. Mastalerz, Dr. V.S. Rao, Dr. E. Ruediger, Mr. J.P. Daris, Mr. P. Lapointe who prepared some of the compounds described. We thank the microbiology department at the Wallingford site for biological data.

REFERENCES

1.   J.S. Kahan, F.M. Kahan, R. Goegelman, S.A. Currie, M. Jackson, E.O. Stapley, T.W. Miller, A.K. Miller, D.Hendlin, S. Mochales, S. Hernandez, H.B. Woodruff and J. Birnbaum, J. Antibiot., 1979, 32, 1.

2.   R. Deziel and M. Endo, Tetrahedron Lett., 1988, 29, 61.

3.   O. Mitsunobu, Synthesis, 1981, 1.

4.   D.H. Shih, F. Baker, L. Cama and B.G. Christensen, Heterocycles, 1984, 21, 29.

5.   M.D. Bachi and J. Vaya, J. Org. Chem., 1979, 44, 4393.

6.   C.U. Kim, P.F. Misco and B.Y. Luh, J. Antibiot., 1987, 40, 1707.

7.   E.H. Ruediger and C. Solomon, J. Org. Chem., 1991, 56, 3183; H. Mastalerz, M. Ménard, E.H. Ruediger and L. Fung-Tomc, J. Med. Chem., 1992, 35, 953.

8.   A. Martel, J.P. Daris, C. Bachand, J. Corbeil and M. Ménard, Can. J. Chem., 1988, 66, 1537.

2

Tribactams: A Novel Class of β-Lactam Antibiotics

A. Perboni, B. Tamburini, T. Rossi, D. Donati, G. Tarzia, and G. Gaviraghi

GLAXO RESEARCH LABORATORIES, VIA FLEMING 4, 37100 VERONA, ITALY

1   INTRODUCTION

The search for novel antibacterial agents has led to the discovery of a new class of compounds containing a tricyclic nucleus as the key structural feature.[1,2]

Characteristic of these tricyclic β-lactam derivatives (Tribactams) **1** is the presence of a reactive β-lactam ring (A), an unsaturated five-membered ring (B), and the final ring (C), which in general could be a five-, six- or seven-membered carbocyclic or heterocyclic ring .

Moreover, ring (C) has given us many opportunities to modify the new tricyclic nucleus by means of the introduction of suitable substituents .

In particular, this paper refers to part of the work performed in our laboratories on the compounds in which ring (C) was an unsubstituted or a 4-substituted carbocyclic six-membered ring as represented by compound **2** .

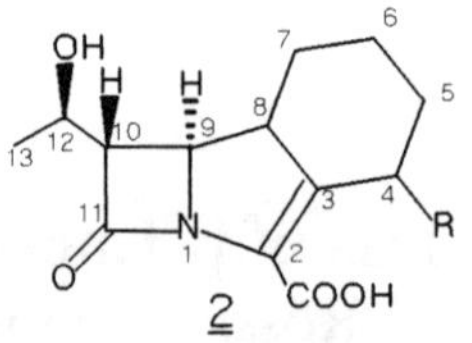

Our investigations began with the preparation and evaluation of the two optically pure unsubstituted compounds **3** and **4** so as to establish which of them was the more active and to identify a synthetic approach capable of giving us predominantly the desired isomer.

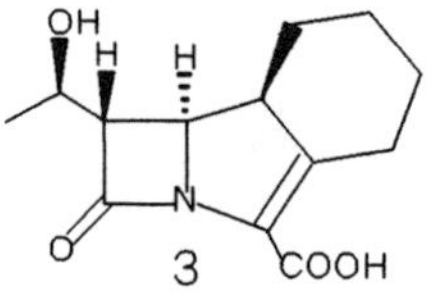

We next turned our attention to the 4-substituted derivatives and again we were able to identify, independent of the substituent, a preferred isomeric configuration .

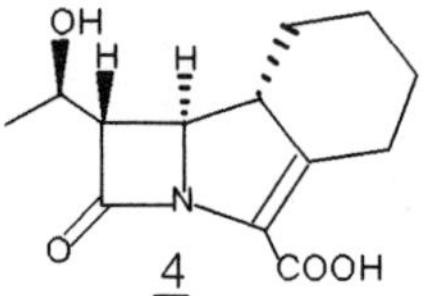

Finally, we focused on preparing the desired isomer by means of a stereocontrolled synthesis with the objective of selecting a general and suitable approach to all the tricyclic derivatives having a side-chain linked through a heteroatom to position 4 .

2   CHEMISTRY AND BIOLOGY

Unsubstituted derivatives

The easier and more general approach to the unsubstituted tricyclic derivatives is described in Scheme 1 . A lithium enolate of a cyclic ketone is reacted with the commercially available chiral acetoxyazetidinone **5** to afford two "open" intermediates **6** and **7** in which two new chiral centres (i.e. position 4 and position 2') are formed .

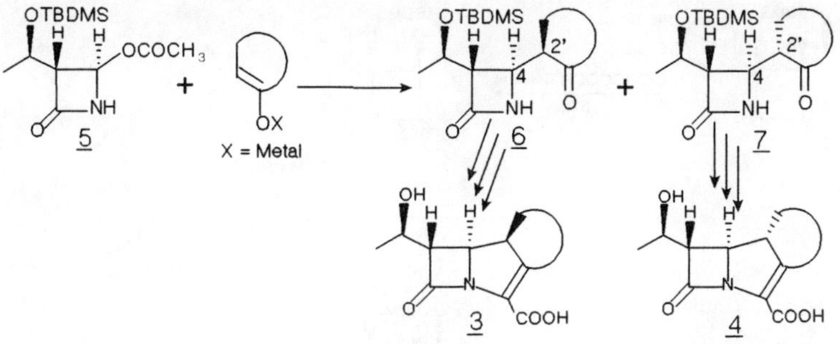

SCHEME 1

The substituent at position 4 is always <u>trans</u> with respect to the hydroxyethyl side-chain[3-5] , whilst the chiral centre in position 2' is usually obtained as a racemic mixture .

The two isomers can usually be separated by chromatography, elaborated, cyclized, and deprotected by means of the previously described methods[6,7] (see Scheme 2 and Scheme 3) .

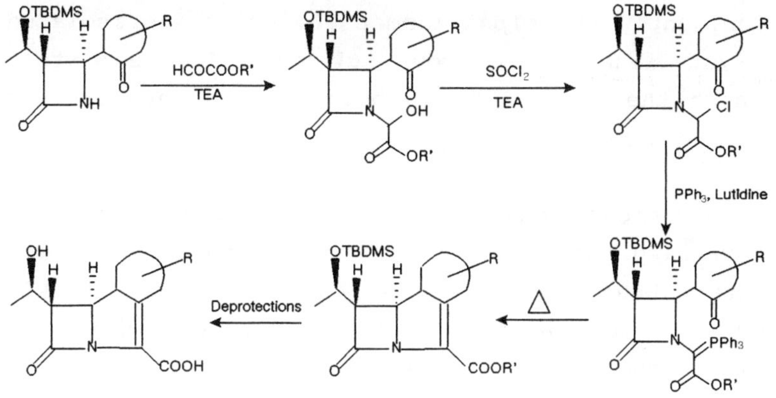

SCHEME 2

When the lithium enolate of cyclohexanone **11** was used the two diasteromeric "open" intermediates **9** and **10** were obtained in around 30% yield each .

In order to increase these yields and to avoid the chromatographic separation of the two "open"

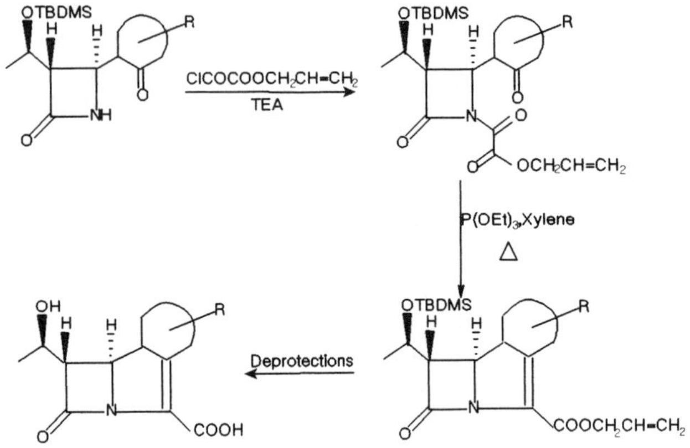

SCHEME 3

intermediates, we looked more closely at the reaction
of the acetoxyazetidinone with 1-trimethylsilyloxy
cyclohexene **12** in the presence of different Lewis
acids .

The use of zinc chloride, as Lewis acid, in
dichloromethane at room temperature for 24hr gave
predominantly the alpha isomer **10**, which was isolated
by crystallization in about 50% yield from a mixture
in which the alpha and the beta isomers were present
in a 7 to 3 ratio.

Concurrent with these investigations, studies
were carried out using an acetoxyazetidinone protected
at the nitrogen with a t-butyldimethylsilyl (TBDMS)
group **8**. In this case the coupling with the
silylenolether in the presence of suitable Lewis acid
(tin tetrachloride) gave mainly the beta isomer **9** ,
which was obtained in 45-50% yield by crystallization
(Scheme 4) after deprotection .

This stereoselectivity was not unexpected in
view of the wide literature precedent regarding the
erythro/threo selectivity in Aldol-type reactions with
cyclic silyl enolethers.[8,9]

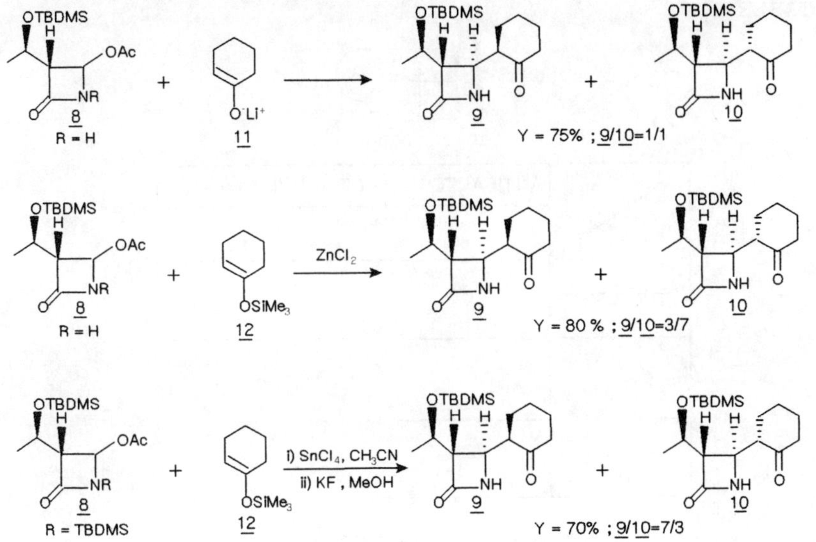

SCHEME 4

Having reached the objective of setting up a chemical approach capable of giving us the "open" intermediates without any chromatographic purifications, we then carried them through to the two final tricyclic derivatives which were submitted for antibacterial testing .

As shown in Table 1 , the antibacterial activities obtained against representative Gram-positive and Gram-negative bacteria proved that the beta isomer (8S) was more active than the alpha isomer (8R) .

4-Substituted derivatives

As in the case of the unsubstituted derivatives, the 4-substituted compounds can easily be prepared by a coupling between the enolate of a 2-substituted cyclohexanone and the acetoxyazetidinone giving the substituted "open" intermediates (Scheme 5) which could be subsequently cyclized .

TABLE 1

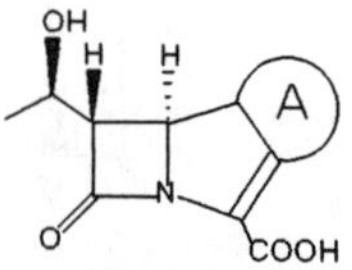

Organism (10⁵CFU/ml)	ANTIBACTERIAL ACTIVITY (MIC=μg/ml)	
Compound A	S.aureus 663	E.coli 851
Imipenem	0.06	0.5
	0.2	0.5
	1	2

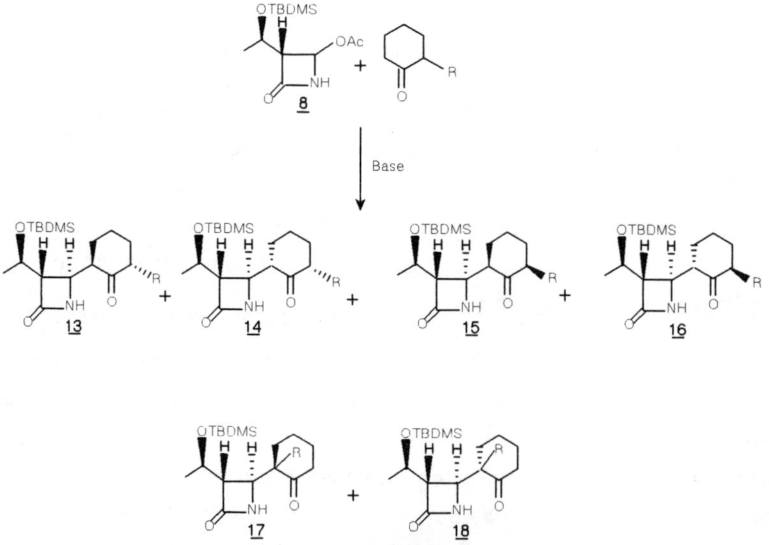

SCHEME 5

This coupling reaction can give six different compounds , four of which (compounds **13**, **14**, **15**, and **16**) are diastereoisomers derived from the desired enolate, whereas the other two (**17** and **18**) derive from the unwanted enolate, which is always present in enolate reactions using alpha-substituted ketones[10-12]. The amount of this undesired enolate is dependent on the substituent , the steric demands of the base used to form the enolate , and the reaction conditions used.

As a result, this approach required a very careful separation of the resultant products with a generally low yield of each of the four desired diastereomeric open intermediates .

In view of the above results and because it was possible to prepare the pure alpha and beta unsubstituted open intermediates (**10** and **9**) it was decided to study the asymmetric introduction of the 4-substituent starting from these intermediates . In particular, the beta isomer **9** was used as a starting point because of the better antibacterial activity of the 8S-tribactam .

The first derivative prepared by means of this two-step reaction was the 4-thiomethyl compound .

The introduction of a thiomethyl group alpha to a ketone by means of the reaction between a lithium enolate and electrophilic methyl methanethiosulfonate is well described[13,14] . Moreover, it has also been shown that the thiomethyl group can be used as nitrogen protecting group on azetidinones[15] . Therefore we reacted the dianion of the open intermediate with two equivalents of electrophile to obtain two protected thiomethyl substituted compounds (Scheme 6) .

The desired substituted derivatives (**19** and **20**) were obtained in high yield after deprotection but in a variable diastereomeric mixture in which the 2'-beta-thiomethyl derivative was always present as the major isomer .

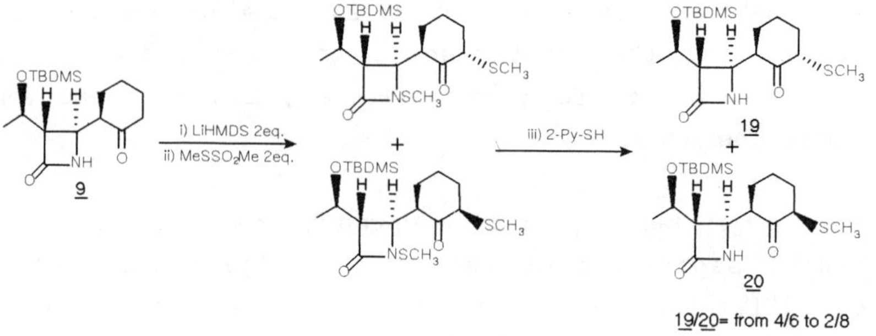

SCHEME 6

This variability was essentially due to the different conditions used when adding the electrophile to the dianion solution at a low temperature .

In particular the slower the rate of addition the higher the percentage of the 2'-beta isomer that was obtained.

Before explaining these observations by analyzing the possible conformations of the dianion and the permitted transition states it was decided to perform the same reaction in two steps by preliminary protection of the nitrogen followed by introduction of the thiomethyl group (Scheme 7) .

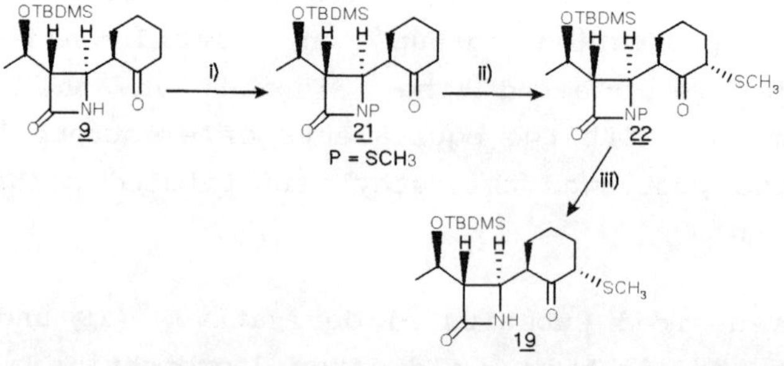

I) LiHMDS 1eq. ; CH₃SSO₂CH₃ 1eq.   II) LiHMDS 1eq. ; CH₃SSO₂CH₃ 1eq.   III) 2-PySH

SCHEME 7

The only compound obtained from this sequence was the 2'alpha-thiomethyl intermediate <u>19</u> .

In view of this result , the variability in the ratio between the two isomers observed in the one step reaction could be explained by assuming an unselective reaction between the dianion and the electrophile.

This unselective reaction was promoted by the presence of an excess of electrophile and produced some protected derivative and consequently the alpha thiomethyl derivative <u>19</u> , whilst the coupling of the unprotected carbanion with the electrophile gave predominantly the beta isomer <u>20</u> .

In conclusion, starting from the unsubstituted beta derivative <u>9</u> , we were able to prepare either the chiral intermediate <u>19</u> by preliminary protection of the nitrogen or compound <u>20</u> by a very slow addition of the electrophile to the dianion solution .

We also proved that it was possible to protect the nitrogen using different protecting groups (e.g. TBDMS) and, more importantly, the alpha unsubstituted derivative proved to have the same reactivity as the beta isomer but with opposite and slightly less efficient stereocontrol (the one step reaction gave mainly the 2'-alpha-thiomethyl derivatives whilst the two-step approach gave the 2'-beta compound) .

The four diastereomeric substituted open intermediates were finally carried through to give the tricyclic derivatives .

One of the isomers was not stable to the deprotection conditions used whilst the others gave the desired final salts and were evaluated <u>in vitro</u> .

The results obtained are shown in Table 2 .

TABLE 2

Organism (10⁵CFU/ml) / Compound A	ANTIBACTERIAL ACTIVITY (MIC=μg/ml)	
	S.aureus 663	E.coli 851
Imipenem	0.06	0.5
	0.2	0.5
	1	2
SMe	0.01	1.5
SMe	2	>32
SMe	2	8

All the compounds showed some antibacterial activity but only one of them (i.e. 4-alpha,8-beta) was as or more active than the unsubstituted 8-beta product .

Consequently, we focused our attention on this isomer with the objective to modify the substituent in position 4, with particular attention to the introduction of substituted amines or alkoxy groups .

The synthetic approach described above (lithium enolate and electrophile) is reported to be useful for the preparation of 4-substituted amines[16] and the 4-hydroxy compounds[17] but not for the synthesis of 4-alkoxy derivatives, which could only be obtained by alkylation of the corresponding hydroxy compound .

In view of this it was decided to study a more general pathway to these derivatives .

Our target was to identify and prepare a key intermediate , which would be reactive in the presence of nucleophiles and which would give us mainly the desired isomer .

One of the possible intermediates that could satisfy the above characteristics if obtained in a chirally pure form was the epoxide **22** .

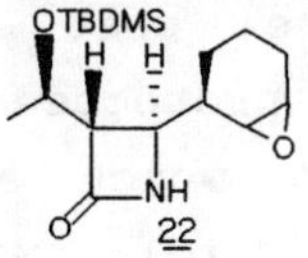

The preparation of this epoxide (see Scheme 8) started from the pure 2'-beta unsubstituted cyclohexanone **9** which was converted to the tosylhydrazone **23** in about 90% yield . This underwent a Shapiro reaction[18] to afford the olefin **24** .

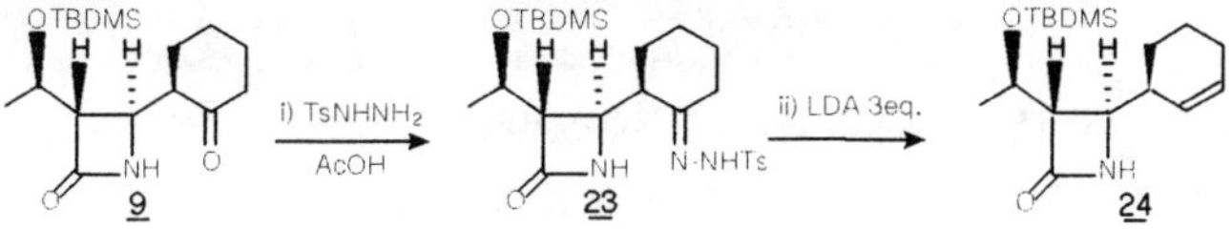

SCHEME 8

The olefin **24** was epoxidized with m-chloroperbenzoic acid and, as in the above thiomethylation, it was possible to prepare in good yield either the beta **25** or the alpha **26** epoxide with high stereoselectivity by means of a TBDMS protection of the nitrogen (Scheme 9) .

In this case the desired isomer **25** was obtained using the unprotected olefin as starting material .

The epoxide was then reacted with methanol and methylamine to give the intermediates **27a** and **27b**

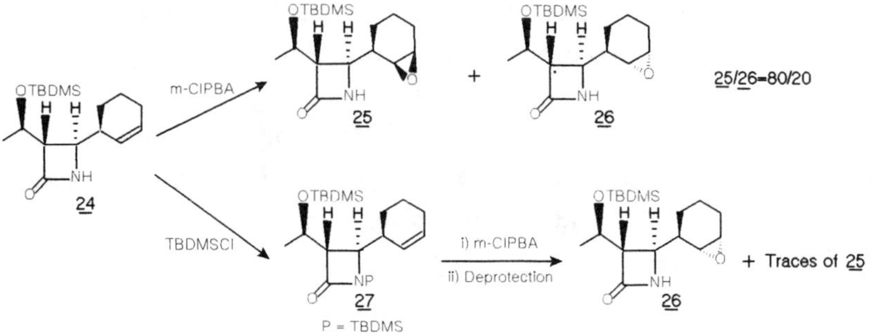

SCHEME 9

respectively (in the case of the methylamine derivative it was also protected as the allylcarbamate), Swern oxidation[19] of which afforded intermediates **28a** and **28b** (Scheme 10), which have a defined stereochemistry at C-4.

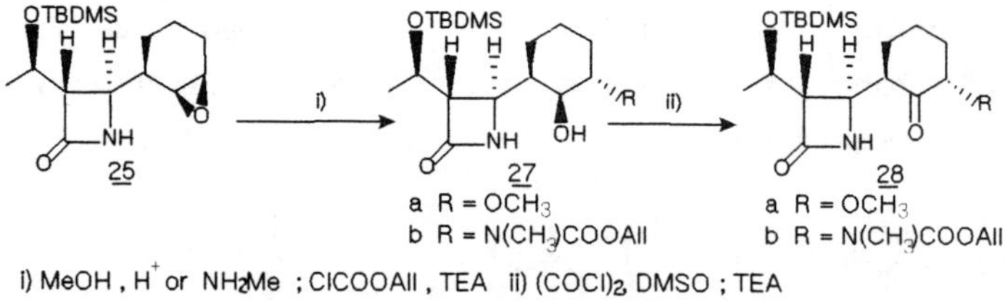

I) MeOH , H$^+$ or NH$_2$Me ; ClCOOAll , TEA   II) (COCl)$_2$ DMSO ; TEA

SCHEME 10

The substituted ketone intermediates were elaborated, cyclized, deprotected (see Scheme 11) and tested giving the results shown in Table 3 .

In conclusion we have discovered a new class of antibacterial agents and developed a flexible, stereoselective synthetic route from the commercially available acetoxyazetidinone .

In the case of the 4-substituted derivatives, of the four possible isomers the 4-alpha,8-beta has provided a series of derivatives with the same level

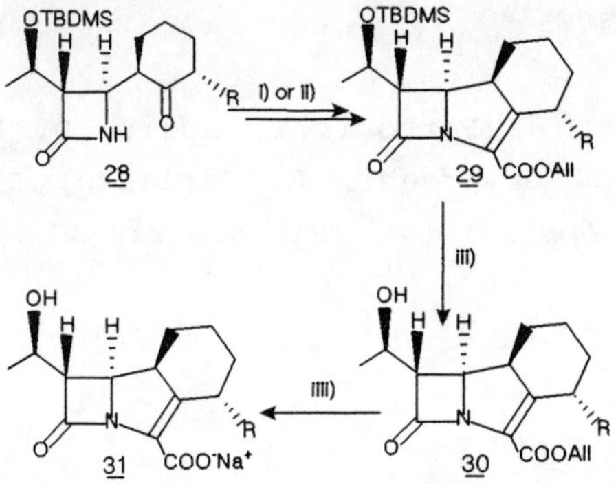

R = OCH₃, NH CH₃ , SCH₃

i) Allylglyoxal ; SOCl₂, Lutidine ; PPh₃ , Lutidine ; heat
ii) Allyloxalylchloride , TEA ; P(OEt)₃, heat   iii) TBAF ,
AcOH IIII) Pd(PPh₃)₄, Sodium-2-ethylhexanoate

SCHEME 11

TABLE 3

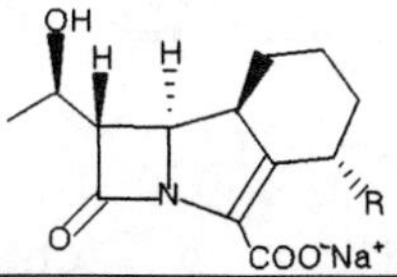

Organism (10⁶CFU/ml) / Compound R	ANTIBACTERIAL ACTIVITY (MIC=µg/ml)		
	S.aureus 663	E.coli 851	P.aeruginosa
Imipenem	0.06	0.5	4
Meropenem	0.25	0.12	2
H	0.2	0.5	>32
SMe	0.01	1.5	>32
OMe	0.06	0.5	>32
NHMe	0.2	2	4

of activity as one of the most potent antibacterial
agents now on the market (Imipenem) , allowing us wide
scope to modify the antibacterial spectrum by means of
a suitable substituent .

3   REFERENCES

1. B. Tamburini, A. Perboni, T. Rossi, D. Donati, D. Andreotti, G. Gaviraghi, R. Carlesso, C. Bismara, Eur. Patent Appl. 1991, EP 0 416 953 A2 .

2. M. Sendai, T. Miwa; Eur. Patent Appl. 1991, EP 0 422 596 A2.

3. M. Alpegiani, A. Bedeschi et al, J. Am. Chem. Soc., 1985, 107, 6398.

4. P.J. Reider, E.J.J. Grobowski, Tetrahedron Lett., 1982, 23, 2293.

5. M. Fuentes, I. Shinkai, T.N. Salzmann, J. Am. Chem. Soc., 1986, 108, 4675.

6. R.N. Guthikonda , L.D. Cama et al, J. Med. Chem., 1987, 30, 871.

7. A. Yoshida , T. Hayashi et al, Chem. Pharm. Bull., 1983, 31, 768.

8. S. Murata, M. Suzuki, R. Noyori, Tetrahedron, 1988, 44, 4259.

9. E. Nakamura, M. Shimizu, I. Kuwajiama et al, J. Org. Chem., 1983, 48, 932.

10. R.D. Miller, D.R. McKean, Synthesis, 1979, 730.

11. S. Ahmad, M.A. Khan, J. Iqbal, Synthetic Comm., 1988, 18, 1679.

12. D.A. Evans, 'Asymmetric Synthesis', Academic Press, Inc. , Orlando, 1984, Vol. 3, Chapter 1, p. 6.

13. D.L. Boger, M.D. Mullican, <u>J. Org. Chem.</u>, 1984, <u>49</u>, 4045.

14. B.M. Trost, G.S. Massiot, <u>J. Am. Chem. Soc.</u>, 1977, <u>99</u>, 4405.

15. N.V. Shah, L.D. Cama, <u>Heterocycles</u>, 1987, <u>25</u>, 221.

16. G. Boche, N. Mayer et al, <u>Angew. Chem.</u>, 1978, <u>90</u>, 733.

17. J. Vedejs, <u>J. Am. Chem. Soc.</u>, 1974, <u>96</u>, 5944.

18. R.H. Shapiro, <u>Org. React.</u>, 1976, <u>23</u>, 405.

19. K. Omura, D. Swern, <u>Tetrahedron</u>, 1978, <u>34</u>, 1651.

3
Crystallography of Penicillin-binding Enzymes

James R. Knox
DEPARTMENT OF MOLECULAR AND CELL BIOLOGY, THE UNIVERSITY OF
CONNECTICUT, STORRS, CT 06269, USA

1 INTRODUCTION

Atomic-level definition of the ß-lactam binding site in bacterial target enzymes of cell wall synthesis, and in the daughter enzymes, the penicillin-destroying ß-lactamases, has been sought for twenty years. Structures of three types of ß-lactamases have been reported since a 1984 RSC symposium on the chemistry of penicillins, but new structures of target enzymes have not been forthcoming. However, expected similarities between the two types of enzymes[1-3] allows modeling the target site from the well-established ß-lactamase sites. A reactive serine common to both types is acylated by the ß-lactam carbonyl group; the enzyme types are distinguishable, however, by very different rates of deacylation:

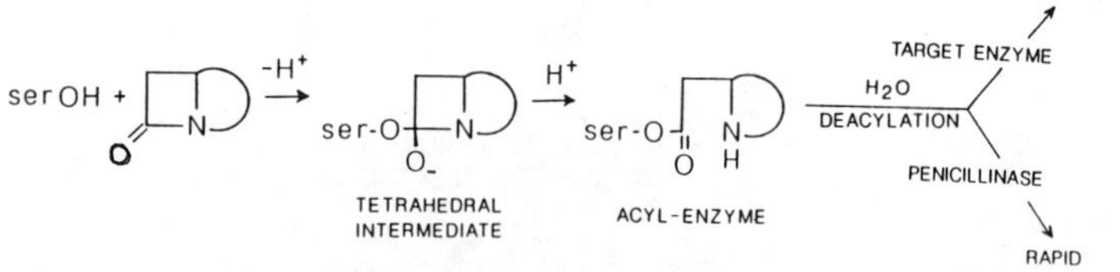

2 ß-LACTAM RECOGNIZING ENZYMES

<u>ß-Lactam Target Enzymes</u> To date, the only crystal structure of a ß-lactam inhibited enzyme having both carboxypeptidase and transpeptidase activity on D-alanyl-D-alanine cell wall peptides is that from *Streptomyces* sp. R61.[4,5] Work by many groups on the structure of the penicillin binding proteins (PBPs) of *Escherichia coli* is hindered by their membrane association and hydrophobicity, and results have not been reported at high resolution. Ghuysen's group in Liege has

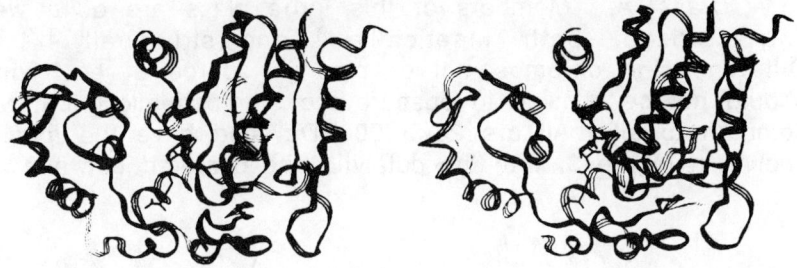

Figure 1. Backbone of the ß-lactam-sensitive DD-peptidase of *S.* R61 with acyl cephalosporin.

demonstrated that the R61 enzyme (Figure 1) is a useful model for PBPs. The reactive serine is located opposite a ß-strand which hydrogen bonds the ß-lactam substrate prior to acylation. Because of the long half-life of such acyl-intermediates, crystallographic mapping of several complexes has been possible.[4] Recently, maps of the cephalothin and cefotaxime complexes have shown the acylamide thiophene substituent of cephalothin is relatively unhindered in the binding pocket and in fact quite exposed (Figure 2). In the map of the branched cefotaxime, the amino-thiazole ring is exposed but the methoxy group is rather crowded beside the ß-strand.[6] Such bulky substituents are synthesized not so much to facilitate binding to the target enzymes, but to reduce binding by the hydrolytic ß-lactamases, which are discussed next.

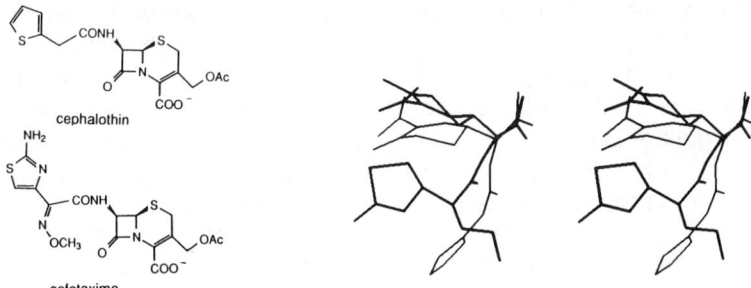

Figure 2. Observed conformations of acylated cephalothin and cefotaxime in the DD-peptidase.

ß-Lactamases These enzymes have been divided into four classes A, B, C and D depending on amino acid sequence and substrate profiles. All are serine-reactive enzymes except those in class B, which contain a catalytic zinc center. X-ray structures of representatives of all except the D class are now known.

Class A.   Members of this large class are quite well characterized, both kinetically[7,8] and structurally.[3,9-12] Differences in chromosomal or plasmid source of each one should not be allowed to obscure the stereochemical features common to all.   All are 28-30,000 Da. and have the folding shown in Figure 3, with the catalytic site centered between

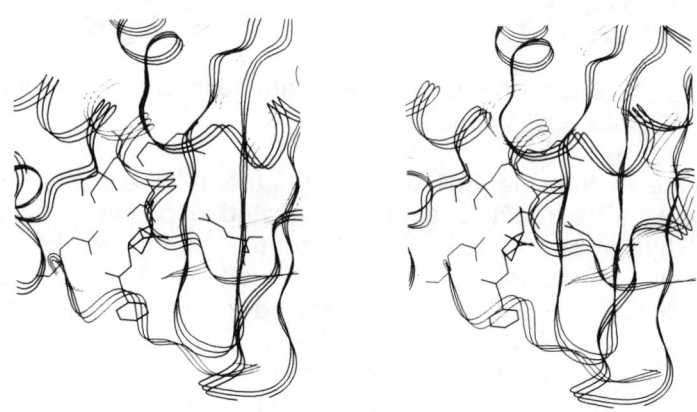

Figure 3. Binding site of a Class A ß-lactamase with penicillin G, showing Serine-70, Lysine-73, Serine-130, Glutamic-166, Lysine-234 and Arginine-244 (on right).

a helical domain on the left and an antiparallel ß-sheet on the right.   The reactive serine (number 70 in the consensus scheme of Ambler et al.[13]) is surrounded by conserved residues Lysine-73 and Glutamic-166, which are thought to deprotonate the serine for catalysis,[14,15] though the latter is 3.6 Å from the serine.   Other conserved residues, Serine-130, Lysine-234 and Arginine-244, are believed to be necessary for substrate and transition-state binding,[16-18] probably by interacting with the carboxylate of ß-lactams.   Because of the high reactivity of ß-lactamase crystals, Michaelis complexes of the enzyme with ß-lactams have not yet been mapped, so the exact steric relationship of these residues to the substrates has not been directly observed.   However, crystallographic observation of acyl-enzyme complexes of the *S.* R61 DD-peptidase mentioned above, a class C acyl complex discussed below, and modeling the binding with energy minimization and dynamics, have resulted in reasonable schemes which agree with kinetics and mutagenesis studies.   Later in this paper we will focus on two of these studies: the inactivation of Class A ß-lactamases by clavulanate-type suicide inhibitors, and second, the use of ß-lactamase mutants to understand the mechanism of the wild-type enzyme.

Class B. The 3.5 Å crystal structure of a 25,000 Da. zinc-containing ß-lactamase from *Bacillus cereus* 569/H/9 has been reported by Sutton at King's College, London.[19] It bears no resemblance to the class A enzymes. No other structures from class B have been determined. Efforts are underway to crystallize the homologous enzyme from *Bacteroides fragilis* and the tetrameric enzyme from *Pseudomonas maltophilia.* As clinical use of carbapenems increases, natural mutations are expected to develop in these class B ß-lactamases, as has been the case with the class A enzymes.

Class C. The 2 Å crystal structure of the class C ß-lactamase from *Citrobacter freundii* was reported in 1990.[20] The molecule is 35 % larger than the class A enzymes, but except for some outer segments of polypeptide, folds very much like the latter. Less complete is crystallographic analysis of the class C enzyme from *Enterobacter cloacae* P99.[6] The relatively stable acyl-complex of the *C. freundii* ß-lactamase with the monobactam aztreonam has been observed crystallographically at 2.5 Å. When this complex is compared with the map of the cefotaxime acyl-complex with the *S.* R61 DD-peptidase (see above), it appears that the rotational conformations of the acylamide substituents differ by almost 180°. Here, the amino-thiazole ring is close to the ß-strand and the larger carboxylate-bearing oxime branch is exposed.

3 INHIBITION OF ß-LACTAMASES BY CLAVULANATE

Examination of the ß-lactam binding site of the class A *Bacillus licheniformis* ß-lactamase (Figure 3) led to the suggestion[16,18] that Arginine-244 on the B4 ß-strand is not only involved in substrate binding but might also inadvertently facilitate the inactivation of ß-lactamases by clavulanate-type inhibitors. This idea has now been further explored by modeling three clavulanate reaction intermediates in the binding site of the enzyme.

Mobashery and coworkers[21] propose that the multi-step inactivation is a non-concerted proton relay process (Figure 4) involving a crystallographic water molecule bridging between Arginine-244 and the carboxylate of clavulanate. Figure 5 shows a schematic of the energy-minimized Michaelis complex modeled from the crystal structure, in which the

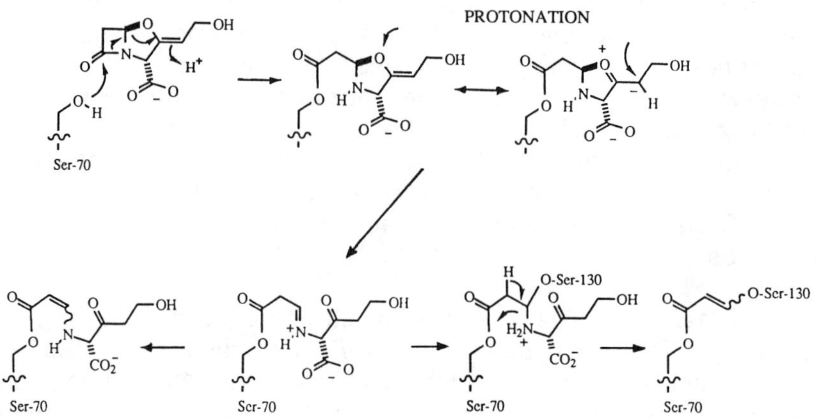

Figure 4. The scheme of Mobashery[21] for the clavulanate reaction.

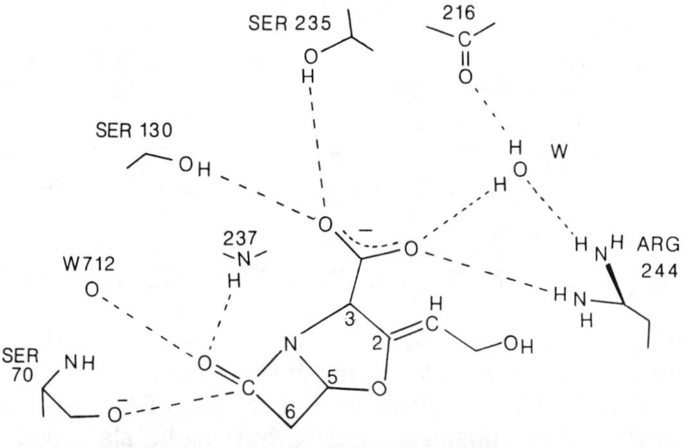

Figure 5. Interactions seen in the minimized Michaelis complex of clavulanaic acid with ß-lactamase.

alkene functionality of clavulanate lies below the water and Arginine-244. After acylation of the Serine-70 by the ß-lactam carbonyl group, the strategically-positioned water molecule (O673) could relay the arginine-derived proton via the carboxylate to the sp2 carbon atom (Figure 6). Support for the involvment of Arginine-244 in this proton-relay scheme was provided by inhibition studies with a poorly-inactivated mutant having serine rather than arginine at the 244 position.[21] It is noteworthy that Serine-130 was found to be well-placed to facilitate the transition state for departure of the incipient amine by hydrogen-bonding with the lone electron pair on the oxazolidine nitrogen, and it may play a similar role in turnover of normal substrates as well.

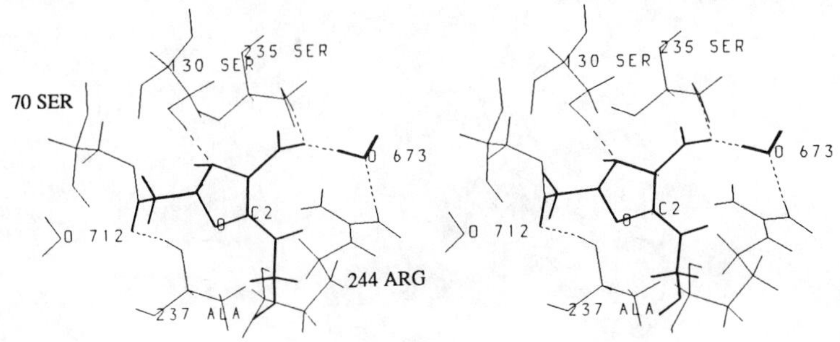

Figure 6. Acylated clavulanate in binding site near the bridging water molecule.

After opening of the oxazolidine ring, one or more protein groups are thought to add to the C5 carbon atom to complete the inactivation.[22] A search of the ß-lactamase binding site for likely nucleophiles, using dynamics simulations for 60 psec. to examine all conformations of the flexible linear intermediate, showed that the expected nucleophiles, Lysine-73 and Lysine-234, remained rather far from the C5 atom. Instead, it was found that the residue closer and better positioned for nucleophilic attack on the intermediate species is the conserved Serine-130.

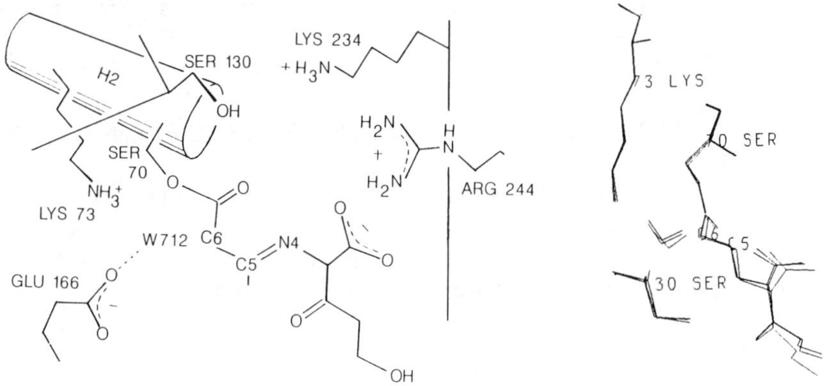

Figure 7. Relation of nucleophiles to the linear intermediate, as seen in dynamics runs (right).

Another unexpected finding from the crystal structure-based modeling is that a Glutamic-166-activated water molecule, also used in hydrolysis of substrates (see below), is well placed for deprotonation of C6 to give, by tautomerization of the linear species, the transiently-inhibited species proposed originally by Knowles.[22]

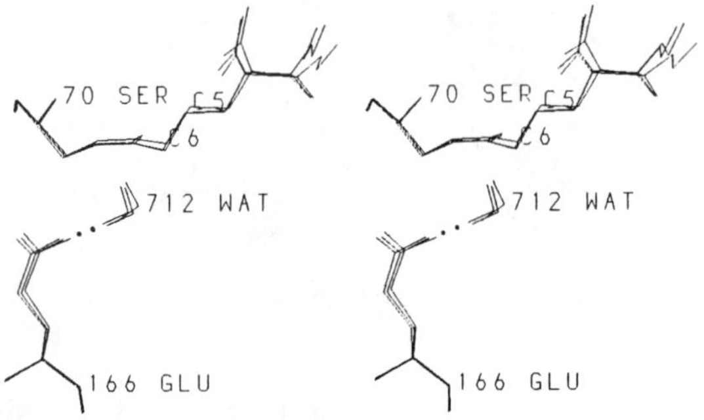

Figure 8. Dynamics results showing a route for deprotonation of C6 via an activated water molecule.

As an extension of the earlier chemical and kinetic studies of many workers, these recent mutagenesis, crystallographic and modeling studies have given a more complete picture of the mechanism of this multi-step inhibition process. The enzyme obviously did not evolve to react efficiently with such an inhibitor; rather the inhibitor happens to have an ideal chemical structure for reacting with the existing functionalities (and bound waters) in the catalytic site of this particular enzyme. Of some concern is a recent clinical report[23] of an ampicillin/clavulanate-resistant *E. coli* strain found to produce a mutant ß-lactamase with Arginine-244 replaced by serine.

4 ß-LACTAMASE MUTANTS, NATURAL AND ENGINEERED

Nature has provided increasing numbers of mutants with which one can study the finer points of ß-lactamase function. Because of heavy antibiotic use by man, the ß-lactamase enzymologist is presented with a spectrum of ß-lactamases, each of which is an effective destroyer of a particular class of ß-lactam. The number of known mutants is approaching one hundred.[24] Most of these are from clinical environments and are characterized only by MICs with a few ß-lactams, and some by mutation position in the DNA sequence. The amino acid changes in many of these mutants can now be rationalized from what we know about the architecture of the binding site.

A Structural Basis for Understanding some Natural
ß-Lactamase Mutants. The class A SHV and TEM-type ß-
lactamases are the best characterized of the plasmid-borne
clinically-derived enzymes.[24] In the early 1980's many third-
generation cephalosporins such as cefotaxime and ceftazidime
became susceptible to hydrolysis by extended-spectrum SHV
ß-lactamases. The new mutants (SHV-2,4..) differ from wild-
type (SHV-1) by one or two changes on the B3 ß-strand at
positions 238 and 240 (which are adjacent in the standard
numbering):

Table 1. Relative activity of SHV ß-lactamases[25]

Enzyme	Position			Hydrolysis of	
	69	238	240	cefotaxime	ceftazidime
SHV-1	Met	Gly	Glu	no	no
SHV-2	Met	Ser	Glu	yes	no
SHV-4	Met	Ser	Lys	yes	yes
SHV-8	Met	Gly	Lys	no	no

As the four published three-dimensional structures of class A
ß-lactamases are very similar, we will assume that the SHV
ß-lactamase and its mutants can be modeled from any of the
class A crystal structures.

Cefotaxime and ceftazidime were docked into the modeled SHV
binding site according to the pre-catalytic binding scheme
proposed for the homologous *B. licheniformis* enzyme.[16] From
crystallographic binding studies of cefotaxime bound to the *S.
R61* DD-peptidase mentioned in Section 2, we assumed that the
discriminating interactions between these ß-lactams and the
SHV enzyme would involve the R1 substituent and amino acids
at the bottom of the binding site (Figure 9).

It was concluded that the size of the amino acid at 238, rather
than its charge or hydrogen-bonding properties, is the factor
which best correlates with binding and catalytic data.[25,26]
Examination of the 3D structure indicates that a large 238
sidechain will push against the interior sidechain at position
69, with a resulting outward movement of the B3 ß-strand.

The consequences of this movement are 1) that the mutant
enzyme is better able to bind a cephalosporin with a rigid R1
sidechain, and 2) that the ß-lactam is shifted slightly away
from the reactive Serine-70. In enzymatic terms, the
Michaelis constant Km will decrease, and the reaction velocity

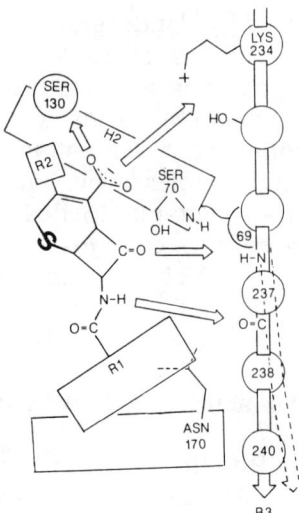

Figure 9. SHV binding site showing 238 and 240 positions on B3 ß-strand relative to position 69 on helix H2.

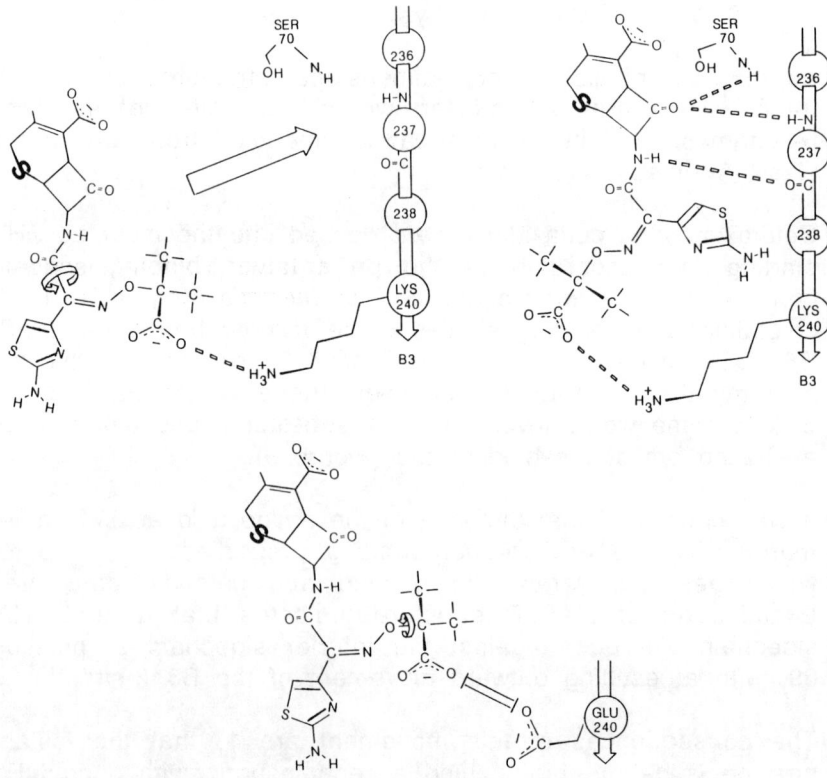

Figure 10. A lysine at 240, rather than a glutamic acid, can assist a needed rotational change in ceftazidime.

(Kcat) may decrease. SHV mutants having sidechains at 238 larger than alanine are better able to bind the newer cephalosporins, especially if the amino acid at position 69 is also large, as it often is in the TEM and SHV ß-lactamases. Therefore, though binding of ß-lactams (Km) by these new broad-spectrun mutants is markedly improved, the catalytic efficiency (kcat/Km) of the mutants is often rather poor, however, because of the low Kcat.

The amino acid at position 240 is very exposed at the bottom of the B3 ß-strand; it is always a polar or charged residue. The data show that only the SHV-2 mutant with Lysine at 240 will measureably hydrolyse ceftazidime. We propose in Figure 10 that a conformational rotation, induced by an electrostatic linkage between Lysine-240 and the oxime side chain, unblocks the acylamide of the ß-lactam to permit better hydrogen-bonding to the B3 ß-strand prior to hydrolysis. This rotation is not readily induced by a glutamic acid at position 240 because an ineffectual rotation about the ester bond is more likely.

A Catalytically-Impaired ß-Lactamase, Glu166Ala. The role of the conserved Glutamic acid 166 (Fig. 3) in ß-lactam hydrolysis by class A ß-lactamases has been under discussion for some time.[27,28] Glutamic-166 has been proposed to function as a general acid/base catalyst in both acylation and deacylation steps, first, by assisting deprotonation of serine, and then, in the deacylation step, to activate a water molecule for hydrolysis of the covalent serine intermediate.

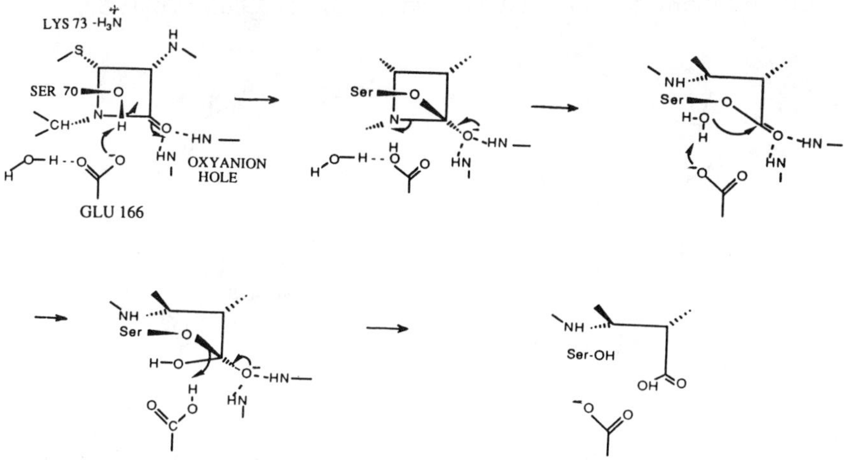

Figure 11. A ß-lactam turnover mechanism involving Glu-166, according to Pratt.[29]

However, recent work[30-32] in which site-directed mutagenesis was used to remove glutamate from the binding site
showed that the requirement for glutamate is limited to the
deacylation step, as acylation can proceed without it.

The kinetic studies of Escobar et al.[32] were definitive in
showing that the binding affinites (Km) and acylation rates
(k1) were essentially unchanged by a glutamate-to-alanine
mutation E166A, but that the rate of deacylation was reduced
by a million-fold for both penicillin and cephalosporin
substrates. They concluded that acylation and deacylation in
ß-lactamase catalysis are not mirror images, as in classical
serine proteases, and must involve different mechanisms.

The E166A mutant of the *B. licheniformis* ß-lactamase
prepared by Fink[17] was subjected to crystallographic
structure determination.[33] Since the wild-type (WT) structure
had also been determined,[11] it was possible to establish
whether the tertiary structure, especially in the penicillin
binding site, had been perturbed by the mutation, and if not, to
explain how only the deacylation step was altered. The mutant
structure was found to be quite equivalent to the WT
structure, with only 0.25 Å rms difference between backbone
atoms (Fig. 12a). Though 30-40 structural water molecules
were seen in their WT positions, two water molecules in the
binding site were significantly displaced. One moved out of
the oxyanion hole by about 1 Å, the second had been hydrogen
bonded to Glu-166, and was uniquely positioned for attack on
the serine-penicilloyl covalent bond of the intermediate (Fig.
12b). Without the carboxylic acid group of Glutamic-166,

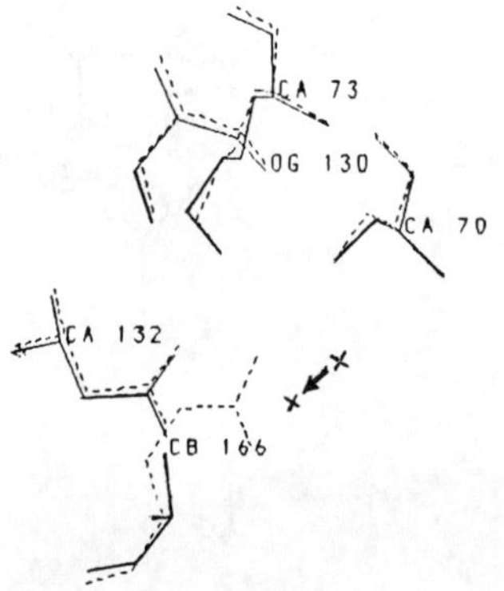

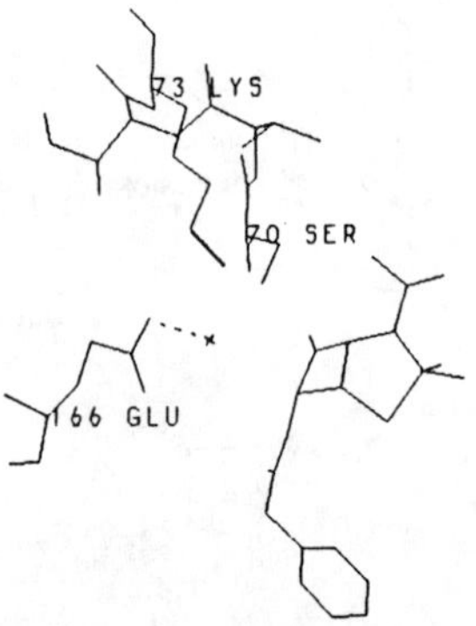

Figure 12. Left: Overlay of E166A mutant with WT (dashed).
Hydrolytic water (x) moves as shown. Right: Activation of
water molecule by Glutamic-166 in WT enzyme with Pen G.

this water shifts by 1.5 Å toward Asparagine-170, which possibly substitutes for Glutamic-166 in this role. As a result, the hydrolytic water is no longer optimally positoned or activated for nucleophilic attack on the ß-lactam intermediate. Because the water is still generally positioned on the α–face of the intermediate, acylation is observed to occur, but only at non-catalytc rates. That acylation continues at catalytic rates without Glutamic-166 supports an alternative hypothesis that the positively-charged Lysine-73, not Glutamic-166, assists the deprotonation of neighboring Serine-70 by simple charge repulsion.

<u>ß-Lactam Binding Affinities and Serine-130</u>. Levesque in Quebec has used site-directed mutagenesis of the ROB-1 plasmid ß-lactamase from *H. influenzae* to investigate a suggestion[16] that the conserved Serine-130 is important in stabilizing the active site structure of ß-lactamases by hydrogen bonding to Lysine-234 (Figure 13). His kinetic measurements of all 19 mutants at position 130, and our modeling of ampicillin and cephalexin Michaelis complexes in the hydrated binding site, revealed that Serine-130 has another role, that of discriminating between penicillin and cephalosporin-type substrates.[34] The cephalosporins as a class are much more affected by the type of amino acid in the 130 position than are the penicillins. Energy minimization placed the carboxylate group of each ß-lactam near conserved amino acids Serine-130, Lysine-234 and Arginine-244 as expected, but interestingly the carboxylate of the cephalosporin was much closer to the Serine-130 than was the carboxylate of the penicillin, which was equidistant between Serine-130 and Arginine-244 (Table 2).

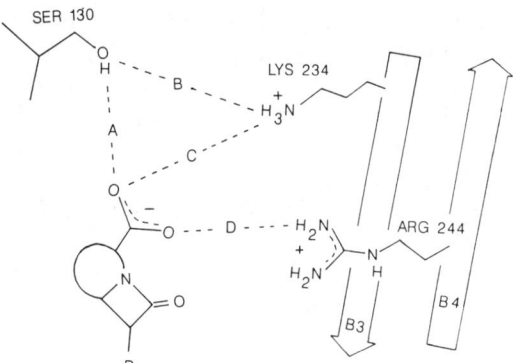

<u>Figure 13</u>. Enzyme environment around carboxylate group of ß-lactam. Distances are given below for WT and S130G.

Table 2. Calculated distances (Å) in ROB-1 complexes

	Ampicillin complex		Cephalexin complex	
Distance in Fig.	Ser130	Gly130	Ser130	Gly130
A	3.35	- -	2.75	- -
B	3.05	- -	3.10	- -
C	4.65	4.35	2.90	2.85
D	3.35	3.75	3.70	4.55

This calculation agrees with the fact that Km for binding of cephalexin is very large in all the mutants, whereas Km for ampicillin binding is reasonable even in the glycine mutant. Apparently the class A ß-lactamase evolved first as a penicillinase, and was confronted with cephalosporins at a later time. As such, it has developed a fail-safe binding site for penicillin-type ß-lactams; but for effective binding of cephalosporins, it cannot tolerate many changes in the stereochemistry of the binding site.

Acknowledgments. This work was suported by Glaxo Group Research, Hoffmann-LaRoche AG, and the Eli Lilly Company. Computations were performed at the University of Connecticut Computing Center and the Pittsburgh Supercomputing Center under grant NIH 1 P41RR04154 and NSF ASC-8500650.

REFERENCES

1. J. R. Knox and J. A. Kelly, 'Molecular Recognition: Chemical and Biochemical Problems', Royal Society of Chemistry (S. M. Roberts, ed.), London, 1989, p. 46.
2. J. A. Kelly, O. Dideberg, P. Charlier, J.-P. Wery, M. Libert, P. C. Moews, J. R. Knox, C. Duez, C. Fraipont, B. Joris, J. Dusart, J.-M. Frere, and J.-M. Ghuysen, Science, 1986, 231, 1429.
3. B. Samraoui, B. J. Sutton, R. J. Todd, P. J. Artymiuk, S. G. Waley, and D. C. Phillips, Nature (London), 320, 378.
4. J. A. Kelly, J. R. Knox, H. Zhao, J.-M. Frere, and J.-M. Ghuysen, J. Mol. Biol. 1989, 209, 281.
5. J. A. Kelly, S. G. Waley, M. Adam and J.-M. Frere, Biochim. Biophys. Acta, 1992 1119, 256.
6. H. Liu, Ph. D. thesis, University of Connecticut, 1991.
7. H. Christensen, M. T. Martin, and S. G. Waley, Biochem. J., 1990, 266, 853.
8. A. Matagne, A.-M. Misselyn-Baudin, B.Joris, T. Erpicum, B. Granier, and J.-M. Frere, Biochem. J., 1990, 265, 131.
9. O. Herzberg, J. Mol. Biol., 1991, 219, 701.

10. O. Dideberg, P. Charlier, J.-P. Wery, P. Dehottay, J. Dursart, T. Erpicum, J.-M. Frere, and J.-M. Ghuysen, Biochem. J.,1987, 245, 5797.

11. J. R. Knox and P. C. Moews, J. Mol. Biol., 1991, 220, 435.

12. C. Jelsch, F. Lenfant, J. M. Masson, and J. P. Samama, FEBS Letters 1992, 299, 135.

13. R. P. Ambler, J.-M. Frere, J.-M. Ghuysen, B. Joris, R. C. Levesque, G. Tiraby, S. G. Waley, and A. F. W. Coulson. Biochem. J. 1991, 276, 269.

14. K. Knap and R. F. Pratt, Biochem. J., 1991, 273, 85.

15. O. Herzberg and J.Moult, Science, 1987, 236, 694.

16. P. C. Moews, J. R. Knox, O. Dideberg, P. Charlier, and J.-M. Frere, Proteins: Struct. Funct. Genet. 1990, 7, 156.

17.L. M. Ellerby, W. A. Escobar, A. L. Fink, C. Mitchinson, and J. A. Wells, Biochemistry, 1990, 29, 5797.

18. G. Zafaralla, E. K. Manavathu, S. A. Lerner and S. Mobashery, Biochemistry, 1992, 31, 3847.

19. B. J. Sutton et al., Biochem. J., 1987, 248, 181.

20. C. Oefner, A. D'Arcy, J. J. Daly, K. Gubernator, R. L. Charnas, I. Heinze, C. Hubschwerfen, and F. K. Winkler, Nature (London), 1990, 343, 284.

21. U. Imtiaz, E. M. Billings, J. R. Knox, E. K. Manavathu, S. A. Lerner, and S. Mobashery, J. Amer. Chem. Soc. (submitted).

22. J. R. Knowles, Accts. Chem. Res., 1985, 18, 97.

23 A. Belaaouaj et al., 31st Interscience Conf. on Antimicrobial Agents and Chemotherapy, Abstr. 944, 1991.

24. G. A. Jacoby and A. A. Medeiros, Antimicrob. Ag. Chemother., 1991, 35, 1697.

25. A. Huletsky, J. R. Knox, and R. C. Levesque, J. Biol. Chem. (in press).

26. J. R. Knox. Fifth ß-Lactamase Workshop, Holy Island, Northumbria, UK, 1992.

27. P. J. Madgwick and S. G. Waley, Biochem. J.1987, 248, 657.

28. R. M. Gibson, H. Christensen and S. G. Waley, Biochem. J.., 1990, 272, 613.

29. R. Pratt, Wesleyan University, personal communication.

30. H. Adachi, T. Ohta and H. Matsuzawa, J. Biol. Chem., 1991, 266, 3186.

31. M. Delaire, F. Lenfant, R. Labia and J.-M. Masson, Protein Engineering, 1991, 4, 805.

32. W. A. Escobar, A. K. Tan, and A. L. Fink, Biochemistry, 1991, 30, 10,783.

33. J. R. Knox, P. C. Moews, W. A. Escobar, and A. L. Fink, Protein Engineering. (submitted).

34. J. M. Juteau, E. M. Billings, J. R. Knox and R. C. Levesque, Protein Engineering (in press).

4

Novel Tetracyclic Carbapenems: Synthesis and Biological Activity

T. Wollmann, U. Gerlach, R. Hörlein, N. Krass, R. Lattrell,
M. Limbert, and A. Markus
SBU ANTI-INFECTIVES, HOECHST AG, 6230 FRANKFURT 80, GERMANY

Within the group of carbapenem antibiotics, the 2-phenylcarbapenems (e.g. **1** and **2**) form an interesting subgroup because they show promising antibacterial activity and, in comparison to imipenem **3**, an improved chemical and dehydropeptidase stability[1-3]. Due to the conjugation of the phenyl group to the β-lactam carbonyl, the biological activity of the system can be influenced by substitution of the aromatic ring with electron-withdrawing or -donating groups.

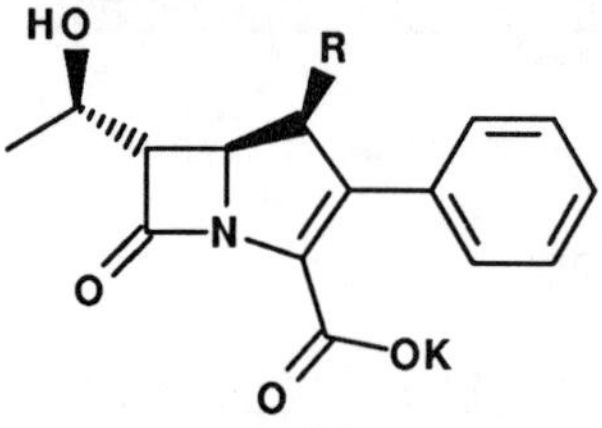

1 R = H

2 R = CH$_3$

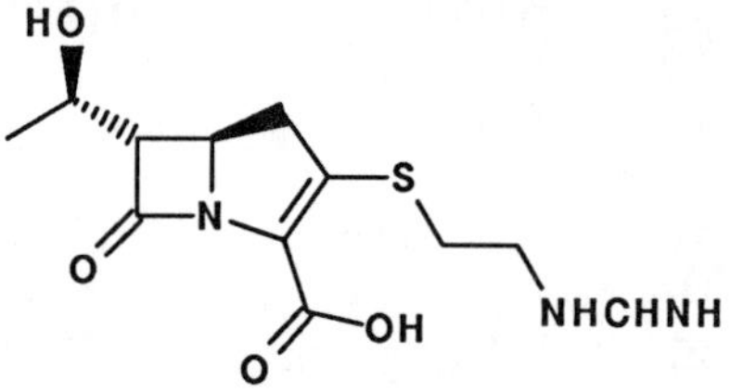

3

The 2-phenylcarbapenems **1** and **2** have been prepared by multistep synthesis (> 10 steps) using an intramolecular Wittig cyclisation for the formation of the carbapenem system[2, 3]. Recently, the introduction of phenyl in the 2-position using a Pd(0) mediated cross-coupling of an enoltriflate with phenyl trimethylstannane was published[4]. As oxalimides like **6** are accessible in two steps from the commercially available azetidinone **4**, we investigated the oxalimide cyclisation[5] as an alternative for the ring closure.

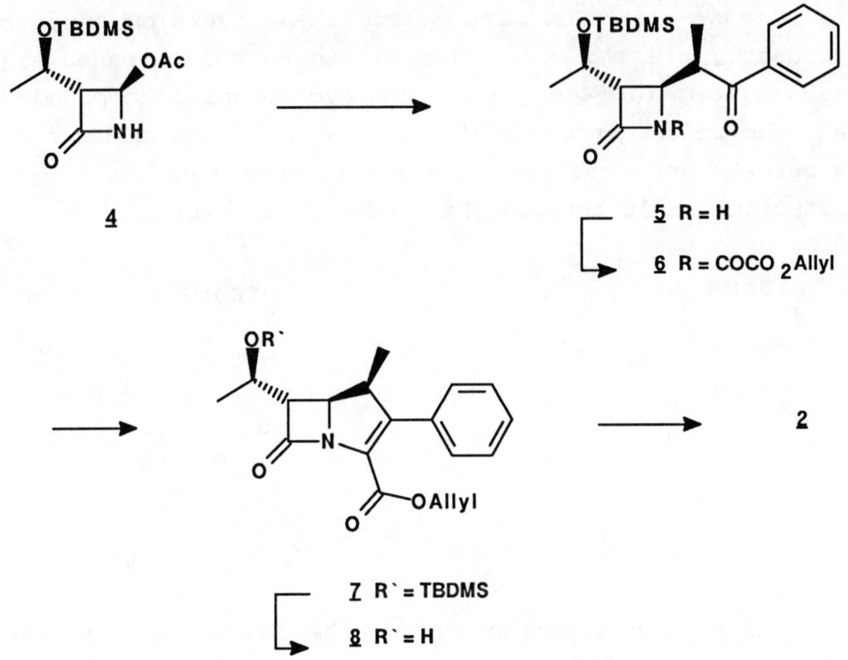

Scheme 1

The reaction of azetidinone **4** with a tin enolate, generated in situ from α-bromopropiophenone and metallic tin, gives the desired β-product **5** after crystallisation of the crude reaction mixture (α/β = 25/75) in high purity (α/β <3/97)[6] . The oxalimide **6** is obtained by acylation of **5** with allyloxalylchloride in the presence of diisopropylethylamine/calcium carbonate[5] .

First attempts to cyclise oxalimide **6** to carbapenem **7** with trimethyl or triethyl phosphite, the standard reagents for the oxalimide cyclisation to penems, in refluxing xylene for up to 24h led to decomposition of the starting material.

As an alternative for alkyl phosphites we next used methylphosphonous acid diethylester for the cyclisation. This compound is a highly active and mild cyclisation reagent and has been successfully used for the synthesis of penems[7] . In refluxing xylene the conversion of **6** to product **7** goes to completion within 6h and carbapenem **7** is isolated in 54 % yield. The cyclopropane compound **9** and the reduced product **10** are formed as byproducts from a carbene intermediate. At lower temperatures (T< 100°C) reduction to **10** becomes the predominant reaction.

   9   **10**

Subsequent deprotection of the silylether and the allylester[3] afforded the carbapenem **2**. Using the oxalimide cyclisation, 2-phenylcarbapenems like **1** and **2** now are accessible in only 5 steps from the commercially available azetidinone **4**.

To study the influence of substituents other than the methyl group in position 1 on antibacterial activity and dehydropeptidase stability, the carbapenems **1** and **11** - **14** were prepared using the oxalimide cyclisation route. The carbapenem **1** exhibits the highest antibacterial activity (table 1). As described by the Merck group[3], introduction of the β-methyl group slightly reduces the activity, but does not improve the dehydropeptidase stability. This is in contrast to thienamycin[8] and its β-methyl derivative, where the β-methyl group significantly improves the dehydropeptidase

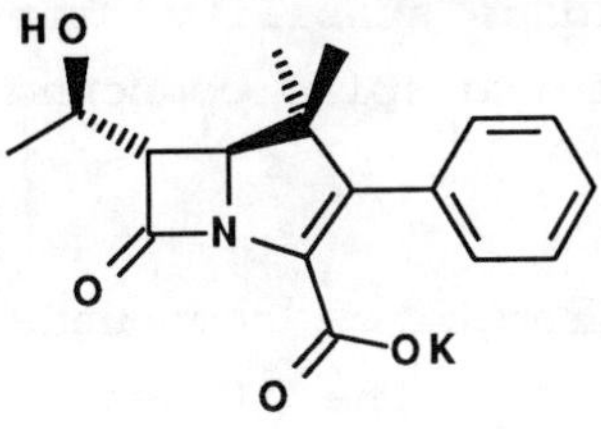

11

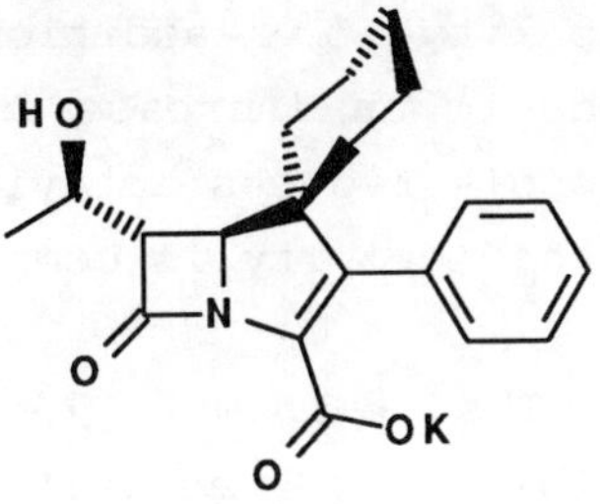

12

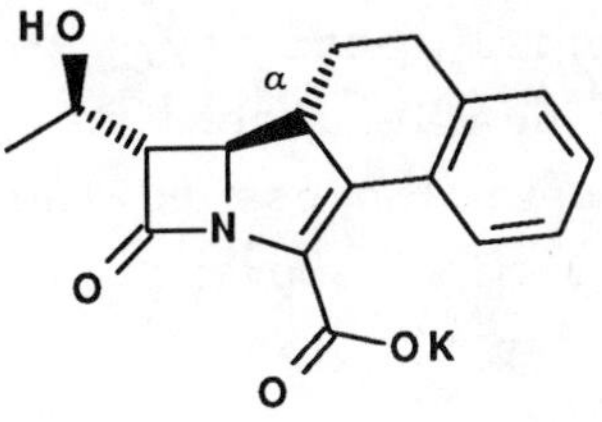

13

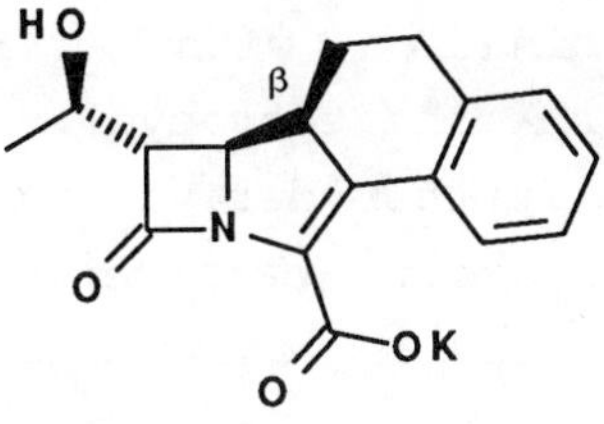

14

Table 1 Antibacterial activity and DHP-stability

Compound	1	2	11	12	13	14
S.a. SG 511[a]	0.025	0.098	1.56	12.5	0.391	0.098
S.a. 503	0.025	0.195	1.56	12.5	0.391	0.098
S.p. 77 A	0.002	0.004	0.195	12.5	0.195	0.004
S.f. D	0.391	3.13	25	12.5	25	1.56
E.c. O 78	0.049	0.391	12.5	>100	6.25	0.391
E.c. TEM	0.098	0.781	100	>100	50	1.56
E.c. 1507 E	0.098	0.781	100	>100	50	1.56
K.a. 1082 E	0.195	1.56	3.13	>100	12.5	0.781
K.a. 1522 E	0.098	0.781	100	>100	25	0.781
E.cl. P 99	3.13	3.13	>100	>100	100	6.25
DHP[b]	6	5	27	>33	>22	37

[a] MIC (µg/ml); agar dilution test; Mueller-Hinton agar; inoculum 5×10^4 cfu/spot; S.a.: Staphylococcus aureus, S.p.: Streptococcus pyogenes, S.f.: Streptococcus faecium, E.c.: Escherichia coli, K.a.: Klebsiella aerogenes, E.cl.: Enterobacter cloacae. [b] Dehydropeptidase stability is given relative to imipenem; enzyme from porcine kidney.

stability. The additional methyl group in compound **11**
leads to an increase in dehydropeptidase stability, but
strongly reduces activity. In case of the spiro compound
12 the activity is completely lost.

The β-isomer **14** of the tetracyclic compounds
exhibits an activity comparable to the β-methyl
derivative **2**, but a very high dehydropeptidase stability. The α-diastereomer **13** is less active than **14**.
Because of the high dehydropeptidase stability in
combination with a good antibacterial activity, we
chose the tetracyclic system as our new lead structure
for further derivatisation. The synthetic access to the
tetracyclic carbapenems via our oxalimide approach is
depicted in scheme 2.

As the β-isomer **14** exhibits higher antibacterial
activity than the α-isomer **13**, the introduction of β-
configuration for the stereocenter in position 1 of the
carbapenem system by alkylation of 4-acetoxyazetidinone
4 with a tetralone enolate is the key step of the
reaction sequence. The reaction of 2-bromotetralone with
azetidinone **4** via tin enolate following Deziel's
procedure[6] (tin, iodine, silver tetrafluoroborate,
dimethylformamide/dichloromethane (2/1)) afforded a
mixture of diastereomers, with α-isomer **15** as the main
product (**15**/**16** = 2/1). When the reaction temperature is
strictly kept between 8 and 12°C, a slight excess of the
desired β-isomer **16** is obained (**15**/**16** = 40/60). This
mixture of isomers is obtained in 87 % yield. When a tin
enolate of tetralone is generated with tin(II) triflate
following Mukaiyama's procedure[9] , the α-isomer **15** is
isolated with 56 % yield in high purity (**15**/**16** = 97/3).
The α- and β-configuration of the stereocenter in
position 2 of tetralone was assigned after cyclisation
to the carbapenem using NOE-experiments. As the
separation of the diastereomers **15** and **16** by
chromatography turned out to be difficult and the
reaction failed completely in case of some substituted

tetralones (e.g. 2-bromo-7-methoxytetralone), alternatives for the alkylation were examined.

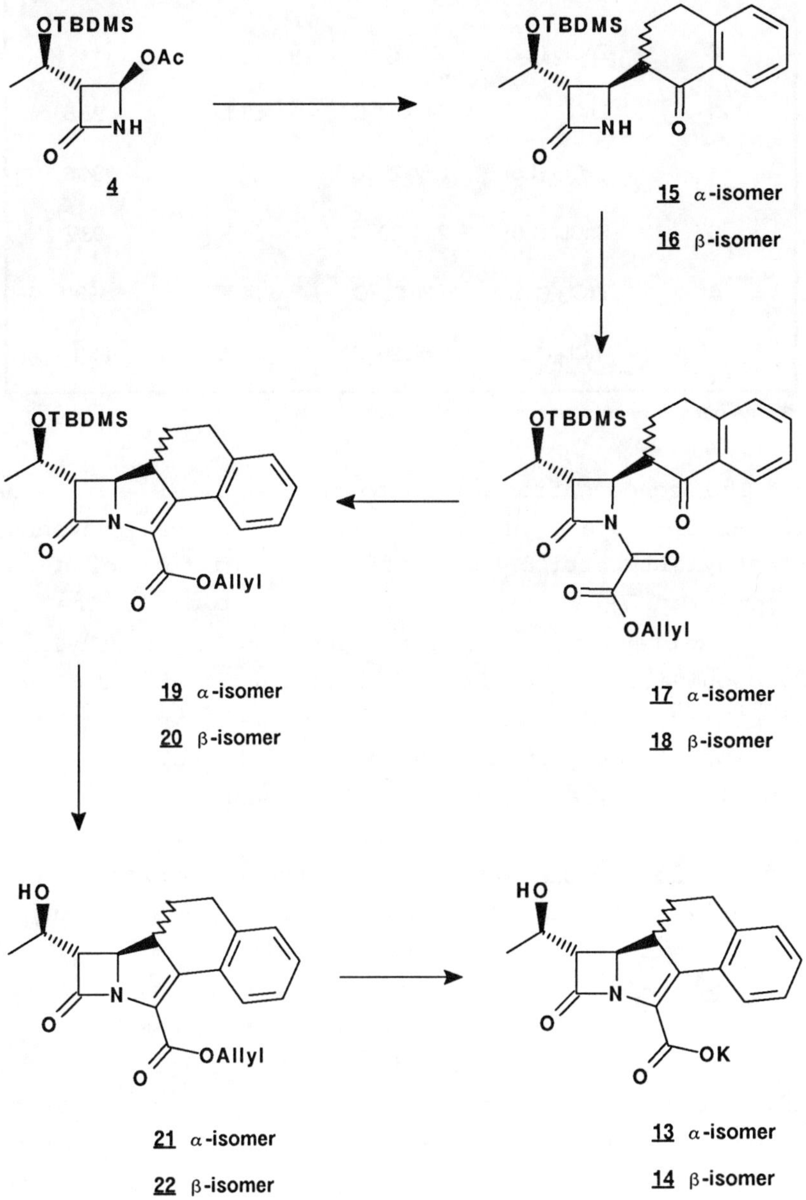

15 α-isomer
16 β-isomer

19 α-isomer
20 β-isomer

17 α-isomer
18 β-isomer

21 α-isomer
22 β-isomer

13 α-isomer
14 β-isomer

Scheme 2

Table 2 Alkylation of **4** with tetralone silyl enol ether

Entry	Lewis acid[a]	Conditions	Yield	15/16 α/β
1	TMSOTf (0.2)	4h/20°C	79 %	50/50
2	$ZnCl_2$ (0.2)	24h/20°C	<5 %	-
3	$ZnCl_2$ (1.5)	24 h/20°C	86 %	95/5
4	$SnCl_2$ (1.5)	2.5h/20°C	57 %	94/6
5	$SnCl_4$ (1.5)	2.5h/20°C	54 %	93/7
6	$ZrCl_4$ (1.5)	2h/20°C	14 %	>99/1
7	$TiCl_4$ (1.5)	2.5h/20°C	24 %	83/17

[a] The number of equivalents is put in brackets.

The condensation of a tetralone silyl enol ether with azetidinone **4** in the presence of a catalytic amount trimethylsilyl triflate[10] afforded a mixture of α- and β-isomers in 79 % yield (entry 1 in table 2). The enol ether is prepared in situ by addition of triethylamine (2.5 equivalents) and trimethylsilyl triflate (2.6 equivalents) to a mixture of azetidinone **4** and tetralone (1.3 equivalents) in dichloromethane at 0°C. Under these conditions silylation of the azetidinone nitrogen occurs and, therefore, after aqueous work-up a mixture of products **15** and **16** together with their N-trimethylsilyl derivatives are isolated. The N-trimethylsilyl group is cleaved with diluted aqueous hydrochloric acid in ethylacetate/dichloromethane. With 2.4 equivalents tri-methylsilyl triflate no product is formed as the reaction stops after formation of the silyl enol ether and, therefore, the influence of other Lewis acids can be studied. The results are shown in table 2. With a catalytic amount of Lewis acid the reaction is very slow and only traces of the product can be detected. In all cases the α-diastereomer **15** is formed as the main product.

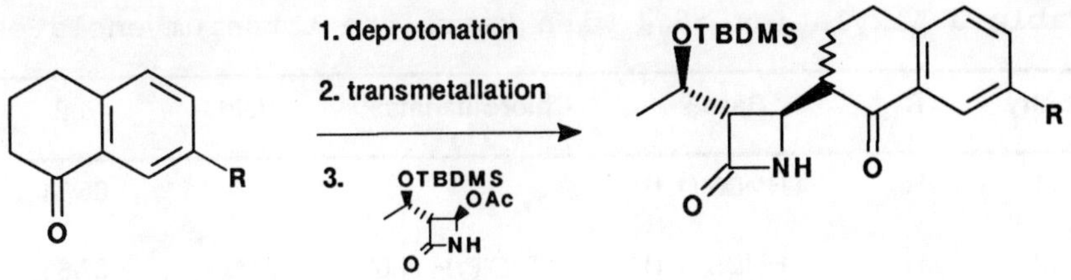

23 R = OCH$_3$

24 R = CO$_2$CH$_3$

4

25 R = OCH$_3$, β-isomer

26 R = CO$_2$CH$_3$, β-isomer

Scheme 3

As the geometry of the enolate in kinetically controlled aldol reactions determines the syn/anti-selectivity[11] , apparently the fixed (E)-geometry of tetralone enolates favours the formation of the α-product **15**.

Reetz[12] showed, that a titanium enolate of cyclohexanone, in contrast to other enolates (e.g. tin, boron), reacts with aldehydes syn-selectively. This prompted us to study these enolates in more detail. The results are summarised in table 3. After deprotonation of tetralone with LHMDS (10 min - 78°C, then for 30 min 0°C) in tetrahydrofuran and transmetallation with chlorotitanium triisopropoxide (1 M solution in hexane, - 78°C for 70 min), the azetidinone **4** is added and the reaction is warmed up to 0°C. After 30 minutes at this temperature the azetidinone **4** is consumed and the reaction is quenched as longer reaction times lead to a lower stereoselectivity. The lithium enolate (entry 1) predominantly forms the α-isomer **15**. The highest β-selectivity is observed when a slight excess of chlorotitanium triisopropoxide in relation to the base is used in the transmetallation step. Replacement of the triisopropoxide by dichlorotitanium diisopropoxide gives the undesired diastereomer **15** as the main product. Although the stereoselectivity now is satisfactory, the

Table 3 Alkylation of **4** with tetralone titanium enolates

Entry	R	Base[a]	Chlorotitanate[a]	Yield	α/β
1	H	LHMDS (1.1)	-	20 %	66/34
2	H	LHMDS (1.1)	$ClTi(O^iPr)_3$ (1.0)	30 %	37/63
3	H	LHMDS (1.1)	$ClTi(O^iPr)_3$ (1.1)	27 %	25/75
4	H	LHMDS (1.1)	$ClTi(O^iPr)_3$ (1.2)	26 %	23/77
5	H	LHMDS (1.1)	$ClTi(O^iPr)_3$ (2.7)	26 %	42/58
6	H	LHMDS (1.1)	$Cl_2Ti(O^iPr)_2$ (1.2)	23 %	62/38
7	H	LDA (1.1)	$ClTi(O^iPr)_3$ (1.2)	56 %	22/78
8	OCH_3	LDA (1.1)	$ClTi(O^iPr)_3$ (1.2)	51 %	23/77
9	CO_2CH_3	LDA (1.1)	$ClTi(O^iPr)_3$ (1.2)	30 %	18/82

[a] The number of equivalents is put in brackets.

yields are low. When LDA instead of LHMDS is used in the deprotonation step, reasonable yields are obtained. Comparable selectivities are found for substituted tetralones like **25** and **26** (entry 9 and 10), too. The β-isomers now can be obtained in high purity (α/β < 5/95) after flash chromatography and recrystallisation.

Acylation of **16** to the oxalimide **18** using standard conditions (allyloxalylchloride, Hünig base, calcium carbonate in dichloromethane, 2h, 0°C) gives a crude product which is purified by recrystallisation. As product **18** exhibits only limited stability to silica gel, only small amounts can be purified by flash chromatography. The crude product contains an impurity, probably **27**[13] , which forms the pyranone **28** during the oxalimide cyclisation.

The cyclisation of oxalimide **18** with methyl-phosphonous acid diethylester (3 equivalents) in refluxing xylene for 4h affords the carbapenem **20** in 52 % yield. Under these conditions the α-isomer **17** cyclises to α-product **19** in 6 % yield.

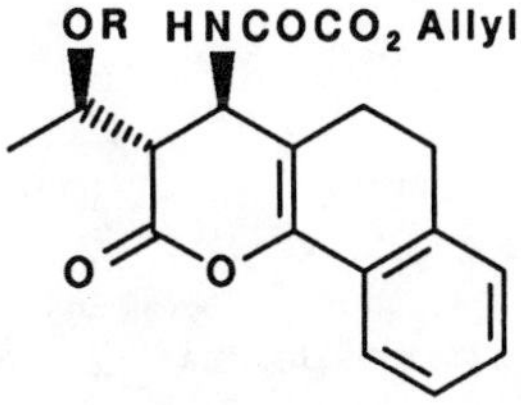

27 R = TBDMS

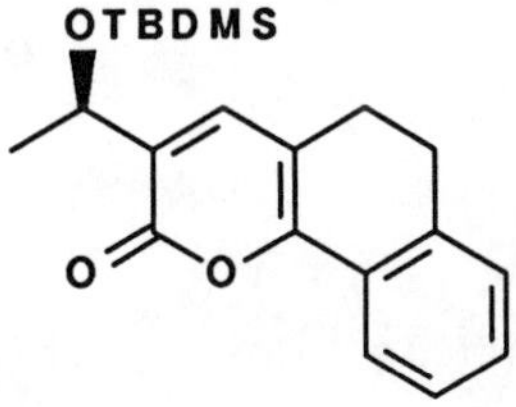

28

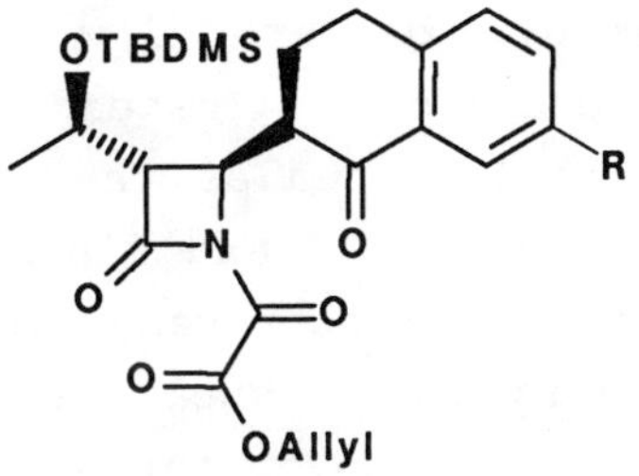

29 R = F

30 R = CO$_2$CH$_3$

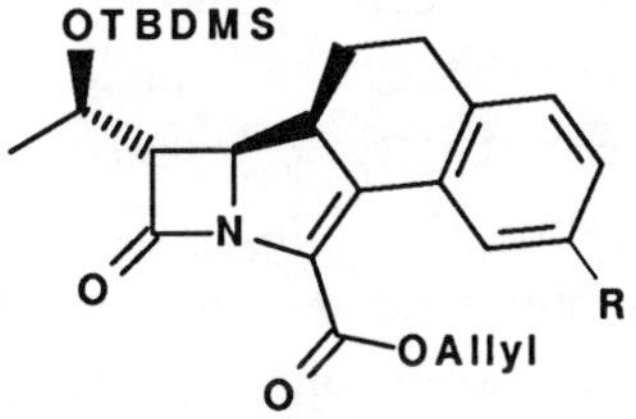

31 R = F

32 R = CO$_2$CH$_3$

In refluxing mesitylene (3h) the yield could be increased to 75 %. As shown in table 4, electron-withdrawing groups like the fluorine and the ester group

Table 4 Influence of substituents on oxalimide cyclisation

Entry	R	Educt	Product	Conditions	Yield
1	H	**18**	**20**	3h/160°C	75 %
2	F	**29**	**31**	0.5h/140°C	44 %
3	CO$_2$C$_6$F$_5$	**30**	**32**	0.2h/140°C	46 %

in **29** and **30**, respectively, significantly increase the cyclisation rate.

While desilylation of 2-phenylcarbapenem **7** was accomplished without problems [3] (tetrahydrofuran, 15 equivalents acetic acid, 5 equivalents tetra-n-butyl-ammonium fluoride as 1M solution in tetrahydrofuran), yields for the deprotection of **20** are low and irreproducible under these conditions. Instead of alcohol **22**, the enol actate **33** is isolated as the main product[14] . When the excess of acetic acid and fluoride is reduced to 5 and 3 equivalents, respectively, and, most importantly, the trihydrate is used instead of the commercially available solution in tetrahydrofuran, the alcohol **22** is reproducibly isolated in 45 - 50% yield. Subsequent deprotection of the allylester with potassium 2-ethylhexanoate/ Pd(0)[15] gives the potassium salt **14**.

Following the oxalimide cyclisation route, the tetracyclic carbapenems **34** - **44** were prepared starting from chromanone, thiochromanone and substituted tetra-lones. The antibacterial activities and dehydropeptidase stabilities are summarised in tables 5 and 6. All derivatives have retained their high stability to renal dehydropeptidase. Against Gram-positive bacteria all compounds show comparable, good activities. When the lipophilic character of the substituents is increased, the activity against Gram-negative bacteria sig-nificantly is diminished as for the dimethyl derivative **44**. Compound **34**, possessing two additional methyl groups in the benzylic position, is devoid of any antibacterial activity. The thia- and oxa-analogues **35** and **36** possess activities comparable to their carba-analogues **13** and **14**, respectively. The fluoro derivative **38** exhibits the highest overall activity. For none of the compounds was activity against Pseudomonas aeruginosa found.

In case of imipenem **3**, a basic group in the 2-position improves penetration of the outer membrane and, therefore, is responsible for the high activity against

34 - 36   **37 - 44**

Table 5 Antibacterial activity and DHP-stability

Compound	13	14	34	35	36
X	CH$_2$	CH$_2$	C(CH$_3$)$_2$	S	O
Stereoisomer	α	β	β	β	α
S.a. SG 511[a]	0.391	0.098	>100	0.049	0.391
S.a. 503	0.391	0.098	>100	0.049	0.391
S.p. 77 A	0.195	0.004	>100	<0.002	0.391
S.f. D	25	1.56	>100	0.781	50
E.c. O 78	6.25	0.391	>100	0.391	25
E.c. TEM	50	1.56	>100	1.56	100
E.c. 1507 E	50	1.56	>100	1.56	100
K.a. 1082 E	12.5	0.781	>100	0.781	12.5
K.a. 1522 E	25	0.781	>100	1.56	50
E.cl. P 99	100	6.25	>100	6.25	>100
DHP[b]	>22	37	-	>24	-

[a] MIC (µg/ml), [b] Dehydropeptidase stability is given relative to imipenem; enzyme from porcine kidney.

Table 6 Antibacterial activity and DHP-stability

Compound	14	37	38	39	40	41	42	43	44
R	H	F	H	H	H	H	H	H	CH$_3$
R'	H	H	H	H	CO$_2$CH$_3$	H	OCH$_3$	H	H
R''	H	H	F	Cl	H	CO$_2$CH$_3$	H	OCH$_3$	CH$_3$
S.a. SG 511a)	0.098	0.049	0.013	0.098	0.025	0.049	0.098	0.195	0.013
S.a. 503	0.098	0.049	0.013	0.195	0.025	0.049	0.098	0.195	0.049
S.p. 77 A	0.004	0.004	<0.002	0.007	<0.002	<0.002	<0.002	0.004	<0.002
S.f. D	1.56	0.781	0.391	6.25	0.781	0.781	1.56	3.13	0.781
E.c. O 78	0.391	1.56	0.195	6.25	1.56	1.56	0.781	3.13	12.5
E.c. TEM	1.56	6.25	0.781	12.5	6.25	3.13	3.13	6.25	25
E.c. 1507 E	1.56	3.13	0.391	12.5	6.25	3.13	3.13	6.25	12.5
K.a. 1082 E	0.781	3.13	1.56	50	12.5	0.781	1.56	3.13	6.25
K.a. 1522 E	0.781	6.25	0.781	25	6.25	1.56	1.56	12.5	50
E.cl. P 99	6.25	3.13	6.25	50	25	25	12.5	25	25
DHPb)	37/-	-/>23	>17/-		-/5	-/13	-/>36	>40/11	>60/5

a) MIC (µg/ml), b) Dehydropeptidase stability is given relative to imipenem; enzyme from porcine kidney/ human kidney.

Gram-negative pathogens, especially Pseudomonas aeruginosa. Thus, we prepared the N-methyl piperazine derivative **50** and the ammonium salt **51**. To differentiate between the influence of steric effects and lipophilicity on one side and effects from the amino group on the other side, we additionally prepared the morpholine derivative **49**.

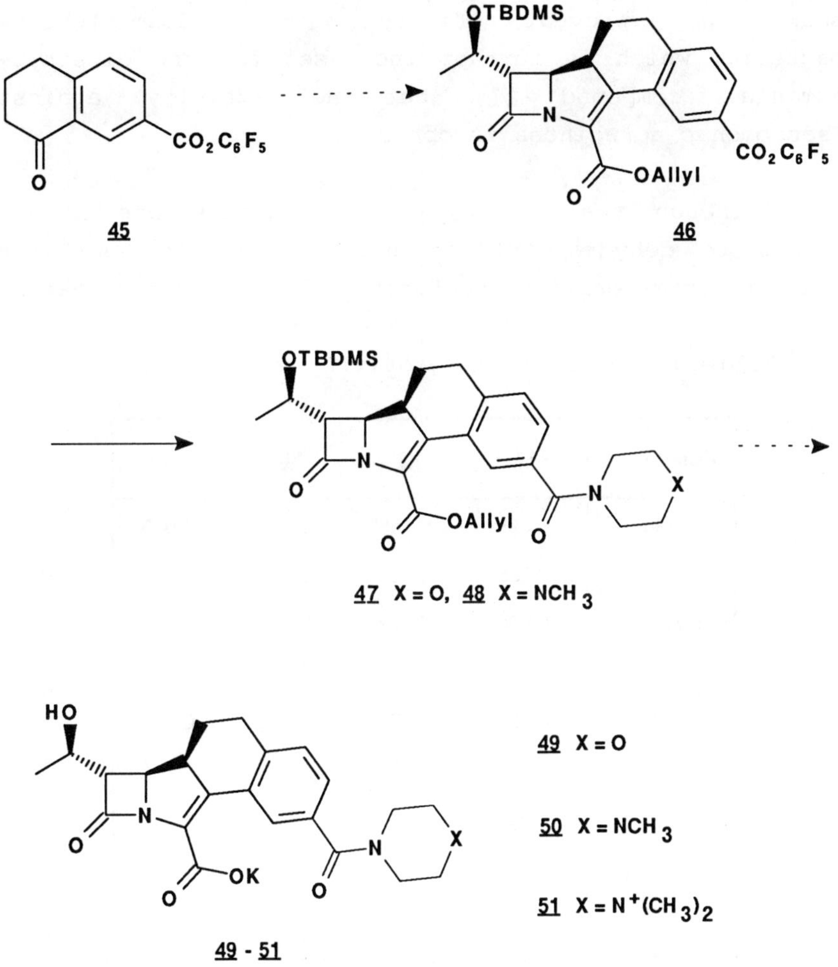
OTBDMS
$CO_2C_6F_5$
OAllyl
45
46
47 X = O, 48 X = NCH_3
HO
OK
49 - 51
49 X = O
50 X = NCH_3
51 X = $N^+(CH_3)_2$

Scheme 4

The amides are prepared from the tetralone derivative **45** as shown in scheme 4. The reaction of pentafluorophenylester **46** with morpholine or N-methyl piperazine (3 equivalents) in dimethylformamide leads to the amides **47** and **48** in 80 - 85 % yield. Cleavage of the protecting groups, as already described, yields the products **49** - **50**. Quaternisation is achieved with methyliodide in dichloromethane after deprotection of the silyl ether. The biological activities are summarised in table 7. Introduction of the morpholine amide leads to a diminished overall activity. Amine **50** shows an improved activity against Gram-negative bacteria, which is further increased for the dimethyl-ammonium compound **51**, but no activity against Pseudomonas aeruginosa is observed.

Although the tetracyclic carbapenems are highly stable to dehydropeptidase and exhibit high activity against Gram-positive bacteria, the activity against

Table 7 Antibacterial activity and DHP-stability

Compound	14	49	50	51
S.a. SG 511[a]	0.098	0.195	0.195	0.049
S.a. 503	0.098	0.195	0.391	0.391
S.p. 77 A	0.004	<0.002	0.025	0.013
S.f. D	1.56	1.56	6.25	6.25
E.c. O 78	0.391	3.13	0.781	0.781
E.c. TEM	1.56	6.25	1.56	0.781
E.c. 1507 E	1.56	6.25	1.56	1.56
K.a. 1082 E	0.781	12.5	3.13	0.781
K.a. 1522 E	0.781	6.25	0.781	1.56
E.cl. P 99	6.25	50	3.13	1.56
DHP[b]	37	-	35	-

a) MIC (µg/ml), b) Dehydropeptidase stability is given relative to imipenem; enzyme from porcine kidney.

Gram-negative pathogens, especially Pseudomonas aeruginosa, needs further improvement.

Acknowledgements

We gratefully thank our coworkers for their
competent technical assistance, Dr. H.-J. Kleiner for
the generous supply with methylphosphonous acid
diethylester, Dr. D. Isert and Dr. N. Klesel for
additional biological testing, and Dr. W. Dürckheimer
and Prof. G. Seibert for stimulating discussions.

References and Notes

1. L.D. Cama and B.G. Christensen, Tetrahedron
    Letters, 1980, 21, 2013.

2. L.D. Cama, K.J. Wildonger, R. Guthikonda, R.W.
    Ratcliffe, and B.G. Christensen, Tetrahedron, 1983,
    39, 2531.

3. R.N. Guthikonda, L.D. Cama, M. Quesada, M.F. Woods,
    T.N. Salzmann, and B.G. Christensen, J. Med. Chem.,
    1987, 30, 871.

4. T.A. Rano, M.L. Greenlee, and F.P. DiNinno,
    Tetrahedron Letters, 1990, 31, 2853.

5. A. Afonso, F. Hon, J. Weinstein, and A.K. Ganguly,
    J. Am. Chem. Soc., 1982, 104, 6138.

6. R. Deziel and M. Endo, Tetrahedron Letters, 1988,
    29, 61.

7. K.H. Budt, W. Dürckheimer, G. Fischer, R. Hörlein,
    R. Kirrstetter, and R. Lattrell, European Patent
    Application 399.228; K.H. Budt, G. Fischer, R.
    Hörlein, R. Kirrstetter, and R. Lattrell, Tetra-
    hedron Letters, submitted

8. A. Andrus, F. Baker, F.A. Bouffard, L.D. Cama, B.G.
    Christensen, R.N. Guthikonda, J.V. Heck, D.B.R.
    Johnston, W. J. Leanza, R.W. Ratcliffe, T.N.
    Salzmann, S.M. Schmitt, D.H. Shih, N.V. Shah, K.J.
    Wildonger, and R.R. Wilkening, "Recent Advances in
    the Chemistry of β-Lactam Antibiotics", The 'Royal
    Society of Chemistry, London, Special Publication
    No 52, 1985, p. 86.

9. T. Mukaiyama, R.W. Stevens, and N. Iwasawa,
    Chemistry Letters, 1982, 353.

10.   A.G.M. Barrett and P. Quayle, J. Chem. Soc., Chem. Commun., 1981, 1076.

11.   C.H. Heathcock, " Asymmetric Synthesis", Academic Press, New York, 1984, Vol.3, Chapter 2, p. 111.

12.   M.T. Reetz, R. Peter, Tetrahedron Letters, 1981, 22, 4691.

13.   For a related intramolecular opening of an acylated azetidinone with an alcohol see: E.J. Thomas and A.C. Williams, J. Chem. Soc., Chem. Commun., 1987, 992.

14.   For a related retro Dieckmann reaction of 3-oxo-carbapenam esters see: J.G. Vries, G. Hauser, and G. Sigmund, Heterocycles, 1985, 23, 1081.

15.   P.D. Jeffrey and S.W. McCombie, J. Org. Chem., 1982, 56, 587.

5

Stereospecific Modifications of Erythromycin Derivatives

Herbert A. Kirst, James P. Leeds, Jonathan W. Paschal, and Jacek Martynow

LILLY RESEARCH LABORATORIES, ELI LILLY AND COMPANY, LILLY CORPORATE CENTER, INDIANAPOLIS, IN 46285, USA

1 INTRODUCTION

A variety of structural modifications of erythromycin have now been identified which improve the antimicrobial spectrum and/or pharmacokinetics of this older but still widely used macrolide antibiotic.[1-3] A common theme of these modifications can be formally regarded as an inhibition of the well known, acid-catalyzed intramolecular cyclization of erythromycin (1) to its 8,9-anhydro-6,9-hemiketal (2) (Figure 1); the latter compound has substantially diminished antimicrobial activity compared to erythromycin.[4] Inhibition of this cyclization reaction leads to greater stability of the molecule and, at least in part, to enhanced antimicrobial action. New modifications which minimize or completely prevent the intramolecular cyclization of erythromycin are still being sought as a means to the discovery of new macrolide antibiotics.

In contrast to their antimicrobial activity, bicyclic compound (2) and some of its derivatives have shown potential utility as gastrointestinal prokinetic agents for the treatment of certain hypomotility disorders of the gastrointestinal tract.[5,6] For this non-antimicrobial application, intramolecular cyclization has been shown to increase biological activity by approximately an order of magnitude.[7]

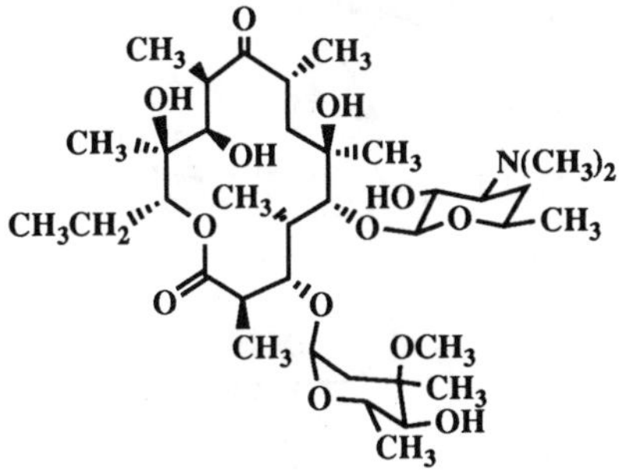

(1)	(2)
Erythromycin	8,9-Anhydroerythromycin-6,9-hemiketal

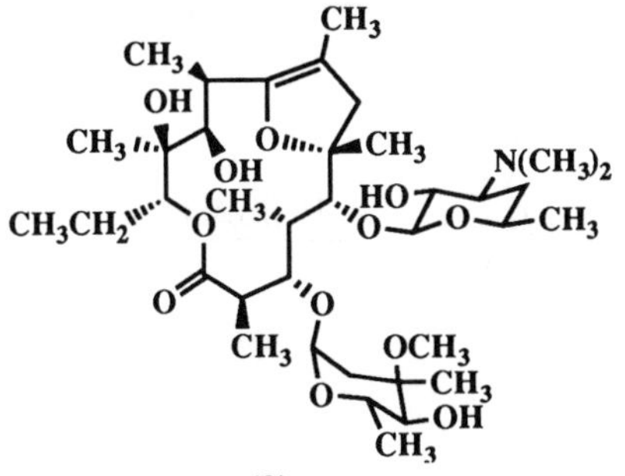

Figure 1 Erythromycin and its intramolecular cyclization product

While pursuing structure-activity investigations for both antimicrobial and non-antimicrobial applications, some stereo- and regiospecific reactions involving the double bond of 8,9-anhydroerythromycin-6,9-hemiketal and its ring-contracted derivative (Figure 2) were observed in which the outcome unexpectedly differed between the 12- and 14-membered ring macrolides. The ring-contracted bicyclic derivative LY267108 (3) had previously been prepared either directly from erythromycin or by base-catalyzed translactonization of 8,9-anhydroerythromycin-6,9-hemiketal.[8,9]

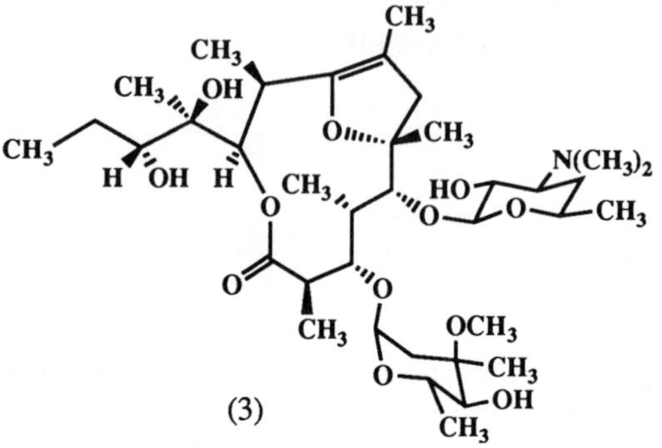

Figure 2   Ring-contracted 8,9-anhydroerythromycin-6,9-hemiketal (LY267108)

2   SYNTHESIS OF 8,9-DIHYDRO DERIVATIVES

Hydrogenation of 8,9-anhydroerythromycin-6,9-hemiketal (2) was conducted under previously described conditions using PtO_2 as catalyst and acetic acid-trifluoroacetic acid as solvent.[10] The structure of the hydrogenated product was determined to be the 8,9-*cis*-α-dihydro derivative (4, Figure 3) on the basis of detailed NMR studies. The stereochemistry that we assigned to this compound proved to be opposite what had previously been depicted for this hydrogenation product.[11]

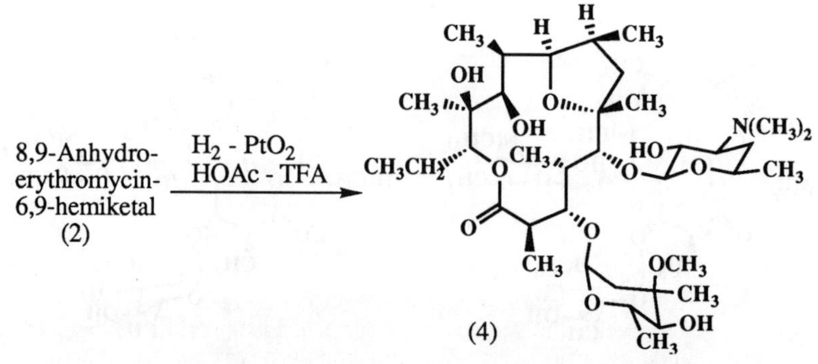

Figure 3   Synthesis of 8,9-*cis*-α-dihydro-8,9-anhydroerythromycin-6,9-hemiketal

In contrast to the hydrogenation of (2), reduction of the ring-contracted LY267108 (3) under identical conditions proceeded by addition of hydrogen to the opposite face of the dihydrofuran ring system, yielding the 8,9-*cis*-β-dihydro derivative (5, Figure 4).

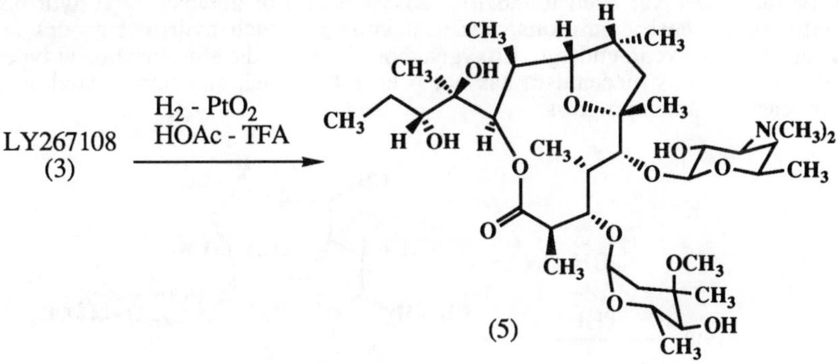

Figure 4 Synthesis of 8,9-*cis*-β-dihydro-LY267108

The addition of hydrogen from opposite faces of (2) and (3) was confirmed by translactonization of 8,9-*cis*-α-dihydro product (4) into the corresponding 8,9-*cis*-α-dihydro-LY267108 (6, Figure 5), which was different from (5), the direct hydrogenation product of LY267108.

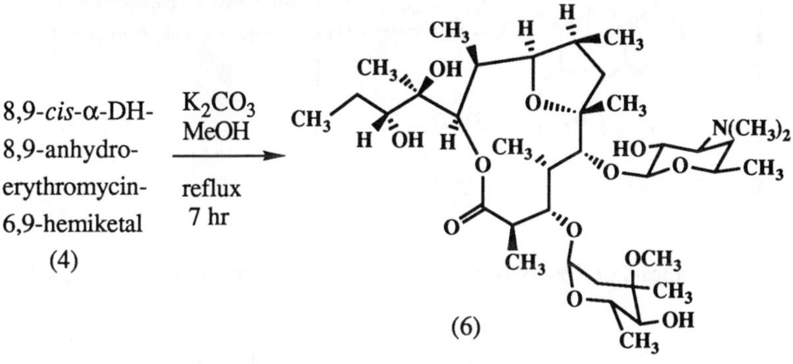

Figure 5 Synthesis of 8,9-*cis*-α-dihydro-LY267108

The final possible *cis*-dihydro compound within this series, 8,9-*cis*-β-dihydro-8,9-anhydroerythromycin-6,9-hemiketal (7, Figure 6), was obtained by treatment of erythromycin with sodium cyanoborohydride and zinc iodide in THF. These reaction conditions have previously been employed for deoxygenation of aryl ketones, benzylic and allylic alcohols, and tertiary alcohols.[12] It had been further reported in this same study that certain aryl enol ethers were stable to reduction under these conditions.[12] These reagents have also been used for dehalogenation of benzylic, allylic, and tertiary halides and reduction of enamines.[13,14]

A possible intermediate in the reduction is erythromycin-6,9-hemiketal (8, Figure 6), which could arise by Lewis acid-catalyzed cyclization and might be present as a complex with zinc and/or borohydride species. This proposed intermediate would retain the stereochemistry of the methyl group at C-8, which has been determined to be the same as that of erythromycin. However, no precedent in the literature has yet been found for deoxygenation of glycosyl-type hydroxyl groups under these conditions, nor is it known if such hydroxyl groups are sufficiently reactive to undergo deoxygenation similar to the aforementioned types. Consequently, this mechanism has not been established, and further studies of these reactions are in progress.

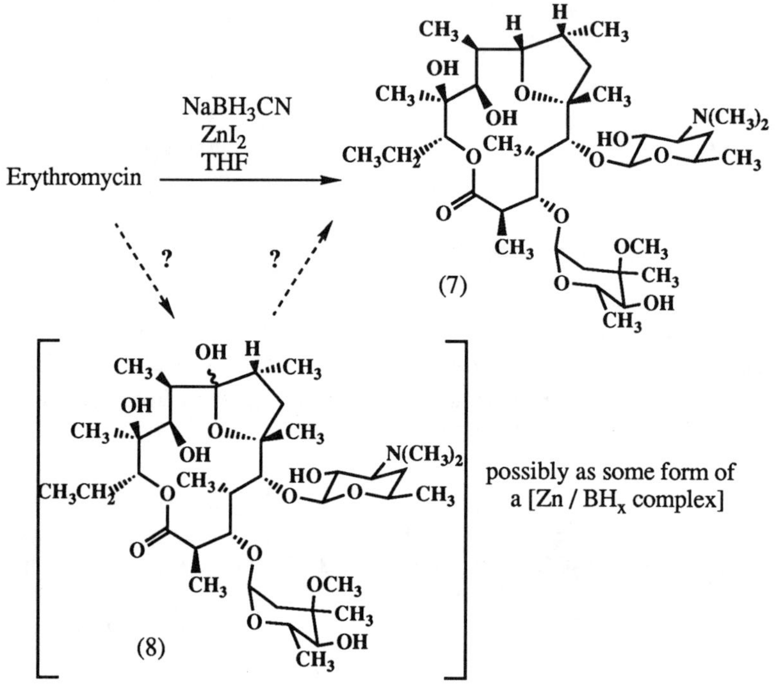

Figure 6 Synthesis of 8,9-*cis*-β-dihydro-8,9-anhydroerythromycin-6,9-hemiketal

3 STEREOCHEMISTRY OF 8,9-DIHYDRO DERIVATIVES

The proton and ^{13}C NMR spectra of the four dihydro derivatives were fully assigned (Tables 1 and 2) and were in complete accord with the structures illustrated above. Stereochemical assignments were made on the basis of coupling constants and results of proton NOE experiments using Difference and NOESY techniques, which were correlated with configurational orientations. The observed NOE effects which contributed most importantly to the stereochemical assignments of the two hydrogenation products are illustrated in Figure 7.

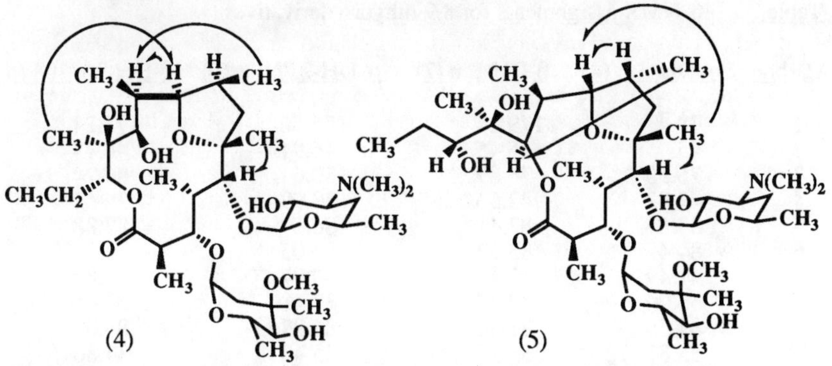

Figure 7 Comparative proton NOE effects of hydrogenation products

Table 1 Proton NMR assignments for 8,9-dihydro derivatives

Proton	α-DH-Em (4)	β-DH-Em (7)	α-DH-267108 (6)	β-DH-267108 (5)
2	2.75	2.71	2.83	2.72
3	4.23	4.11	4.48	4.28
4	1.91	1.97	1.79	2.08
5	3.75	3.73	3.54	3.70
7	1.49/2.37	1.48/1.89	1.32/2.57	1.43/1.88
8	2.32	2.48	2.08	2.36
9	3.71	3.77	3.46	3.91
10	2.16	1.92	2.21	2.17
11	3.30	3.86	5.13	5.53
13	4.79	4.78	2.84	2.82
14	1.49/1.91	1.47/1.86	1.31/1.66	1.32/1.68
15	0.89	0.91	0.99	0.99
2-Me	1.17	1.18	1.28	1.24
4-Me	1.09	1.07	1.09	1.12
6-Me	1.49	1.35	1.54	1.36
8-Me	1.11	0.95	1.02	1.12
10-Me	0.98	0.91	1.15	1.36
12-Me	1.13	1.11	1.17	1.14
1'	4.35	4.38	4.23	4.35
2'	3.23	3.22	3.24	3.21
3'	2.51	2.52	2.51	2.72
4'	1.20/1.71	1.25/1.70	1.22/1.65	1.24/1.85
5'	3.50	3.51	3.49	3.50
6'	1.23	1.22	1.21	1.23
NMe	2.32	2.33	2.30	2.42
1"	5.20	5.13	4.84	4.80
2"	1.58/2.41	1.56/2.38	1.56/2.36	1.55/2.36
4"	3.04	3.01	3.01	3.01
5"	4.08	4.03	4.03	4.00
6"	1.32	1.27	1.31	1.32
3"-Me	1.25	1.23	1.22	1.24
3"-OMe	3.34	3.32	3.24	3.27

Table 2 ^{13}C NMR assignments for 8,9-dihydro derivatives

Carbon	α-DH-Em (4)	β-DH-Em (7)	α-DH-267108 (6)	β-DH-267108 (5)
1	178.47	179.24	176.56	177.24
2	45.35	45.45	46.91	46.73
3	76.50	76.89	82.61	80.74
4	43.92	42.56	38.99	36.62
5	83.66	82.41	86.59	82.88
6	84.45	83.37	84.05	na
7	41.49	42.41	40.97	41.87
8	36.04	32.77	37.48	33.66
9	84.56	81.30	84.97	84.08
10	33.81	33.25	35.45	33.86
11	73.47	69.80	79.91	74.89
12	74.67	76.02	76.54	na
13	79.47	80.47	76.85	76.10
14	21.48	22.00	22.41	22.58
15	10.98	11.28*	11.89	11.98
2-Me	13.57	13.66	15.12	14.00
4-Me	10.07	9.60	9.72	9.78
6-Me	29.33	24.97	30.93	22.28
8-Me	18.12	15.94	15.95	17.60
10-Me	10.61	11.21*	9.57	16.07
12-Me	17.56	17.18	16.33	16.99
1'	103.13	103.09	104.94	103.76
2'	70.70	70.70	70.82	70.99
3'	65.98	65.82	65.49	65.36
4'	28.93	28.86	28.83	29.72
5'	68.93	68.80	69.12	68.87
6'	21.26	21.22	21.07	21.03
NMe	40.36	40.30	40.27	40.38
1"	94.65	94.66	98.52	98.20
2"	34.76	34.76	35.51	35.28
3"	73.02	72.90	72.40	72.44
4"	78.19	78.09	78.24	78.20
5"	65.58	65.59	65.36	65.51
6"	18.00	18.20	18.00	18.10
3"-Me	21.60	21.56	21.47	21.51
3"-OMe	49.53	49.49	49.18	49.25

Proton and ^{13}C spectra were measured in CDCl$_3$ solution. Chemical shifts are given in δ values from internal TMS or CHCl$_3$. The primed ring is the amino sugar (desosamine); the double primed ring is the neutral sugar (cladinose).
* means assignments may be interchanged.
na means not able to be assigned; attempts to assign these two resonances using QUATD were not successful.

Since both (4) and (5) are hydrogenation products, their protons at C-8 and 9 are *cis* to each other. In dihydro derivative (5), the observed NOE effect from the methyl group at C-6 to the methine proton at C-9 indicates that these two substituents lie on the same side of the tetrahydrofuran ring, thereby establishing the β orientation of the proton at C-9. Consequently, the 12-membered macrolide LY267108 is hydrogenated from the β face to yield a 8,9-*cis*-β-dihydro derivative.

Since derivatives (4) and (6) are related via translactonization, and the two 12-membered macrolides (5) and (6) differ from each other, the stereochemistry at C-8 and C-9 in (5) must be opposite that in (4) and (6). Consistent with this conclusion, an NOE effect from the methyl group at C-6 to the methine proton at C-9 is not observed in dihydro isomer (4), indicating that its proton at C-9 has the α configuration. In this compound, an NOE effect is instead observed from the methyl group at C-8 to the methine proton at C-10. Such an effect is consistent with the methyl group at C-8 and the C-9,10 bond being *cis* to each other. This result confirms that hydrogenation of the 14-membered macrolide (2) occurs from its α face, yielding its 8,9-*cis*-α-dihydro derivative. Its ring-contracted isomer (6) also possesses this stereochemistry since the tetrahydrofuran ring is not altered by the translactonization process. Consistent with this assignment, no NOE effect was observed in the spectrum of (6) between the methyl group at C-6 and the proton at C-9.

The dihydro derivative (7) produced NOE effects similar to those observed in compound (5), with the most important effect being that between the methyl group at C-6 and the proton at C-9. In compound (7), the coupling constant between H-8 and H-9 is 8.5 Hz, which agrees closely with the value (9.5 Hz) found for this coupling in (5). As a result of the close similarity of their NOE effects and coupling constants, this second pair of compounds, (5) and (7), were assigned the same stereochemistry (8,9-*cis*-β).

4 MOLECULAR MODELING OF 8,9-DIHYDRO DERIVATIVES

Conformational analyses of both starting materials and the four dihydro derivatives were conducted by molecular modeling using MacroModel 3D (version 3.5X).[15] Standard procedures were employed to perform Monte Carlo conformational searches and to derive conformational energies using the MacroModel MM2* force field.[15-19] Energy minimizations were conducted in vacuum and in chloroform and water as solvents. By using a Cray-2 supercomputer for the calculations, the entire molecules were modeled without simplifying them. The lowest energy structures initially found were then further refined by additional Monte Carlo searches to yield new sets of lower energy conformers, which were analyzed for experimental consistency with recorded NMR spectra. After satisfactory conformations of starting materials (2) and (3) had been derived, conformations for the four dihydro derivatives were generated in an analogous manner.

The calculated "minimum conformational energies" derived for the 14- and 12-membered starting materials (2) and (3) were 27.23 and 24.78 kcal/mol, respectively. This result is consistent with the observed direction of translactonization, which proceeds from (2) to (3). Inspection of the lowest energy conformations of 8,9-anhydroerythromycin-6,9-hemiketal (2) revealed that the dihydrofuran ring is oriented in a moderately concave manner relative to the larger macrolide ring, an orientation which causes the back (α) face of the 8,9-double bond to be more accessible to approach. In addition, the positions of the methyl groups at C-6 and C-10 sterically hinder approach from the front (β) face of (2). In contrast, the shorter, 12-membered macrolide ring in LY267108 (3) is constrained to lie largely under the dihydrofuran ring, thereby hindering the approach of reagents from the lower (α) side, but leaving the top (β) face much more exposed. This analysis satisfactorily explains the formation of the 8,9-*cis*-α-dihydro derivative from the hydrogenation of (2) and formation of the opposite 8,9-*cis*-β-dihydro derivative from the hydrogenation of (3).

Table 3   Comparison of calculated and experimental coupling constants[*]

Parameter	α-DH-Em (4)	β-DH-Em (7)	α-DH-267108 (6)	β-DH-267108 (5)
Values calculated from lowest energy conformation derived by molecular modeling				
Energy[+]	40.6	36.0	40.5	36.9
$J_{8,9}$	4.5	6.5	3.8	8.3
$J_{9,10}$	0.6	10.6	0.6	0.9
$J_{10,11}$	10.5	0.3	2.0	0.5
Experimentally observed values extracted from NMR spectra				
$J_{8,9}$	5.2	8.5	4.9	9.5
$J_{9,10}$	0	8.5	0	1.0
$J_{10,11}$	9.2	0	3.0	0

[*]   values for coupling constants measured in Hertz
[+]   minimum conformational energy (kcal/mol) found by Monte Carlo searching
      in vacuum

In order to assess the lowest energy conformers of the dihydro derivatives, their coupling constants in the region of C-8 to C-11 were calculated and compared with the values measured from their proton NMR spectra. Generally good agreement was found within these three sets of model-derived and experimentally measured values (Table 3). In addition to the lowest energy conformer, other low energy states of these derivatives were also found which would be accessible under NMR conditions. Since NMR values are time averaged, contributions from other low energy accessible conformations probably account for the small variations between the calculated and measured coupling constants. Although a complete analysis of all coupling constants was not performed, the relative agreement between the calculated and measured values in the C-8 to C-11 region indicates that molecular modeling produced plausible low energy conformers that were consistent with the stereochemistry assigned to the dihydro derivatives by NMR.

4 SYNTHESIS OF BROMOHYDRINS

Other examples of differential reactivity between the 14- and 12-membered macrolide derivatives (2) and (3) have also been noted. Additions of halogen from electropositive halogen donors such as N-bromosuccinimide (NBS), NaOBr, or CF_3OF to 8,9-anhydroerythromycin-6,9-hemiketal (2) have been shown to yield the corresponding 8β-halo derivatives (Figure 8).[20-22] These reactions most likely proceed by a mechanism in which the halogen adds to the β face of (2), with the hydroxyl group on C-11 sufficiently close to help stabilize a partial positive charge at C-9 and perhaps also to influence the stereochemistry of addition. Some intramolecular influence on the stereochemistry of addition is to be expected, since the analysis of hydrogenation results discussed above had indicated that the α face was the more accessible face of the double bond. Computer-assisted molecular modeling of the intermediate bromonium ion (10) has initially been explored using the 8β,9β-epoxide as an approximate model. This study indicated that the C-11 hydroxyl group was in a suitable position for such intramolecular interactions.

Subsequent reaction of the bromonium ion may then occur by either intramolecular or intermolecular attack. The factors controlling the choice of attack are not yet understood, although spiroketalization with the hydroxyl group at C-11 apparently does not occur.[20-22] In addition to the hydroxyl group at C-11, solvation of the bromonium ion and interactions with the hydroxyl group at C-12 or the dimethylamino substituent of the amino sugar may also play a role under different reaction conditions. The intramolecular stabilization by the hydroxyl group at C-11 in the bromonium ion (10) may also assist attack at C-9 by a molecule of water, thereby producing the 8β-bromo-6,9-hemiketal (9).

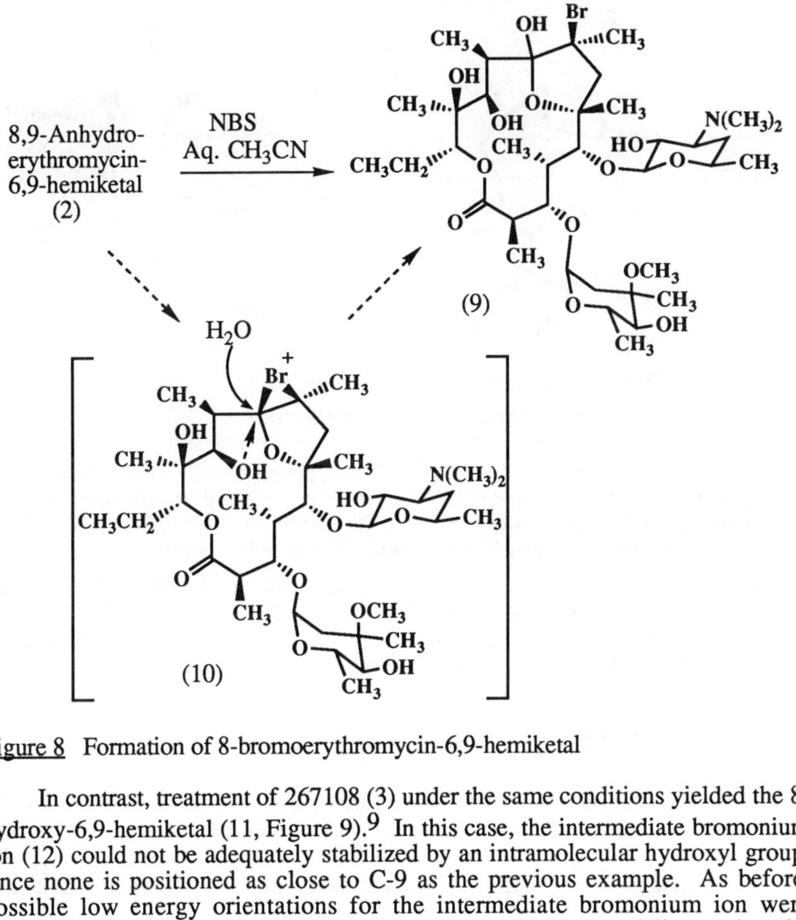

Figure 8 Formation of 8-bromoerythromycin-6,9-hemiketal

In contrast, treatment of 267108 (3) under the same conditions yielded the 8-hydroxy-6,9-hemiketal (11, Figure 9).[9] In this case, the intermediate bromonium ion (12) could not be adequately stabilized by an intramolecular hydroxyl group, since none is positioned as close to C-9 as the previous example. As before, possible low energy orientations for the intermediate bromonium ion were investigated with the aid of computer-assisted molecular modeling, using the 8β,9β-epoxide of LY267108 as a model. Lacking intramolecular stabilization at C-9, the bromonium ion (12) may be attacked intermolecularly by a molecule of water at the less sterically hindered C-8 position, producing the intermediate 8-hydroxy-9-bromo derivative (13). However, under aqueous conditions, this intermediate, which is analogous to a glycosyl bromide, is readily hydrolyzed to the 9-hydroxyl-6,9-hemiketal (11). Although the stereochemistry of the newly introduced hydroxyl group at C-8 has not yet been established, this mechanism indicates that it should have the 8α configuration. Additional studies to establish its orientation are being conducted.

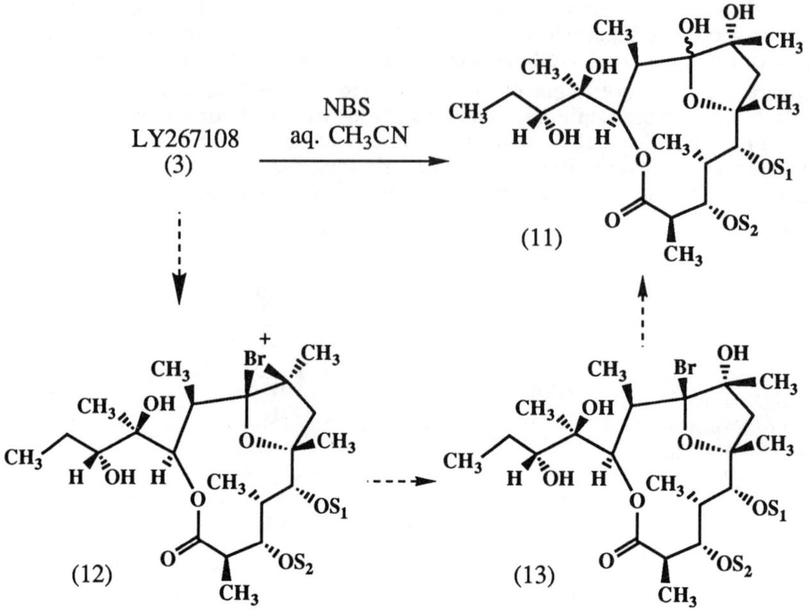

Figure 9   Formation of 8-hydroxy-267108-6,9-hemiketal
(S_1 = β-desosaminyl, S_2 = α-cladinosyl)

A third illustration of the differences between the 12- and 14-membered bicyclic macrolides (2) and (3) was their propensity to ring-open to the corresponding 6-hydroxy-9-keto monocyclic tautomers. The initial evaluation of the ring-contracted 8-hydroxy-6,9-hemiketal (11) indicated that it existed entirely as the bicyclic structure illustrated in Figure 9.[9] In contrast, 8-hydroxyerythromycin-6,9-hemiketal has been well established to open to its 9-keto tautomeric structure.[23-24] These reactions and their products are being examined further to provide more complete explanations for these observed differences.

5 ANTIMICROBIAL ACTIVITY

Table 4   Minimum inhibitory concentrations (μg/ml) of derivatives against *Staphylococcus aureus*, *Streptococcus pyogenes*, and *Streptococcus pneumoniae*

Compound	*S. aureus* X1	*S. pyogenes* C203	*S. pneumoniae* Park
1	0.25	0.06	0.06
2	8	1	1
3	128	32	16
4	2	0.25	0.25
5	64	64	64
6	16	32	8
7	1	0.5	0.25
9	2	1	1
11	128	16	8

All of the compounds were tested for *in vitro* antimicrobial activity by agar-dilution methodology. All of the intramolecular cyclization products possessed antimicrobial activity substantially diminished from that of erythromycin (Table 4), a result previously recognized for some of these compounds by others.[4,11] Another trend was seen in that the 14-membered bicyclic compounds (2), (4), (7), and (8) were significantly more active than the ring-contracted 12-membered analogs (3), (5), (6), and (11). Interestingly, saturation of the 8,9-double bond actually increased *in vitro* activity somewhat; furthermore, this activity did not depend on the stereochemistry at C-8 and C-9.

6 CONCLUSIONS

The 14- and 12-membered bicyclic macrolide derivatives (2) and (3) have shown unexpected differences in the stereochemistry of hydrogenation of their 8,9-double bonds and in the regiochemistry of formation of their 8,9-bromohydrins, which governs their subsequent reactivity. Supercomputer-assisted molecular modeling studies suggest that the steric environments and three-dimensional orientations of the dihydrofuran rings in compounds (2) and (3) are significantly different and that such differences can satisfactorily explain the observed stereo- and regiochemical results. The inability of 8-hydroxy-hemiketal (11) to ring-open to its 9-keto tautomer is another example of the differences between the 14- and 12-membered ring bicyclic macrolides. Finally, a novel application of the reducing complex of $NaBH_3CN$ and ZnI_2 has been found.

The authors gratefully acknowledge the excellent assistance provided by many of their colleagues, especially Ms. J. A. Wind, Mr. J. B. Campbell, T. K. Elzey, L. L. Huckstep, J. A. Occolowitz, J. D. Snoddy, and R. J. Thomas, and Drs. F. T. Counter and J. S. Gidda.

REFERENCES

1. H. A. Kirst and G. D. Sides, <u>Antimicrob. Agents Chemother.</u>, 1989, <u>33</u>, 1413.
2. H. A. Kirst and G. D. Sides, <u>Antimicrob. Agents Chemother.</u>, 1989, <u>33</u>, 1419.
3. H. A. Kirst, <u>Annu. Rep. Med. Chem.</u>, 1990, <u>25</u>, 119.
4. P. Kurath, P. H. Jones, R. S. Egan, and T. J. Perun, <u>Experientia</u>, 1971, <u>27</u>, 362.
5. N. Inatomi, H. Satoh, Y. Maki, N. Hashimoto, Z. Itoh, and S. Omura, <u>J. Pharmacol. Exp. Ther.</u>, 1989, <u>251</u>, 707.
6. H. A. Kirst, B. Greenwood, and J. S. Gidda, <u>Drugs of the Future</u>, 1992, <u>17</u>, 18.
7. S. Omura, K. Tsuzuki, T. Sunazuka, S. Marui, H. Toyoda, N. Inatomi, and Z. Itoh, <u>J. Med. Chem.</u>, 1987, <u>30</u>, 1941.
8. I. O. Kibwage, R. Busson, G. Janssen, J. Hoogmartens, H. Vanderhaeghe, and J. Bracke, <u>J. Org. Chem.</u>, 1987, <u>52</u>, 990.
9. H. A. Kirst, J. A. Wind, and J. A. Paschal, <u>J. Org. Chem.</u>, 1987, <u>52</u>, 4359.
10. P. Kurath, R. S. Egan, and P. H. Jones, U. S. Patent 3,681,323, Aug. 1, 1972.
11. K. Tsuzuki, T. Sunazuka, S. Marui, H. Toyoda, S. Omura, N. Inatomi, and Z. Itoh, <u>Chem. Pharm. Bull.</u> 1989, <u>37</u>, 2687.
12. C. K. Lay, C. Dufresne, P. C. Belanger, S. Pietre, and J. Scheigetz, <u>J. Org. Chem.</u>, 1986, <u>51</u>, 3038.

13. S. Kim, Y. J. Kim, and K. H. Ahn, Tetrahedron Lett., 1983, 24, 3369.
14. S. Kim, C. H. Oh, J. S. Ko, K. H. Ahn, and Y. J. Kim, J. Org. Chem., 1985, 50, 1927.
15. F. Mohamadi, N. G. J. Richards, W. C. Guida, R. Liskamp, M. Lipton, C. Caufield, G. Chang, T. Hendrickson, and W. C. Still, J. Comput. Chem., 1990, 11, 440.
16. G. Chang, W. C. Guida, and W. C. Still, J. Amer. Chem. Soc., 1989, 111, 4379.
17. M. Saunders, K. N. Houk, Y.-D. Wu, W. C. Still, M. Lipton, G. Chang, and W. C. Guida, J. Amer. Chem. Soc., 1990, 112, 1419.
18. MacroModel V3.5X User Manual, Columbia University, April, 1992.
19. MacroModel V3.5X Batchmin User Manual, Columbia University, April, 1992.
20. L. Toscano, G. Fioriello, S. Silingardi, and M. Inglesi, Tetrahedron, 1984, 40, 2177.
21. S. Auricchio, G. Fronza, S. V. Meille, A. Mele, and D. Favara, J. Org. Chem., 1991, 56, 2250.
22. S. Auricchio, G. Fronza, A. Mele, and D. Favara, J. Org. Chem., 1992, 57, 452.
23. J. Tadanier, P. Kurath, J. R. Martin, J. B. McAlpine, R. S. Egan, A. W. Goldstein, S. L. Mueller, and D. A. Dunnigan, Helv. Chim. Acta, 1973, 56, 2711.
24. K. Krowicki and A. Zamojski, J. Antibiotics, 1973, 26, 575.

6
Mechanism-based Dual-action Cephalosporins

D. D. Keith,* H. A. Albrecht, G. Beskid, K. K. Chan,
J. G. Christenson, R. Cleeland, K. Deitcher, M. Delaney,
N. H. Georgopapadakou, F. Konzelmann, M. Okabe,
D. Pruess, P. Rossman, A. Specian, R. Then,[1] C.-C. Wei,
and M. Weigele[2]
ROCHE RESEARCH CENTER, HOFFMANN-LA ROCHE INC., NUTLEY, NJ, USA
[1]F. HOFFMANN-LA ROCHE & CO., LTD., BASEL, SWITZERLAND
[2]PRESENT ADDRESS: ARIAD PHARMACEUTICALS, INC., CAMBRIDGE, MA,
USA

1 INTRODUCTION

The first dual-action cephalosporin appeared in the literature in 1976, when O'Callaghan, Sykes, and Staniforth (and in an accompanying paper Greenwood and O'Grady) described the properties of the cephalosporin MCO.[1,2] This cephalosporin incorporated as its 3'-substituent pyrithione (which is 2-mercaptopyridine N-oxide), a substance with antibacterial and antifungal properties, which is used today as a disinfectant in shampoo. The expanded spectrum of activity of MCO, compared to that of traditional first-generation cephalosporins,

MCO

seemed to reflect a contribution from pyrithione. The exciting concept of the dual-action cephalosporin, as elucidated in these papers, seemed to present an invitation for medicinal chemists to custom-design cephalosporins with superior properties.

2 THE DUAL-ACTION MECHANISM

A cell wall of peptidoglycan surrounds most bacteria, providing rigid support and protection against environmental changes in osmotic pressure. This structure is unique in nature. The biochemistry of cell-wall synthesis is highly specific, and provides selective targets for useful chemotherapeutic agents. One lethal target is the final crosslinking step, through which rigidity is introduced into the peptidoglycan. β-Lactams inhibit this step by acylating active-site serine residues of the transpeptidases responsible for carrying out the crosslinking.[3,4] Similarly, acylation of active-site serine residues occurs as a key step in the mechanism by which most β-lactamases inactivate these antibiotics.[5-7] In either case, when a cephalosporin containing a potential leaving group, X, at the 3'-position reacts with a bacterial enzyme, that group is eliminated (Scheme 1).

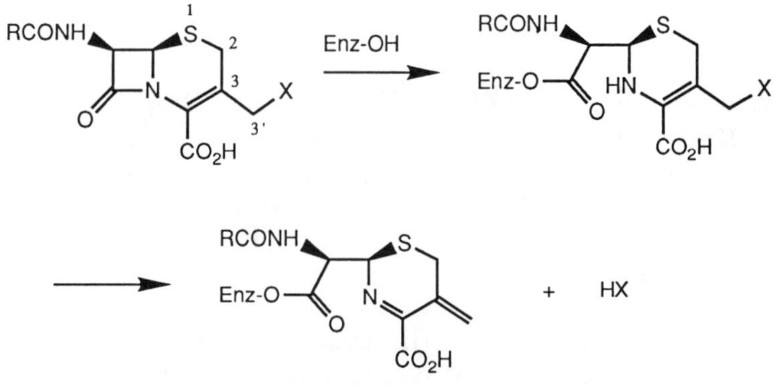

Scheme 1. Dual-Action Mechanism

Published evidence suggests that opening of the β-lactam ring correlates with elimination of the leaving group, although the reaction is probably not concerted.[8-14] Quantitative release of such 3'-substituents as acetate, azide, and pyridine can be brought about by treatment with β-lactamases, and this reaction appears to be quite general.[9,15] If the leaving group possesses intrinsic antibacterial activity, then the cephalosporin should exhibit a dual mode of action.[1,2,16-18] In addition to providing its own β-lactam activity, the cephalosporin should also act as a targeted prodrug for the second antibacterial agent, delivering it close to its site of action. From today's point of view, this mechanism continues to present opportunities to expand the antibacterial spectrum to include organisms which are resistant to the third-generation cephalosporins.

3 CEPHALOSPORIN-QUINOLONES

A variety of antibacterials could have been selected as the second agent for incorporation into dual-action cephalosporins. We chose to explore the use of quinolones for a number of reasons. (1) The antibacterial spectra of the two classes are complementary, with quinolones being active against β-lactam-resistant strains, while cephalosporins are more active against streptococci. (2) The mechanism of action of the quinolones, which inhibit DNA gyrase, seems compatible with that of cephalosporins; i.e. there should be no antagonism.[19,20] (3) Most quinolones are not very soluble under physiological conditions, and might benefit by incorporation into a more soluble prodrug form. Cephalosporins often have excellent solubility and pharmacokinetic properties. (4) Toxic effects of quinolones might be diminished by incorporation into a prodrug form. Central nervous system (CNS) toxicity as well as the arthropathic effects which have discouraged use of quinolones in pediatrics can possibly be overcome. (5) Although choices of quinolones were limited prior to 1976, the years immediately thereafter witnessed spectacular discoveries in the fluoroquinolone field, which provided a selection of extremely potent new examples.

Among the earliest cephalosporin-quinolones to be prepared by our team was Ro 23-5068 (**1**).[16] This target was specifically designed to test the dual-action hypothesis. The 7-phenoxyacetylamino group on a traditional cephalosporin, such as **2**, characteristically limits the useful antibacterial spectrum to gram-positive organisms. Oxolinic acid is a second generation quinolone with broad-spectrum

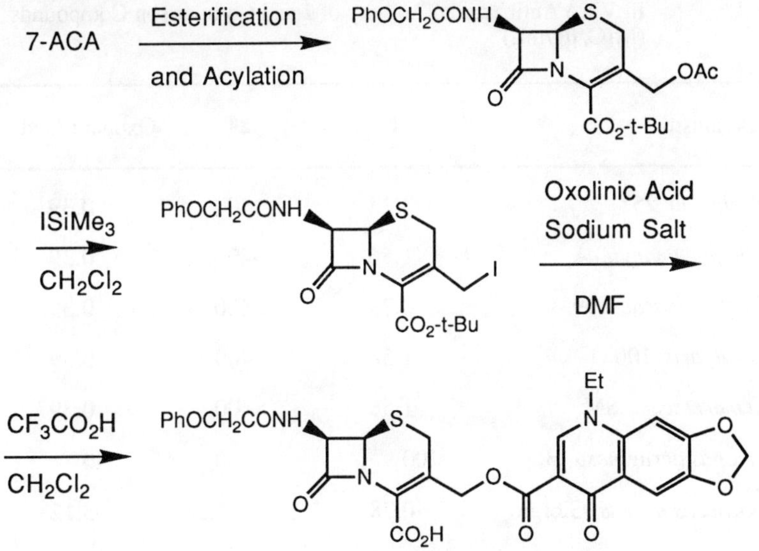

Scheme 2. Synthesis of Ro 23-5068

activity, which is quite potent against gram-negative bacteria. Oxolinic acid was ester-linked to the 3'-position using the synthetic approach outlined in Scheme 2. The resulting dual-action cephalosporin, **1**, demonstrated broad-spectrum in vitro activity (Table 1). The activity of **1** against gram-negative bacteria obviously seemed to derive from a contribution of the quinolone component. Since quinolone esters do not have antibacterial activity, except as a consequence of hydrolysis to the free acid,[21,22] this clearly supported the dual-action thesis. Furthermore, β-lactamases (*Enterobacter* cephase and *Proteus* cephase) had dramatic effects upon the activity of **1**. When in vitro testing was done in their presence, potency against the β-lactam-susceptible but oxolinic acid-resistant *Micrococcus luteus* was greatly reduced, while potency against *E. coli* 2721B, which is much more susceptible to oxolinic acid than to **1**, was greatly enhanced.[23]

1 was reasonably stable under physiological conditions, with a degradation half-life at 37 °C of 6 hours in human plasma.

Oxolinic acid itself has an unfortunate toxic liability, being a CNS stimulant. In a study of spontaneous locomotor activity, mice treated with oxolinic acid demonstrated significantly greater activity than mice treated at a comparable dose level with **1**, suggesting that toxic effects of the quinolone might indeed be avoided through this approach.[16]

Encouraged by these results, we optimized activity in a series of ester-linked dual-action cephalosporins to arrive at Ro 23-9424 (**3**), the first clinical candidate to emerge from this work.[16] When **3** showed outstanding activity (Table 2) against selected bacterial strains both in vitro and in vivo, the need for large quantities of this product stimulated a search for an improved synthesis. Initially, **3** had been prepared by the same general method as **1**, but using a trityl protecting group on the aminothiazole function. While establishing the 3'-ester bond by means of a

Table 1.   In Vitro Antibacterial Activity of **1** and Comparison Compounds (MIC, μg/mL)

Organism	1	2[a]	Oxolinic Acid
Escherichia coli 257	3.12	400	0.39
Klebsiella pneumoniae A	1.56	>400	0.39
Enterobacter cloacae P99	0.78	>400	0.39
Proteus vulgaris 100	1.56	>400	0.39
Serratia marcescens SM	1.56	>400	0.39
Pseudomonas aeruginosa B	100	>400	50
Staphylococcus aureus Giorgio	0.78	1.56	3.12
Staphylococcus aureus 95	6.25	400	6.25

a)

2

nucleophilic displacement reaction had advantages in the synthesis of a series of compounds of diverse structures, it was not well suited to large-scale preparation of **3**. Later, after West and Wei[24] demonstrated the feasibility of forming cephalosporin 3'-esters in good yield through acylation of a 3'-alcohol, Okabe and Sun[25] developed a practical synthesis capable of providing kilogram quantities of **3** in 35% yield from 7-ACA (Scheme 3). Because the carboxyl of the cephalosporin is not protected by esterification in this scheme, the double-bond migration often observed in reactions of cephalosporins under basic conditions is not a problem.

Activity and Mechanism of Action of Ro 23-9424

Ro 23-9424 showed impressive broad-spectrum activity (Table 3) against both gram-positive and gram-negative bacteria, with potent antistreptococcal activity characteristic of the cephalosporin moiety, and activity against strains normally resistant to the third-generation cephalosporins which can be attributed to quinolone activity.[16,26,27] The antistreptococcal activity was especially good against penicillin-resistant strains (Table 4),[27] and may indicate that **3** could have a special role in the treatment of pneumonia.

However, **3** has a chemical degradation half-life of just 3 hours, in 0.067 molar pH 7 phosphate buffer at 37 °C. During the prolonged incubation of the in vitro susceptibility assay, this hydrolytic instability becomes a complicating factor in

Table 2. Antibacterial Activity of Ro 23-9424 (**3**)

Organism	MIC (μg/mL)	ED$_{50}$ (mice, mg/kg, s.c.)
Staphylococcus aureus Smith	1	5.6
Streptococcus pneumoniae 6301	≥ 0.008	<1
Escherichia coli 257	0.063	2
Pseudomonas aeruginosa 5712	32	60

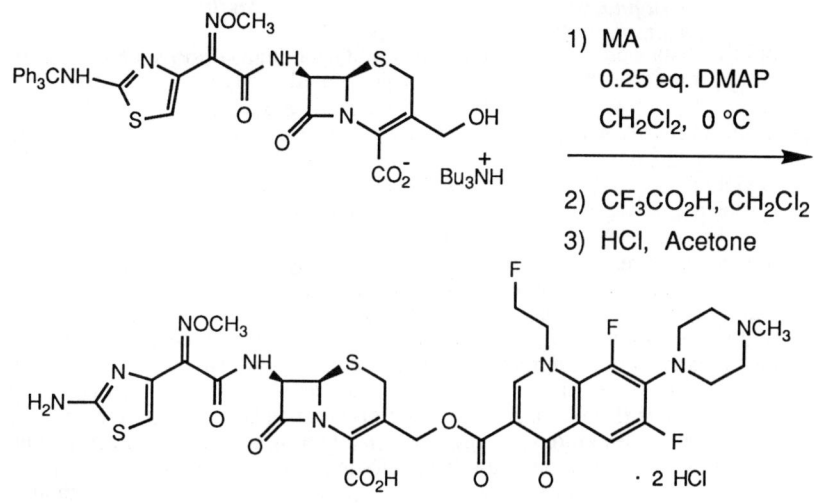

Scheme 3. A Practical Synthesis of Ro 23-9424

Table 3.   Spectrum of Activity of Ro 23-9424 (**3**)

Strains >90 % Susceptible

Enterobacteriaceae

E. coli	S. enteritidis
K. pneumoniae	Shigella spp.
K. oxytoca	Y. enterocolitica
E. aerogenes	P. mirabilis
E. agglomerans	P. vulgaris
E. cloacae	P. morganii
S. marcescens	P. stuartii
C. diversus	P. alcalifaciens
C. freundii	P. rettgeri

Other Gram-Negative Organisms   Gram-Positive Organisms

A. anitratus	S. aureus (methicillin-susceptible and resistant)
M. catarrhalis	
H. influenzae	S. epidermidis (methicillin-susceptible and resistant)
N. gonorrhoeae	
N. meningitidis	S. saprophyticus
	S. pneumoniae
	Streptococci Lancefield groups A, B, C, and G

Susceptibility 50-90%   Susceptibility <50%

Other Gram-Negative Organisms   Other Gram-Negative Organisms

P. aeruginosa (70-85%)   B. fragilis
X. maltophilia (50%)

Other Gram-Positive Organisms

E. faecalis
E. faecium

mechanistic interpretations. In addition to the enzyme-mediated events which generate quinolone according to the proposed dual-action mechanism (Scheme 1), chemical hydrolysis undoubtedly can occur. Neither process precludes the other. If bioactive degradation products generated by non-enzyme mediated hydrolysis accumulate early enough in the test, some effect on MICs might be anticipated. Since many, though not all, aspects of the in vitro activity can be explained by the hydrolysis products fleroxacin and desacetylcefotaxime (**4**), a variety of techniques have been applied to demonstrating the mechanism of action of this compound. These studies were designed to discriminate between the behavior of the bifunctional cephalosporin and its components (or hydrolysis products). Although it has proven difficult to unequivocally demonstrate the dual-action mechanism, a substantial body of evidence suggests that this is at least one mechanism by which **3** and the other dual-action cephalosporins act. Since the release of quinolone was

Table 4. In Vitro Activity of Ro 23-9424 (**3**) against Penicillin-Resistant
Streptococcus pneumoniae

Compound	MIC_{50} (µg/mL)	MIC_{90} (µg/mL)	Range
3	0.12	0.25	≥0.06-0.5
4[a] + FLX[b]	0.5	2	0.25 - 4
4	0.5	4	0.25 - 8
Cefotaxime	0.25	0.5	0.12 - 2
FLX	8	8	4 - 16

a)

Desacetylcefotaxime (**4**)

b)

Fleroxacin (FLX)

obvious from the in vitro results, the demonstration that the observed cephalosporin-like activity was, in fact, due to the intact bifunctional compound provided powerful support for the dual-action mechanism. This evidence is reviewed in the following sections.

In Table 5, the activity of the dual-action cephalosporin, **3**, against several bacterial strains is clearly superior to that of either of the two potential hydrolysis products, **4** and fleroxacin, and to that of a one-to-one molar mixture of the two. Especially notable is the potency against *E. coli* TE-18-28-1, a strain resistant to both desacetylcefotaxime and fleroxacin.[28] In these cases, it is clearly the intact bifunctional cephalosporin which is responsible for the activity.

When the effects of fleroxacin, **4**, and a one-to-one combination were compared to the effect of **3** on the viable count of exponentially growing *E. coli* B, only **3** was rapidly lethal at 0.044 µM concentration. The killing curve clearly distinguished the dual-action cephalosporin from its separate components and from a one-to-one equimolar mixture thereof.[29]

Moreover, **3** has greater affinity for essential PBPs of *E. coli* and *S. aureus* than does desacetylcefotaxime.[30] **3** is also a more potent inhibitor of peptidoglycan synthesis in permeabilized *E. coli* cells than is desacetylcefotaxime.[31] Membrane penetration studies provided further support for the thesis that intact **3** permeates the outer membrane of *E. coli* via porins. Cephalosporins are dependent upon porins

Table 5.   MICs of Ro 23-9424 (3) and Related Compounds (μM)

Organism	3	4[a]	1:1 FLX + 4	FLX[b]
M. luteus PCI	0.0746	0.597	0.597	38.2
S. pneumoniae 6301	0.0187	0.0746	0.0746	9.56
E. coli TE 18-28-1	0.57	242	>50	134

a, b see Table 4

to penetrate the outer membrane, while quinolones can also penetrate by diffusion through exposed lipid domains on the outer membrane.[32] The effects of porin mutations in *E. coli* on the growth-inhibiting activity of **3** were similar to the effect on cefotaxime, and different from that on fleroxacin.[30]

Both quinolones and cephalosporins cause filamentation in growing *E. coli*; the former also disrupt nucleoid segregation, while the latter do not.[33] At concentrations causing filamentation in *E. coli,* **3** did not affect nucleoid segregation during the first hour of incubation. Thus, early in the assay the activity of **3** was cephalosporin-like. After two hours, however, quinolone effects were noted, and were attributed to *in situ* release of fleroxacin. In similar studies with an *E. coli* strain producing TEM-3, an expanded spectrum β-lactamase, only the quinolone response was seen, presumably due to rapid enzyme-mediated hydrolysis of **3** with concomitant release of fleroxacin.[34]

The N-oxide of fleroxacin is nearly devoid of antibacterial activity. When compound **5**, the N-oxide of Ro 23-9424, was prepared, it showed a cefotaxime-like spectrum (Table 6). Activity against β-lactam susceptible strains remained very similar to that of Ro 23-9424. However, against such strains as *E. cloacae* P99, *C. freundii* BS-16, and *P. aeruginosa* 5712, the N-oxide **5** was markedly less active than **3** due to the lack of a significant quinolone contribution to its spectrum.

Thus, there is an impressive array of evidence which supports biochemical aspects of the dual-action mechanistic concept.

Pharmacokinetic profiles of **3** were determined in rats, dogs, and baboons after a single intravenous dose of 20 mg/kg.[35,36] Elimination half-lives were 17, 36, and 75 minutes, and peak plasma levels were 160, 72, and 124 μg/mL, respectively. At these concentrations of **3**, fleroxacin equivalents equal to several times the normal therapeutic level are made available *via* the dual-action cephalosporin to counter bacterial infections. Interestingly, prolonged low plasma concentrations of about 1 μg/mL of free fleroxacin were observed in all species. Low but persistent fleroxacin levels were still detectable after 8 hours in the rat and 24 hours in the dog and baboon. Free quinolone levels have implications both for therapy, since this concentration is well above the MIC for many bacteria, and for toxicity, since one reason for designing dual-action cephalosporins had been to avoid concerns about quinolone toxicity. Although the possibility of mammalian enzymes playing a role in mediating hydrolytic release of quinolone has not been

Table 6.   In Vitro Activity of **3** Compared to its N-Oxide **5** (MIC μg/mL)

Organism	3	5[a]
Staphylococcus aureus Smith	1	1
Streptococcus pneumoniae 6301	0.0625	0.0078
Escherichia coli B	0.0156	0.0313
E. coli DCO	0.25	0.125
Klebsiella pneumoniae A	0.125	0.125
Enterobacter clocae P99	0.125	8
Citrobacter freundii BS-16	0.25	32
Pseudomonas aeruginosa 5712	16	>128

Ro 23-9424 N-Oxide (**5**)

ruled out, chemical instability is one property of **3** with obvious room for improvement. In an attempt to avoid the observed low plasma levels of free quinolone, other more chemically stable types of dual-action cephalosporins were designed and prepared. Their properties are described in the following section.

Second-Generation Dual-Action Cephalosporins

As **3** moved into clinical trial, a search for second-generation dual-action cephalosporins with superior properties began. Although **3** had excellent antibacterial activity, it had some shortcomings. Improvements in such properties as solubility, stability, and potency against specific troublesome organisms such as *P. aeruginosa* were highly desirable. It also became a goal to eliminate the persistent low plasma levels of quinolone observed with **3**, but to do so without compromising antibacterial activity.

When compound **6**, the thiol-ester analogue of Ro 23-9424 was prepared, it proved to have interesting properties. The quinolone component itself, "thiofleroxacin", was not a very active quinolone, and so **6**, like **5**, was a somewhat crippled dual-action cephalosporin. In this case, the tertiary amine function of the quinolone component was still present. The thioester bond was found to be surprisingly stable. Since compound **6** was not soluble enough at more

Table 7.   Antibacterial Activity of **6**[a]

Organism	MIC (µg/mL)	ED$_{50}$ (mg/kg, s.c., mice)
S. aureus Smith	2	43
S. pneumoniae 6301	≥0.0157	-
E. coli 257	0.0625	<2
P. aeruginosa 5712	>128	-

Thiol-Ester Analogue of Ro 23-9424 (**6**)

neutral pH, comparative stability studies were run at pH 10, at 37 °C. Under these conditions, **6** ($t_{1/2}$ = 200 min) proved about four times as stable as **3** ($t_{1/2}$ = 55 min). In vitro, against β-lactam susceptible strains, **6** showed activity rather similar to that of **3** and **5**, but like **5** it lacked activity against β-lactam-resistant bacteria; in vivo, **6** showed good activity in infections caused by susceptible strains (Table 7).

The design of other new target compounds focused on linking the quinolone through functionality other than its carboxyl group. The first and second generation quinolones, such as nalidixic and oxolinic acids, possessed only a single chemically reactive functionality, a carboxylic acid, which could be used to establish an ester bond to the cephalosporin nucleus. However, most of the third-generation quinolones also contain primary, secondary or tertiary amine functionalities which can be used to establish other, more stable linkages. Accordingly, bifunctional cephalosporins were synthesized in which the cephalosporin 3'-position and the quinolone were joined by a bond through a quaternary ammonium function[17] or a carbamate link.[18]

The quaternary-linked compound **7** was prepared,[17] and demonstrated broad-spectrum activity in vitro. It had a degradation half-life at 37 °C of 29 hours, nearly ten times that of **3**. However, in vivo **7** was disappointing, as it failed to show activity in a murine infection caused by *P. aeruginosa* (Table 8). It did show excellent activity in the mouse-protection test against infections due to *S. pneumoniae* and *E. coli*, and demonstrated that chemical stability was compatible with broad-spectrum antibacterial activity.[17]

Table 8.   Antibacterial Activity of **7**[a]

Organism	MIC (µg/mL)	ED$_{50}$ (mg/kg, s.c., mice)
S. aureus Smith	4	-
S. pneumoniae 6301	0.063	12
E. coli 257	0.125	<0.5
P. aeruginosa 5712	16	>250

a)

7

The ciprofloxacin-derived carbamate-linked compound **8** proved more interesting. The synthesis of **8** was accomplished as described.[18] Product **8** showed excellent broad-spectrum activity in vitro, and also showed good activity against selected infections in the mouse-protection test (Table 9). The half-life of **8** in pH 7.4 buffer at 37 °C was 10 hours, significantly better than that of **3**. The pharmacokinetic profile of **8** in the rat was rather similar to that of **3**, with the exception that no low level background concentration of quinolone was detected.[29] Thus, it appears that the carbamate-linked dual-action cephalosporins may have the potential to provide both β-lactam and quinolone activities without the risk of quinolone-type toxicity.

The synthesis of an ester-linked cephamycin-type compound, **9**, led to some surprising results. **9** had broad-spectrum activity in vitro, and demonstrated activity in murine models of infection (Table 10). It also proved to be exceptionally stable. In pH 7.4 phosphate buffer at 37 °C, **9** showed a degradation half-life of 20 hours. In a pharmacokinetic study in the rat, **9** also produced a prolonged background level of ciprofloxacin, but the concentration was much lower than the level of fleroxacin produced by **3**. This result confirms that increasing stability of a bifunctional cephalosporin can result in lower levels of free quinolone. Yet, barely detectable quinolone levels persisted. Thus, there are other factors at work, as well. The nature of the chemical linkage may prove to be critical.

Table 9. Antibacterial Activity of **8**[a]

Organism	MIC (μg/mL)	ED$_{50}$ (mg/kg, s.c., mice)
S. aureus Smith	0.063	12
S. pneumoniae 6301	$\leq$0.016	10
E. coli 257	0.031	1.4
P. aeruginosa 5712	4	67

a)

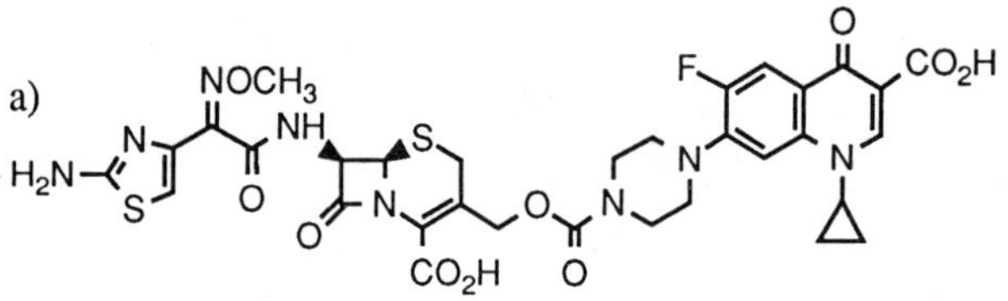

8

Table 10. Antibacterial Activity of **9**[a]

Organism	MIC (μg/mL)	ED$_{50}$ (mg/kg, s.c., mice)
S. aureus Smith	1	-
S. pneumoniae 6301	0.0625	1
E. coli 257	0.125	-
P. aeruginosa 5712	4	41

a)

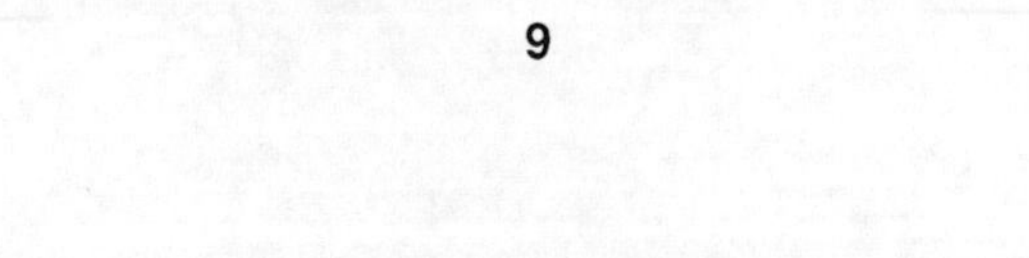

9

4 CONCLUSIONS

The dual-action mechanism, in which bacterial enzyme-mediated opening of the β-lactam ring of a cephalosporin results in the liberation of a second antibacterial agent, provides a fascinating rationale for drug design, and has led to the synthesis of antibacterials with interesting and unusual properties. The bifunctional cephalosporins in which quinolones are incorporated as ester-, quaternary-, or carbamate-linked second agents behave in many respects as dual-action cephalosporins would be expected to. Much of the research into establishing the essential features of the mechanism of action of these compounds has been done with **3**, the first clinical candidate to emerge from this project. Although preliminary studies suggest that compounds with chemically more stable linkages act similarly, some mechanistic details remain elusive. The specific dual-action mechanism of Scheme 1 is difficult to prove unequivocally, and may not be the exclusive mechanism by which these compounds act. However, our investigations have provided convincing evidence that it is indeed realistic to expect a dual mechanism of action to be operative in vivo. While the initial clinical candidate **3** has many fine characteristics as an antibacterial, it is less than ideal in some respects. Therefore, an optimistic search continues for a "second-generation" dual-action cephalosporin.

REFERENCES

1. C. H. O'Callaghan, R. B. Sykes and S. E. Staniforth, *Antimicrob. Agents Chemother.*, 1976, *10*, 245.
2. D. Greenwood and F. O'Grady, *Antimicrob. Agents Chemother.*, 1976, *10*, 249.
3. D. J. Waxman and J. L. Strominger, *J. Biol. Chem.*, 1980, *255*, 3964.
4. J. M. Frere and B. Joris, *CRC Crit. Rev. Microbiol.*, 1985, *11*, 299.
5. R. P. Ambler, *Phil. Trans. R. Soc. Lond. (Biol.)*, 1980, *289*, 321.
6. A. L. Fink, *Pharm. Res.*, 1985, *2*, 55.
7. V. Knott-Hunziker, S. G. Waley, B. S. Orlek, and P. G. Sammes, *FEBS Lett.*, 1979, *99*, 59.
8. J. M. T. Hamilton-Miller, G. G. F. Newton and E. P. Abraham, *Biochem. J.*, 1970, *116*, 371.
9. C. H. O'Callaghan, S. M. Kirby, A. Morris, R. E.Waller and R. E. Duncombe, *J. Bacteriology,* 1972, *110,* 988.
10. A. D. Russell and R. H. Fountain, *J. Bacteriology,* 1971, *106*, 65.
11. M. I. Page and P. Proctor, *J. Am. Chem. Soc.,* 1984, *106*, 3820.
12. W. S. Faraci and R. F. Pratt, *J. Am. Chem. Soc.*, 1984, *106,* 1489.
13. D. B. Boyd, *J. Org. Chem.*, 1985, *50*, 886.
14. E. J. Grabowski, A. W. Douglas and G. B. Smith, *J. Am. Chem. Soc.*, 1985, *107,* 267.
15. L. D. Sabath, M. Jago and E. P. Abraham, *Biochem. J.,* 1965, *96*, 739.
16. H. A. Albrecht, G. Beskid, K.-K. Chan, J. G. Christenson, R. Cleeland, K. H. Deitcher, N. H. Georgopapadakou, D. D. Keith, D. L. Pruess, J.Sepinwall, A. C. Specian, R. L. Then, M. Weigele, K. F. West and R. Yang, *J. Med. Chem.*, 1990, *33*, 77.
17. H. A. Albrecht, G. Beskid, J. Christenson, J. Durkin, V. Fallat, N. H. Georgopapadakou, D. D. Keith, F. M. Konzelmann, E. R. Lipschitz, D. H. McGarry, J. Siebelist, C.-C.Wei, M. Weigele and R. Yang, *J. Med. Chem.*, 1991, *34,* 669.
18. H. A. Albrecht, G. Beskid, J. Christenson, N. H. Georgopapadakou, D. D. Keith, F. M. Konzelmann, D. L. Pruess, P. L. Rossmann and C.-C.Wei, *J. Med. Chem.,* 1991, *34*, 2857.
19. I. Haller, *Arzneim.-Forsch.*, 1986, *36*, 226.

20. J. S. Wolfson and D. C. Hooper, *Antimicrob. Agents Chemother.*, 1985, *28*, 581.
21. R. Albrecht, *Prog. Drug Res.*, 1977, *21*, 9.
22. D. Kaminsky and R. E. Meltzer, *J. Med. Chem.*, 1968, *11*, 160.
23. J. Christenson, H. Albrecht, N. Georgopapadakou, D. Keith, D. Pruess, M. Talbot and R. Then, 28th Intersci. Conf. Antimicrob. Agents Chemother., 1988, Abstract No. 450.
24. K. F. West and C. C. Wei, *J. Org. Chem.*, 1992, *57*, in press.
25. M. Okabe and R. Sun, *Synthesis,* 1992, in press.
26. G. Beskid, J. Siebelist, C. M. McGarry, R. Cleeland, K.-K. Chan, D. D. Keith, *Chemotherapy,* 1990, *36*, 109.
27. R. N. Jones, A. L. Barry and C. Thornsberry, *Antimicrob. Agents Chemother.* 1989, *33*, 944-950.
28. J. Pace, A. Bertasso and N. H. Georgopapadakou, *Antimicrob. Agents Chemother.,* 1991, *35*, 910.
29. J. G. Christenson, G. Beskid, R. Cleeland, H. H. Farrish, Jr., B. D. Holzknecht, D. D. Keith, D. L. Pruess, and M. K. Talbot, 30th Intersci. Conf. Antimicrob. Agents Chemother., 1990, Abstract No. 405.
30. J. S. Chapman and N. H. Georgopapadakou, *Antimicrob. Agents Chemother.*, 1988, *32*, 438.
31. J. G. Christenson, V. Brocks, K.-K. Chan, D. D. Keith, D. L. Pruess, F. F. Schaefer, M. K. Talbot, 28th Intersci. Conf. Antimicrob. Agents Chemother., 1988, Abstract No. 444.
32. N. H. Georgopapadakou, A. Bertasso, K.-K. Chan, J. S. Chapman, R. Cleeland, L. M. Cummings, B. A. Dix and D. D. Keith, *Antimicrob. Agents Chemother.*, 1989, *33*, 1067.
33. N. H. Georgopapadakou and A. Bertasso, *Antimicrob. Agents Chemother.*, 1991, *35*, 2645.
34. A. Bertasso and N. H. Georgopapadakou, 92nd Annual Meeting, American Society for Microbiology, 1992, Abstract No. A-102.
35. J. G. Christenson, K.-K. Chan, H. H. Farrish, I. H. Patel and A. Specian, 28th Intersci. Conf. Antimicrob. Agents Chemother., 1988, Abstract No. 449.
36. J. G. Christenson, K.-K. Chan, R. Cleeland, B. Dix-Holzknecht, H. H. Farrish, I. H. Patel and A. Specian, *Antimicrob. Agents Chemother.*, 1988, *34*, 1895.

7

Novel Quinolone Antibacterial Agents: Synthesis and Biological Activity of 6H-6-Oxopyrido[1,2-a] pyrimidine-7-Carboxylic Acids

D. T. W. Chu,* Q. Li, C. M. Lee, K. Raye, K. Tanaka, J. Alder, and J. J. Plattner

ANTI-INFECTIVE RESEARCH, ABBOTT LABORATORIES, ABBOTT PARK, IL 60064, USA

1 INTRODUCTION

Nalidixic acid **1**[1], a synthetic antibacterial agent active against Gram-negative organisms, was introduced into clinical practice in early 1960s. Since then, a large number of related analogs have been synthesized. However, it was not until the discovery of norfloxacin **2**[2] that a large number of fluorinated compounds having a 1-substituted-1,4-dihydro-4-oxopyridine-3-carboxylic acid moiety collectively known as quinolones have successfully been developed as clinically useful antibacterial agents.[3] Examples of these clinically important quinolones are norfloxacin **2**, temafloxacin **3**,[4] ciprofloxacin **4**,[5] lomefloxacin **5**,[6] fleroxacin **6**,[7] ofloxacin **7**,[8] and tosufloxacin **8**.[9]

2 R=C_2H_5, R_1=R_2=R_3=H

3 R=o,p-$C_6H_3F_2$, R_1=R_3=H, R_2=CH_3

4 R=c-C_3H_5, R_1=R_2=R_3=H

5 R=C_2H_5, R_1=H, R_2=CH_3, R_3=F

6 R=C_2H_4F, R_1=CH_3, R_2=H, R_3=F

These new quinolones have been shown to be very effective against a number of bacterial infections. Hence, there is a considerable interest in the synthesis of 1,4-dihydro-4-oxoquinoline-3-carboxylic acid derivatives. Detailed structure-activity relationships (SAR) within this class have been reviewed.[3,10] Clinically useful members invariably contain a condensed N-substituted 4-pyridone-3-carboxylic acid moiety. Bioisosteres having N-1 nitrogen replaced with a carbon atom or oxygen atom were reported to be inactive.[11,12] Current structure-activity effort has been focused on novel substitutions on N-1, C-5, C-7, and C-8. In search for broad spectrum quinolone antibacterials with novel ring skeletons, we synthesized a series of quinolone nucleus modified analogs, 6H-6-oxopyrido[1,2-a]pyrimidines, in which the N-1 nitrogen atom of the naphthyridine nucleus was replaced by a carbon atom (Figure 1).

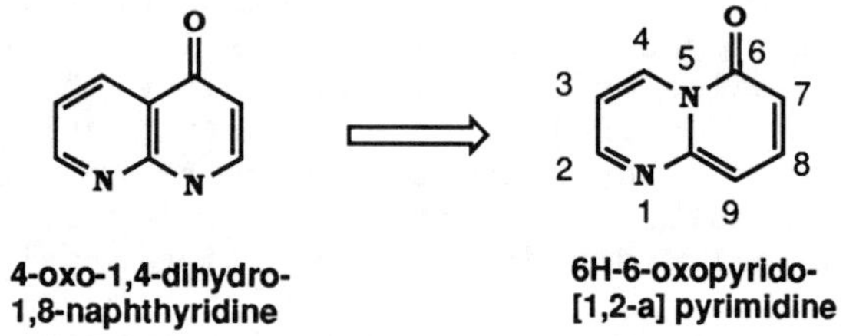

4-oxo-1,4-dihydro-
1,8-naphthyridine

6H-6-oxopyrido-
[1,2-a] pyrimidine

Figure 1   6H-6-oxopyrido[1,2-a]pyrimidine skeleton

In this paper, we report the synthesis and antibacterial activity of 2-substituted amino-3-fluoro-9-(2,4-difluorophenyl)-6H-6-oxopyrido[1,2-a]pyrimidine-7-carboxylic acid derivatives **9**. These compounds are nuclear modified analogs of tosufloxacin **8**, which is a potent clinically useful antibacterial agent. This study shows that in order to retain antibacterial activity, replacement of the nitrogen atom with carbon atom at the 1 position of naphthyridine nucleus is permissible if a nitrogen atom is placed at the ring juncture at C-4a. This investigation provides further insights into the SAR of quinolone antibacterial agents.

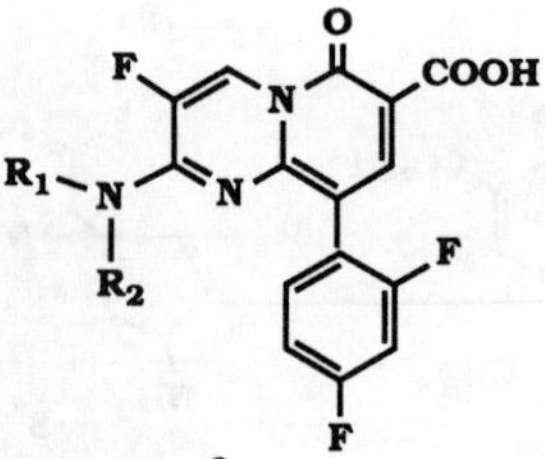

9

2 CHEMISTRY

In order to have a rapid assessment of our concept that
the 6H-6-oxopyrido[1-2a]pyrimidines may possess potent
antibacterial activity, the 2-N-methylpiperazin-1-yl-3-
fluoro-9-(2,4-difluorophenyl)-6H-6-oxopyrido[1,2-
a]pyrimidine-7-carboxylic acid (**10**) was chosen as our
initial synthetic target based upon known SAR consideration as well as that no protection of the 2-N-methyl-
piperazinyl group is required during its synthesis. The
retrosynthetic analysis of **10** indicated that it could be
derived from compound **11** upon disconnecting the N_5-C_6
amide bond. Compound **11** could be generated by condensation of diethyl ethoxymethylene malonate **12** and 2-(2,4-
difluorobenzyl)-5-fluoro-6-N-methylpiperazin-1-yl pyrimidine (**13**). The compound **13**, in turn, could be derived
from condensation of α-amidino-2,4-difluorotoluene **14** and
ethyl 2-fluoro-2-formyl acetate (**15**) (Figure 2).

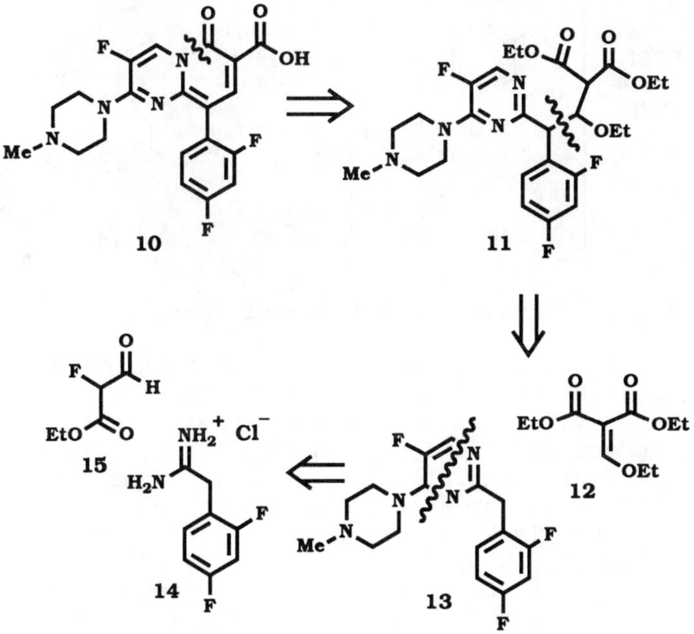

Figure 2 Retrosynthetic analysis for the synthesis of **10**

The key intermediate **13** was synthesized by a route
as illustrated in Scheme 1. Treatment of 2,4-difluo-
rophenyl acetonitrile (**16**) with hydrogen chloride in dry
ethanol and then ammonia yielded the α-amidino-2,4-diflu-
orotoluene (**14**) (98%). Condensation of **14** with sodio
ethyl-2-fluoro-2-formylacetate in methanol in the presence of triethylamine at reflux gave the pyrimidine **17**

(59%). Reaction of **17** with phosphorus oxychloride in
dimethylformamide at 0° yielded the 6-chloro-5-fluoro-2-
(2,4-difluorobenzyl)-pyrimidine (**18**) which, without
purification, was converted to the pyrimidine **13** upon
treatment with N-methylpiperazine in methylene chloride
at room temperature (92%).

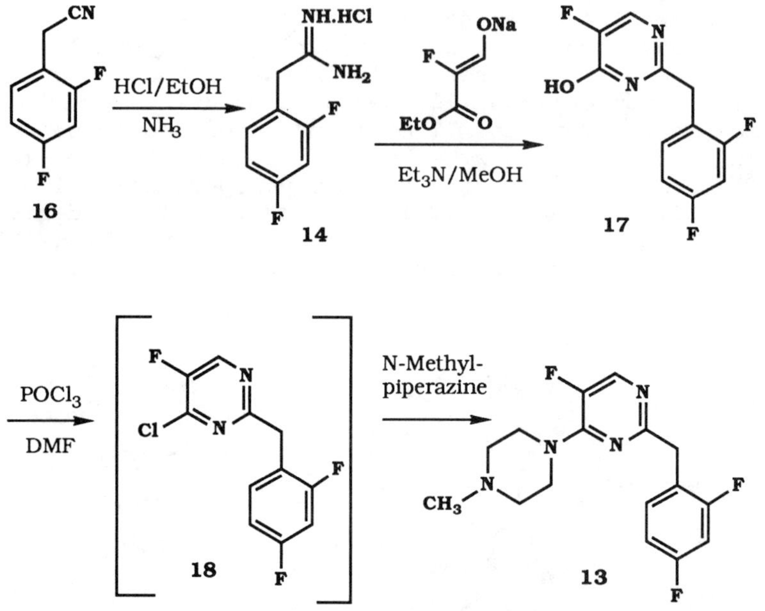

Scheme 1   Synthetic route for pyrimidine **13**

Treatment of pyrimidine **13** with lithium diisopropyl
amide at -78° in tetrahydrofuran followed by reaction
with ethoxymethylene malonate yielded the Michael addition product **19** (62%). Intramolecular cyclization of **19**
with piperidine in refluxing ethanol in the pressure of
catalytic amount of acetic acid gave the ethyl 2-N-
methyl-piperazin-1-yl-3-fluoro-9-(2,4-difluorophenyl)-6H-
6-oxopyrido[1,2-a]pyrimidine-7-carboxylate (**20**)
(Scheme 2). Treatment of **20** with sodium hydroxide solution did not provide the desired acid **10**, yielding the
ethyl 2-hydroxy-3-fluoro-9-(2,4-difluorophenyl)-6H-6-ox-
opyrido[1,2-a]pyrimidine-7-carboxylate (**21**) instead.
Apparently, the 2-N-methylpiperazinyl group was replaced
by the hydroxyl group faster than the hydrolysis of the
C_7 ester group. Although we did not obtain the desired
target compound **10**, compound **21** served as an important
correlative compound to establish the structure assignment for the 6H-6-oxopyrido[1,2-a]pyrimidine **20**.

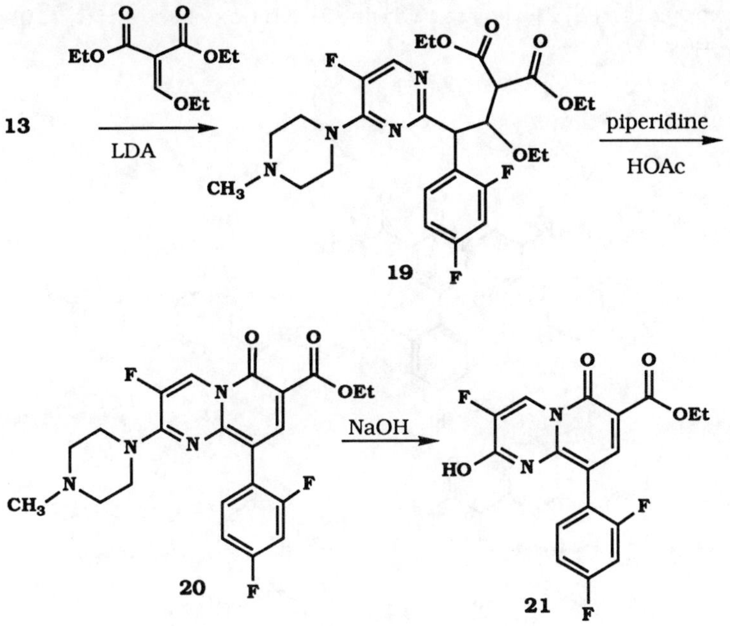

Scheme 2   Synthetic route for **20**

Intramolecular cyclization of **19**, in theory, may produce two regioisomers **20** and **22** (Figure 3). Since the experimental result indicated the production of only one regioisomer, the intramolecular cyclization of **19** was a regiospecific reaction. Because of the lack of the other isomer, the ^{1}H NMR spectrum for the cyclization product could not be assigned with certainty as **20**. The structural assignment for **20** was initially based upon reactivity consideration since the production of **22** could be avoided due to the steric bulk of the adjacent N-methylpiperazinyl group. The NMR spectrum of **21** showed the presence of a broad singlet at 9.3 ppm corresponding to the 2-hydroxyl proton. This confirmed the structure assignment of **21** since the corresponding proton in the regioisomer **23** would be expected to have a chemical shift of >10 ppm due to the intramolecular hydrogen bonding with the neighboring carbonyl group. Thus, upon correlation with **21**, the structure assignment for the cyclization product of **19** as **20** was expected to be correct.

Titanate-mediated transesterification[13] of the ethyl ester **20** in benzyl alcohol in the presence of catalytic amount of titanium tetraethoxide gave a mixture of the desired acid **10** (23%) and the benzyl ester **24** (67%). Deprotection of the benzyl ester **24** in methanol and formic acid with palladium on charcoal yielded the desired 2-N-methylpiperazin-1-yl-3-fluoro-9-(2,4-difluorophenyl)-6H-

6-oxopyrido[1,2-a]pyrimidine-7-carboxylic acid (**10**) (72%) (Scheme 3).

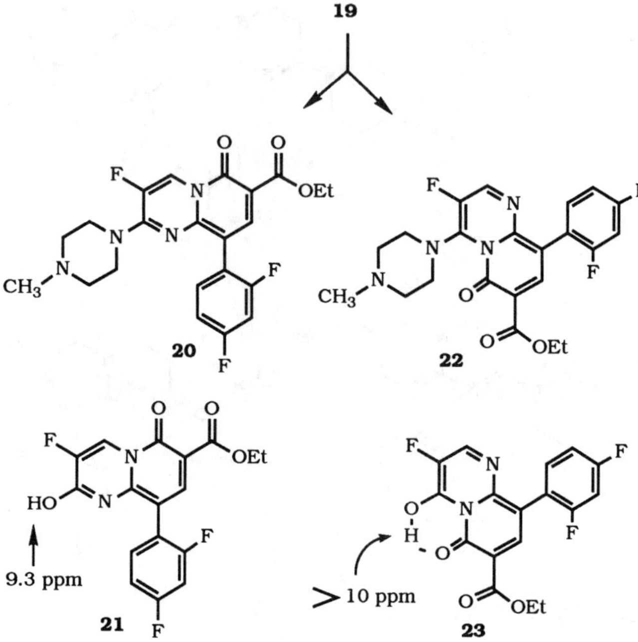

Figure 3   Structures for **21** - **23**

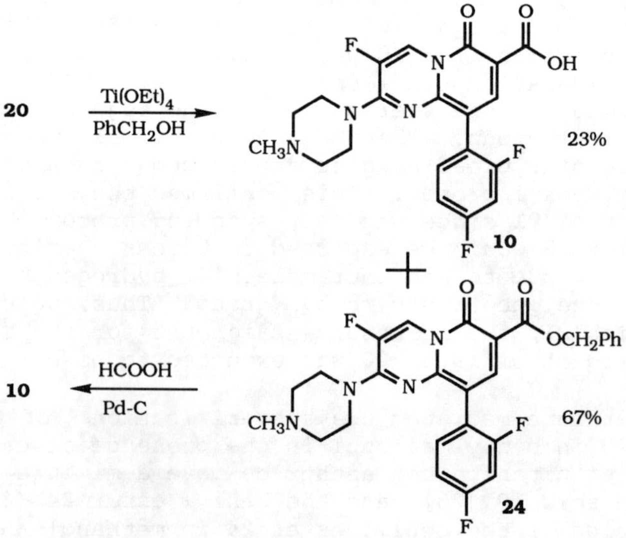

Scheme 3   Synthesis of **10** from **20**

The ^{1}H NMR spectrum of the benzyl ester **24** possessed a doublet at 9.17 ppm with a J value of 10 Hz for the H$_4$ proton. These data are identical to the H$_4$ proton (9.16 ppm, d, J=10 Hz) for benzyl 2-N-methylpiperazin-1-yl-3-fluoro-9-cyclopropyl-6H-6-oxopyrido[1,2-a]pyrimidine-7-carboxylate (**25**) whose structure had been confirmed by X-ray analysis. Hence, the structure of **20** as the regiospecific cyclization product of **19** was unequivocally confirmed (Figure 4).

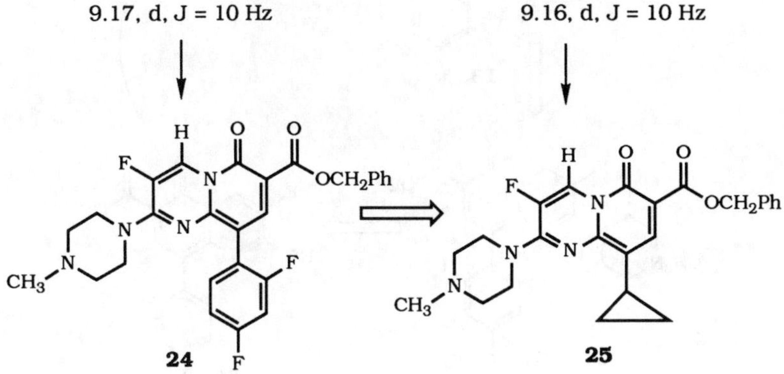

Figure 4   Structures for **24** and **25**

In order to prepare a large number of analogs with different amino substituents at the C$_2$ position, an intermediate having the 6H-6-oxopyrido[1,2-a]pyrimidine ring system with a leaving group at C$_2$ would be required. In this way, displacement at C$_2$ with different amines would produce many derivatives for biological evaluation. The 2-chloro derivative **28** was chosen as this intermediate. It has been shown previously that the 2-hydroxy derivative **26** could be made from **17** involving **13** and **20** as intermediates as shown in Scheme 4. The N-methylpiperazinyl group was used as a bulky substituent to ensure the right regiocyclization reaction. However, this route for the preparation **26** is redundant since it involved the insertion of an amino group at C$_2$ and subsequently removal of the same amino group. Hence, we searched for an alternate and efficient synthesis of **26**. The desired benzyl ester **28** could be prepared from **26** by a simple chlorination reaction.

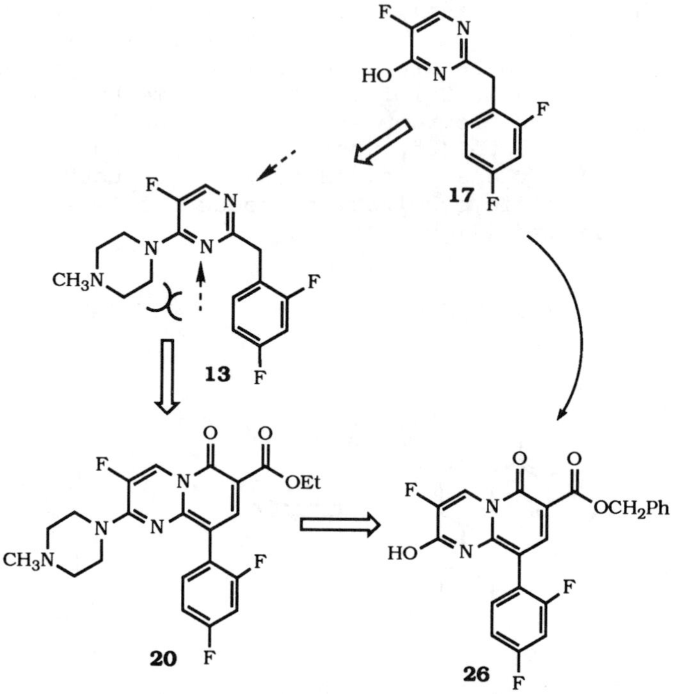

Scheme 4   Synthesis of **26**

Treatment of the pyrimidine **17** with 2 mole equivalents of n-butyl lithium at -78° in tetrahydrofuran followed by reaction with ethoxymethylene malonate yielded
the Michael addition product **27**. Without purification **27**
was subjected to intramolecular cyclization with piperidine in refluxing ethanol in the presence of a catalytic
amount of acetic acid to give the 2-hydroxy derivative **21**
(96%). This compound was identical to the one obtained
previously by sodium hydroxide treatment of **20**. Hence,
the intramolecular cyclization reaction for **27** went
through the same reaction pathway as the regiospecific
cyclization of **19** and was regiospecific. Titanate-mediated transesterification of the ethyl ester **21** in benzyl
alcohol in the presence of catalytic amount of titanium
tetraethoxide yielded the benzyl ester **26** (81%). Chlorination of the benzyl ester **26** with phosphorus oxychloride
in dimethylformamide at 0° gave the 2-chloro derivative **28**. Without purification, displacement of the
2-chloro group of **28** with 3-N-t-butoxycarbonylpyrrolidine
in methylene chloride at room temperature yielded the
2-substituted amino derivative **29** (94%). Deprotection of
the benzyl ester **29** in methanol and formic acid with
palladium on charcoal gave the corresponding acid **30**
(75%). Removal of the t-butoxycarbonyl group of **30** with
trifluoroacetic acid yielded the 3-aminopyrrolidinyl

derivative **31** (A-79527) (Scheme 5). Similarly, displacement of **28** with other appropriate amines followed by removal of protection groups yielded **32, 33** and **34**.

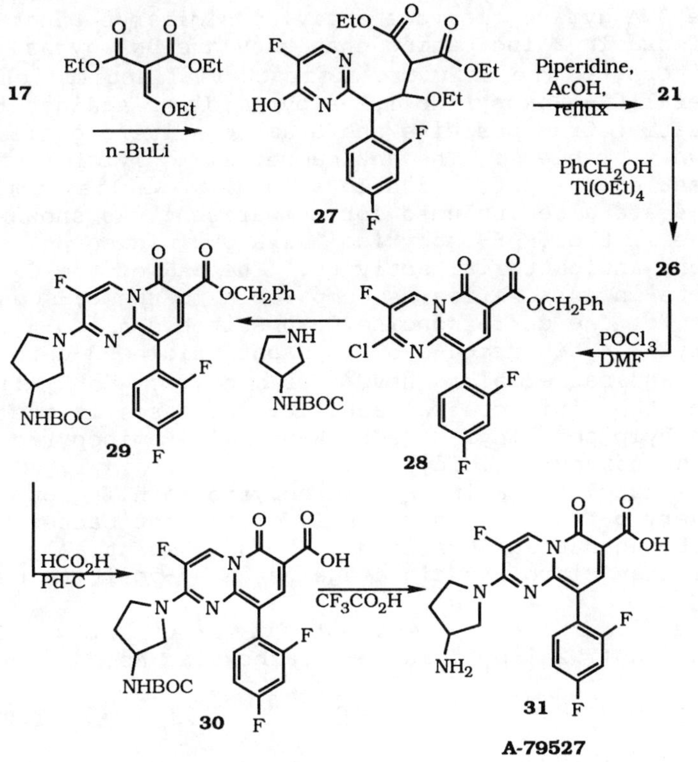

Scheme 5   Synthesis of **31**

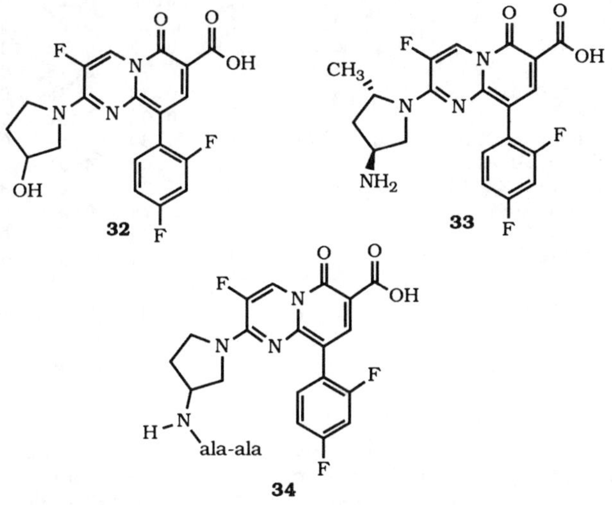

3   BIOLOGICAL RESULTS AND DISCUSSION

The novel 6H-6-oxopyrido[1,2-a]pyrimidine-7-carboxylic acid derivatives prepared in this study were found to have DNA gyrase cleavage activity similar to that of norfloxacin **2** indicating that they are DNA gyrase inhibitors.[14] The *in vitro* antibacterial activities of several 6H-6-oxopyrido[1,2-a]pyrimidines against representative Gram-positive and Gram-negative organisms are shown in Table 1. The antibacterial activities of the parent counterpart, tosufloxacin **8**, as well as ciprofloxacin **4** are also included for comparison. As shown in Table 1, the 6H-6-oxopyrido[1,2-a]pyrimidine possess potent antibacterial activity. The SAR of the C_2-substituents in this series is comparable to the C_7-substituents for the corresponding 7-substituted amino-6-fluoro-1-aryl-1,4-dihydro-4-oxo[1,8]naphthyridine-3-carboxylic acid antibacterial agents.[9] With respect to C_2-substitution, the antibacterial activity increases in the order 4-methylpiperazinyl < (s,s)-2-methyl-4-aminopyrrolidinyl ≤ 3-hydroxypyrrolidinyl ≤ 3-aminopyrrolidinyl. Compound **31** (A-79527), bearing a 3-aminopyrrolidinyl group at C_2, is very potent having slightly better antibacterial activity than tosufloxacin **8**. It is substantially more potent than ciprofloxacin **4** against Gram-positive bacteria.

Table 1   *In Vitro* Antibacterial Activity of 6H-6-Oxopyrido[1,2a]-pyrimidine-7-carboxylic Acids

Organism	10	31	32	33	Tosu	Cipro
S. aureus ATCC6538P	0.2	0.1	0.05	0.1	0.1	0.2
S. aureus A5177	0.39	0.1	0.1	0.1	0.2	0.39
S. aureus 5278	0.2	0.1	0.1	0.1	0.1	0.39
S. aureus NCTC10649	0.2	0.1	0.05	0.1	0.1	0.39
S. epidermidis 3519	0.39	0.2	0.2	0.2	0.2	0.39
S. faecium ATCC8043	6.2	0.78	0.78	0.78	0.78	0.39
S. agalactiae CMX508	3.1	0.78	0.78	0.78	0.78	0.78
S. pyogenes EES61	3.1	0.39	0.78	0.78	0.78	0.39
E. coli Juhl	0.39	0.02	0.02	0.02	0.05	0.01
Ent. aerogenes ATCC13048	0.78	0.05	0.39	0.1	0.39	0.05
Kleb. pneumoniae ATCC8045	0.2	0.02	0.05	0.01	0.01	0.02
Prov. stuartii CMX640	25	1.56	6.2	6.2	3.1	0.78
P. aeruginosa BMH10	3.1	0.2	0.78	0.2	0.39	0.1
P. aeruginosa A5007	6.2	0.39	0.78	0.39	0.78	0.1
Acine. calcoaceticus CMX669	0.78	0.1	0.2	0.05	0.1	0.39

Efficacy in systemic infections due to *S. aureus* NCTC 10649 and *E. coli* Juhl in mice of several selected compounds and of ciprofloxacin **4** is shown in Table 2. The *in vivo* efficacy on the experimental infection due to *S. aureus* NCTC 10649 of **31** is similar to ciprofloxacin **4**, when administered subcutaneously (sc) or orally (po). Compound **34**, prepared as a prodrug of **31**, was found to be more potent than ciprofloxacin **4** when administered

orally. Both compounds **31** and **34** were found to be
slightly less potent as ciprofloxacin **4** when tested
against *E. coli* Juhl infection model.

As a result of this study, we have found a series of
potent quinolone antibacterial agents having a novel ring
system, namely, 6H-6-oxopyrido[1,2-a]pyrimidine. The
common feature of 1-substituted-1,4-dihydro-4-oxo-pyri-
dine-3-carboxylic acid moiety found in the current clinically used quinolone antibacterials can be replaced by
1,5-disubstituted-2H-2-oxopyridine-3-carboxylic acid
without loss of biological activity. The 2-substituted
amino-3-fluoro-9-(2,4-difluorophenyl)-6H-6-oxopyrido[1,
2a]pyrimidine-7-carboxylic acids possess potent antibacterial activity as well as DNA gyrase inhibitory activity. This study also provides further insights into the
investigation of alternate ring systems in quinolone
antibacterial agents.

<u>Table 2</u> Mouse Protection Test of 6H-6-Oxopyrido[1,2a]-
pyrimidine-7-carboxylic Acids

Test Organism (Dose)	Compound	Route	ED_{50}, mg/kg (95% confidence limits)
S. aureus NCTC10649 ($100xLD_{50}$)	**31**	sc	4.4 (3.2-5.9)
		po	12.0 (7.6-19.0)
	34	sc	2.7 (1.5-4.9)
		po	3.0 (0.4-23.2)
	Cipro	sc	1.6 (1.0-3.5)
		po	15.5 (9.9-24.1)
E. coli Juhl ($100xLD_{50}$)	**31**	sc	1.0 (0.6-1.8)
		po	5.6 (3.6-8.9)
	34	sc	1.7 (0.7-4.3)
		po	14.3 (8.1-25.3)
	Cipro	sc	0.2 (0.1-0.2)
		po	1.9 (1.2-3.0)

4 EXPERIMENTAL SECTION

The 2-substituted amino-3-fluoro-9-(2,4-difluorophenyl)-
6H-6-oxopyrido[1,2a]pyrimidine-7-carboxylic acid derivatives were synthesized at Abbott Laboratories by the
route described in the chemistry section.

The *in vivo* antibacterial activity of the test compound was determined in a side-by-side comparison with

ciprofloxacin **4** and tosufloxacin **8** by conventional agar
dilution procedures. The organisms were grown overnight
in brain-heart infusion (BHI) broth (Difco 0037-01-6) at
36°C. Twofold dilutions of the stock solution
(2000 µg/ml) of the test compound were made in BHI agar
to obtain a test concentration ranging from 200 to
0.005 µg/ml. The plate was inoculated with approximately
10^4 organisms. It was then incubated at 36°C for 18 h.
The minimal inhibitory concentration (MIC) was the lowest
concentration of the test compound that yielded no visi-
ble growth on the plate.

The *in vivo* antibacterial activity of the test compounds was determined in CF-1 female mice weighing approximately 20 g. Aqueous solutions of the test compounds were made by dissolving the hydrochloride salt in
distilled water or by dissolving the compound in dilute
NaOH and diluting it with distilled water to the desired
volume. The median lethal dose of the test organism was
determined as follows.

After 18 h incubation, the cultures of the organism
in BHI broth were serially diluted using 10-fold dilutions in 5% (w/v) hog gastric mucin. Cultures (0.5 ml),
dilution from 10^{-1} to 10^{-8}, were injected intraperitoneally into mice. The LD_{50} for the test organism was
calculated from the cumulative mortalities on the sixth
day using the Reed and Muench procedure.[15]

The 18-h culture of the above was diluted in 5%
(w/v) hog gastric mucin to obtain 100 times the LD_{50} and
0.5 ml was injected intraperitoneally into mice. The
mice were treated subcutaneously or orally with a specific amount of the test compound divided equally to be
administered at 1 and 5 h after infection. A group of
10 animals each for at least three dose levels were thus
treated, and the deaths were recorded daily for 6 days.
Ten mice were left untreated as infection control. Fifty
percent effective dose values (ED_{50}) were calculated from
the cumulative mortalities on the sixth day after infection using the trimmed version of the Logit method.[16]

REFERENCES

1. G.Y. Lesher, E.J. Froelich, M.D. Gruett, J.H.
    Bailey, P.R. Brundage, <u>J. Med. Chem.</u>, 1962, <u>5</u>, 1063.
2. H. Koga, A. Itoh, S. Murayama, S. Suzue, T. Irikura,
    <u>J. Med. Chem.</u>, 1980, <u>23</u>, 1358.
3. D.T.W. Chu and P.B. Fernandes, <u>Antimicrob. Agents
    Chemother.</u>, 1989, <u>33</u>, 131.
4. D.J. Hardy, R.N. Swanson, D.N. Hensey, N.R. Ramer,
    R.R. Bower, C.W. Hanson, D.T.W. Chu and P.B.

Fernandes, Antimicrob. Agents Chemother., 1987, 31, 1768.

5. R. Wise, J.M. Andrews and L.J. Edwards, Antimicrob. Agents Chemother., 1983, 23, 559.

6. Hokuriku Pharmaceutical Co. Ltd. (Japan), Drugs of the Future, 1986, 11, 578.

7. G. Shantz, Drugs, News and Perspectives, 1988, 1, 297.

8. K. Sata, Y. Matsuura, M. Inoue, T. Une, Y. Osada, H. Ogawa and S. Mitsuhashi, Antimicrob. Agents Chemother., 1982, 22, 548.

9. D.T.W. Chu, P.B. Fernandes, A.K. Claiborne, E.H. Gracey and A.G. Pernet, J. Med. Chem., 1986, 29, 2363.

10. D.T.W. Chu and P.B. Fernandes, Adv. Drug Res., 1991, 21, 39.

11. T. Hogberg, I. Khanna, S.D. Drake, L.A. Mitscher and L.L. Shen, J. Med. Chem., 1984, 27, 306.

12. T. Hogberg, M. Vora, S.D. Drake, L.A. Mitscher and D.T.W. Chu, Acta Chem. Scand., 1984, B38, 359.

13. D. Seebach, E. Hungerbühler, R. Naef, P. Schnurrenberger, B. Weidmann and M. Zugar, Synthesis, 1982, 138.

14. Private communication with Dr. Linus Shen.

15. L.J. Reed and H. Muench, Am. J. Hyg., 1938, 27, 493.

16. M.A. Hamilton, R.C. Russo and R.V. Thurston, Environ. Sci. Technol., 1977, 11, 714.

8

New Antibacterial Agents: Synthesis and Antibacterial Activity of Heterocyclic Derivatives of Pseudomonic Acid

G. Walker, P. Brown, M. J. Crimmin, A. K. Forrest, P. J. O'Hanlon, and J. E. Pons

SMITHKLINE BEECHAM PHARMACEUTICALS, BROCKHAM PARK, BETCHWORTH, SURREY, RH3 7AJ, UK

Pseudomonic acid (**1**) is the major component of a group of five closely related antibiotics isolated from <u>Pseudomonas fluorescens</u> NC1B 10586.[1] It is active against Gram-positive bacteria and mycoplasma, and in addition some Gram-negative bacteria such as Haemophilus and Pasteurella are also susceptible.[2]

Pseudomonic acid (1)

Organism		MIC (μg ml^{-1})
Staphylococcus aureus	Oxford	0.25
Staphylococcus aureus	Russell	0.50
Streptococcus pyogenes		0.50
Streptococcus pneumoniae		0.25
Haemophilus influenzae		0.13
Pasteurella multocida		0.50
Escherichia coli		>128
Pseudomonas aeruginosa		>128
Mycoplasma pneumoniae		2.5

MIC = minimum inhibitory concentration

The antibiotic is bacteriostatic and pseudomonic acid is now used clinically as a topical agent for the treatment of skin infections. The mechanism of action is competitive inhibition of isoleucyl tRNA synthetase (IRS). This is an essential enzyme found in all cells which ligates isoleucine to its cognate tRNA prior to incorporation of the isoleucine into protein. The lack of toxicity of pseudomonic acid in man and other mammalian species is due to the very low affinity of pseudomonic acid for mammalian IRS compared to bacterial IRS.

Enzyme from	K_m Isoleucine	K_i Pseudomonic Acid
E.coli B	11×10^{-6} M	3×10^{-9} M
Human liver	24×10^{-6} M	10×10^{-6} M

We have now also shown that the lack of activity against Gram-negative bacteria is not due to poor binding to the IRS enzyme from these organisms, but due to poor penetration through the Gram-negative outer membrane. This mode of action also explains the bacteriostatic action of the compound: inhibition of IRS leads to a build-up of uncharged tRNA which produces the stringent response protecting the bacterium from damage.

This paper describes a series of semisynthetic analogues of pseudomonic acid investigating structure-activity relationships (SAR) around C-1. Some earlier investigations had shown that a conjugated double bond at C2-3 was essential for the retention of antibacterial activity, and that a wide range of esters were acceptable at C-1. Of these esters, some of the most active were the substituted benzyl esters (**2**). We have now prepared a range of heterocycles (**3**), where R is as defined herein, as ester isosteres to probe the SAR in this part of the molecule.[3]

(**2**)   (**3**)

We began by introducing the intact heterocycle and our first approach used an ozonolysis/ Wittig-Horner olefination sequence <u>via</u> ketone (**4**) which is shown in Scheme 1. Protection of the three hydroxyl groups was obviously necessary and we have found the TMS group to be most useful in this series because of ease of protection and deprotection.

A number of heterocycles were prepared by this method, for example the two isoxazoles and the oxazole and oxadiazole shown in Table 1. The reaction always gave predominantly the desired E double bond, and in these and all subsequent cases, the Z isomers were found to lack any antibacterial activity. The Z isomers could, however, be equilibrated to a 2:1 E:Z mixture by either irradiation or fluoride catalysis. Of the four compounds shown only (**5b**) and (**5d**) had significant antibacterial activity and this led us to look at other compounds with ring nitrogen adjacent to the vinylic substituent.

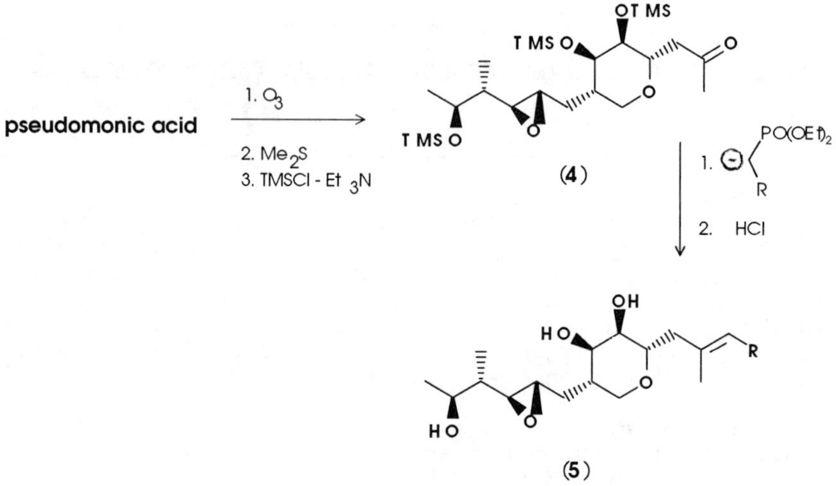

Scheme 1

(5)	R	Yield %	E:Z
(a)	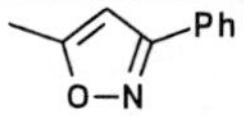 Ph, O–N	36	6:1
(b)	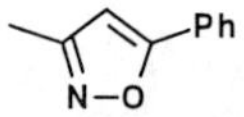 Ph, N–O	18	3:1
(c)	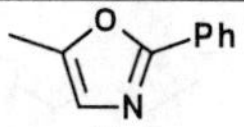 Ph, O...N	10	3:1
(d)	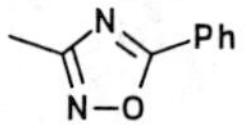 Ph, N–O	29	3:1

Table 1

The Wittig-Horner reaction however proved to be limited in scope because the very slow reaction between phosphonates of this sort and hindered ketone (**4**) gave generally poor yields.

A second approach used the more reactive Peterson reagents to prepare the heterocycles shown in Table 2.

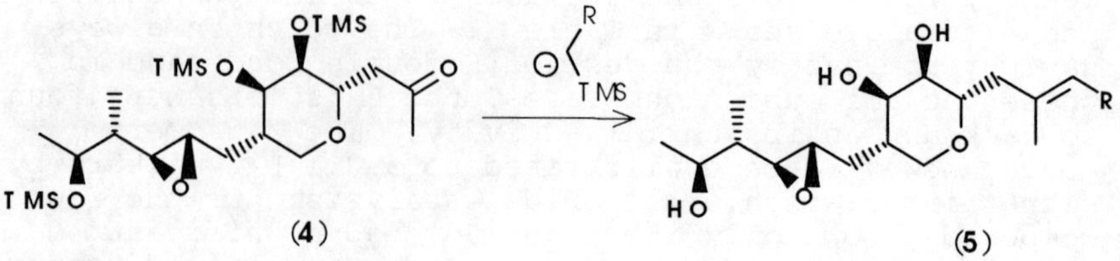

Those containing a nitrogen adjacent to the vinyl substituent were significantly more active and are shown in the table: the E-isomers were predominantly produced but yields were only moderate. A further problem with the Peterson reaction proved to be competitive ring metallation when the ring protons were more acidic than those in the side-chain.

(5)	R	Yield %	E:Z
(e)	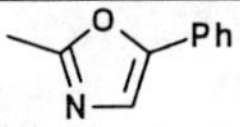	38	92:8
(f)	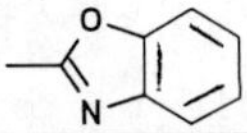	39	60:40
(g)	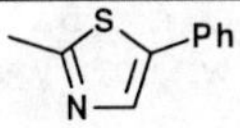	37	80:20
(h)	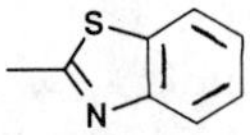	21	52:48
(i)	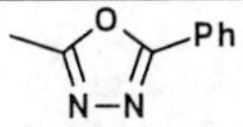	6	90:10
(j)	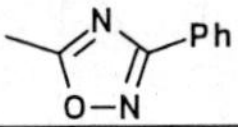	38	80:20

Table 2

Of the compounds shown in Table 2 phenyloxazole (**5e**) was the most active, and so the above difficulties led us on to investigate dehydrative cyclisations for the preparation of such oxazoles, and also diazoles and triazoles, which were difficult to prepare by olefination.

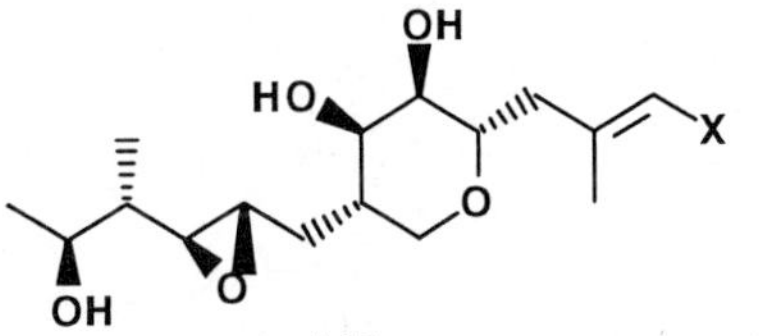

(6)

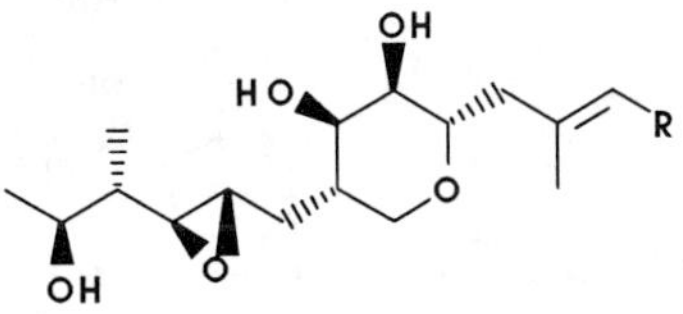

(5)

X

(e)

(i)

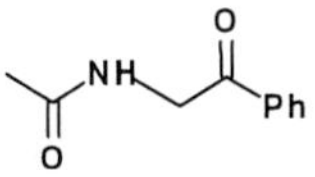

(j)

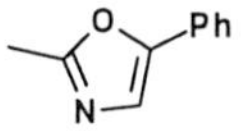

R

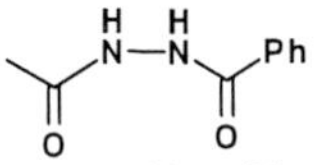

(e)

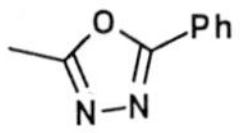

(i)

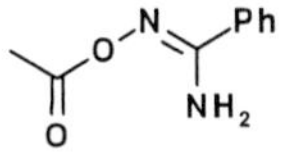

(j)

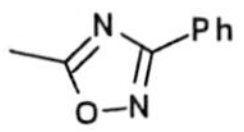

(k)   (k)

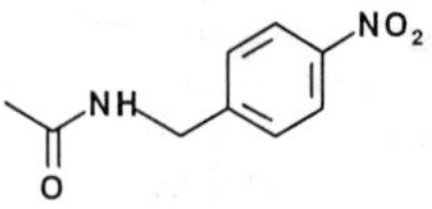

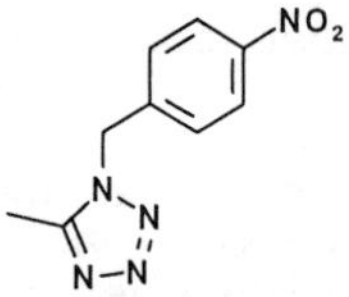

The necessary precursors were easily prepared by
hydrolysis of the ester in pseudomonic acid and amide
formation *via* a mixed anhydride. Because of the
chemical sensitivity of pseudomonic acid and its derivatives, many of the usual cyclisation reagents used in
heterocyclic synthesis proved unsuitable in this
series.[4] However, some standard routes could be used:
for instance oxadiazole (**5j**) could be prepared by
thermolysis of (**6j**),[5] and tetrazole (**5k**) was prepared
from (**6k**) *via* the iminochloride which was generated with
phosgene.[6]

However, in other cases phosgene was unsuitable.
For instance amide (**6e**) gave an oxazolone instead of the
expected oxazole (**5e**). Of many alternative cyclisation
reagents tried by far the most useful proved to be
PPh_3-CCl_4 and PPh_3-C_2Cl_6, especially the former. In most
cases protection of the three hydroxyl groups was not
necessary, as they react very slowly with these
reagents. For example, PPh_3-CCl_4 cleanly produced 1,3,4-
oxadiazole (**5i**) from (**6i**), 1,2,4-triazole (**5l**) from
(**6l**), and oxazoline (**5m**) from (**6m**). Bicyclic derivatives could also be prepared, such as pyridoimidazole
(**5n**) from (**6n**) and benzimidazole (**5o**) from (**6o**).

(6)   (5)

(l)   (l)

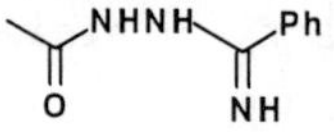

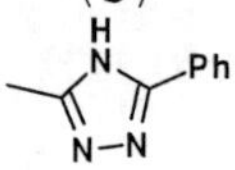

(m) 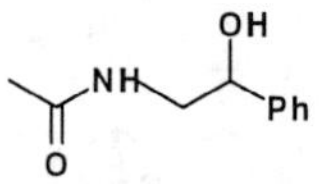   (m)

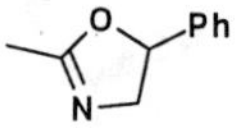

(n) 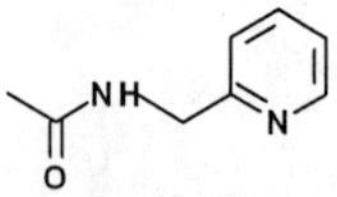   (n)

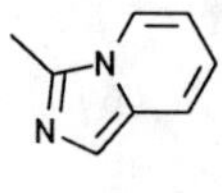

(o) 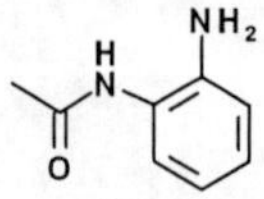   (o)

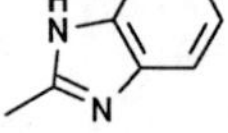

(p) 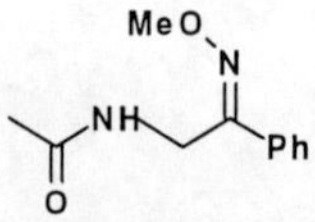   (p)

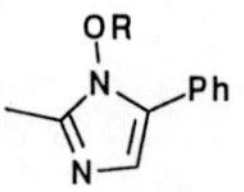

(q) 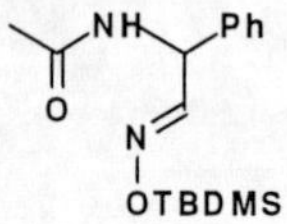   (q)

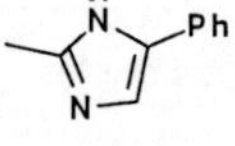

Problems were encountered however in the preparation of phenylimidazole (**5p**) which was intended to be
the precursor of the NH imidazole (**5q**). Oxime (**6p**)
appeared to dimerise on attempted PPh_3-CCl_4 cyclodehydration, presumably because the required oxime Z-isomer is
not generated under the reaction conditions. To test
this hypothesis we prepared the silyl-protected aldoxime
(**6q**), in which the oxime nitrogen lone pair is now
correctly positioned for cyclisation, starting from α-
azidophenylacetaldehyde.[7] Under standard conditions
(**6q**) cleanly cyclised and subsequent deprotection and
deoxygenation ($TiCl_3$) gave (**5q**)[8]. The preparation of
aldoxime (**6q**) was rather tedious and a second route to
imidazole (**5q**) was therefore developed, illustrated by
the preparation of methylthiophenyl imidazole (**5r**).
Thus treatment of nitrile (**6r**) with excess diisobutyl-
aluminium hydride both reduced the nitrile to an imine
and deprotonated the amide. *In situ* cyclisation with
PPh_3-Br_2 then converted the above intermediate directly
to imidazole (**5r**), providing a simple two-pot conversion
of an acid to the corresponding phenylimidazole.

Two final compounds of interest in this series were
the pyrazoles (**5s**) and (**5t**). Acylation of acetophenone
with the isobutyl carbonic anhydride of monic acid gave
diketone (**6s**) which gave pyrazole (**5s**) with hydrazine.
1-Phenyl pyrazole (**5t**) was prepared <u>via</u> cycloaddition:
treatment of hydrazide (**6t**) with PPh_3-CCl_4 in the
presence of acrylonitrile gave cyanopyrazoline (**5u**) and
elimination of hydrogen cyanide with copper (I) gave
(**5t**). Any copper (II) present led instead to oxidation
giving a cyanopyrazole and so oxygen-free conditions
needed to be maintained in this reaction.

MIC and enzyme inhibition results for the above
compounds are shown in the Table 3. In particular the
MIC values for staphylococci and streptococci show that
the oxazole (**5e**) was the most active of the heterocycles
prepared, a result confirmed by enzyme inhibition
measurement.

We have therefore prepared a series of such oxazoles. For range of applicability and large scale preparations, the above PPh_3-CCl_4 cyclisation proved less than ideal in the oxazole series, and we have therefore examined a range of other cyclodehydrating reagents

(5)	R	MIC µg/ml			I_{50} nM
		S.aureus Oxford	*H. influenzae*	*S. pneumoniae*	(*S.aureus* Oxford IRS)
(e)	Ph	2	0.06	0.25	2.7
(m)	Ph	10	1	25	–
(q)	Ph	64	2	64	–
(l)	Ph	>128	0.5	32	>100
(o)		>128	1	128	>100
(n)		16	2	8	5.8
(s)	Ph	64	32	64	27.3
(t)		>128	2	128	>100

Table 3

specifically for the preparation of oxazoles.[9] The most useful reagents proved to be electron deficient acid chlorides, and of these trichloroacetyl chloride in the presence of pyridine was the best. This is the reagent that we now routinely use for the conversion of amidoketones to oxazoles. It has the added advantage in this series of also protecting the three hydroxyl groups as trichloroacetate esters which can easily be removed by treatment with potassium carbonate in methanol. One un-expected disadvantage of the reagent is that it caused *E/Z* isomerisation of the C2/3 double bond when used for the synthesis of 5-alkyl, but not 5-aryl substituted oxazoles. Presumably acylation on the oxazole nitrogen makes the allylic methyl group sufficiently acidic for base-catalysed isomerisation to occur.

A variety of compounds could be prepared in this way. For example, acid (9), a direct analogue of pseu-domonic acid was prepared by enzymatic hydrolysis of ester (**8**).[10] Monosubstituted oxazole (**11**) was prepared by decarbonylation of aldehyde (**10**) using Wilkinson's catalyst.[11]. The antibacterial properties of some representative oxazoles are shown in Table 4.

	R	
(8)	-(CH$_2$)$_8$CO$_2$Me	X = COCCl$_3$
(9)	-(CH$_2$)$_8$CO$_2$H	X = H
(10)	-CHO	
(11)	-H	

R	MIC µg/ml			I$_{50}$nM
	S.aureus Oxford	*H. influenzae*	*S. pneumoniae*	
-H	16	2	128	8.1
-(CH$_2$)$_8$CO$_2$H	4	1	4	5.2
⬡	2	0.06	0.25	2.6
⬡-NO$_2$	0.5	0.5	1	1.3
⬡-SO$_2$Me	1	0.25	0.25	1.8
⬡	8	4	2	9.1
⬡-NH$_2$	8	2	2	7.0

Table 4

The results show that best activity is obtained
when R is an electron-withdrawing aryl group, and that
antibacterial activity and enzyme inhibition equal to or
better than pseudomonic acid can be obtained.

Interestingly for the most electron deficient R
groups, the Cl_3CCOCl-pyridine cyclodehydration proved
unsuitable, not because of difficulties with the
cyclisation, but because the intermediate α-aminoketones
required to prepare the precursor amides were unstable.
To avoid these unstable intermediates a cycloaddition
route was developed for these compounds via the key
tosylmethyl amide (**13**).

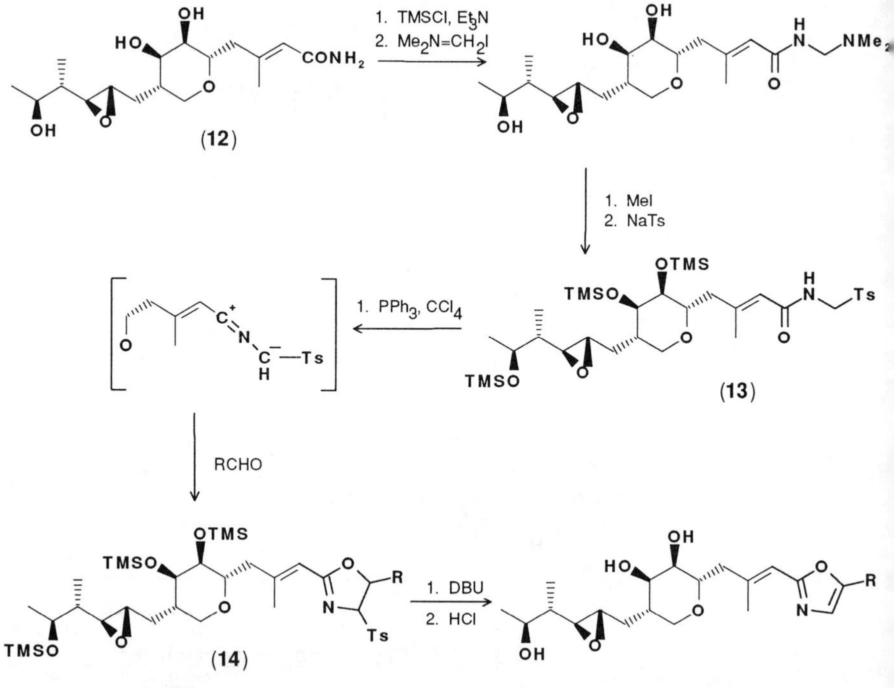

Preparation of (**13**) by direct Mannich reactions on the
primary amide (**12**) proved not to be possible and so the
four step route shown above was developed and shown to
be suitable for kilogram scale preparations. Dehydration of amide (**13**) in the presence of electron-deficient
aldehydes gives oxazoline (**14**) which loses toluenesulphinic acid on base treatment to give the oxazole. Conversely this route is unsuitable for electron-rich aldehydes, in which ring opening of the oxazoline takes
place in preference to elimination of toluenesulphinic
acid.[12]

Although the electron-deficient aryl oxazoles were
clearly the best isosteric replacement for the ester,
the reason for this is not fully understood. At first

we thought that the ester and oxazole might be isoelec-
tronic about C-1, and so we calculated partial charges
using the CNDO method.[13] However, this showed that the
partial charges were considerably different in the oxazoles and esters, indicating that they are not isoelec-
tronic. Furthermore, the oxazole substituent did not
significantly affect the partial charge on the nitrogen,
oxygen, or C-2 carbon of the oxazole ring, showing that
this was not the cause of the different activities of
the variously substituted oxazoles.

Likewise we found no correlation between antibacterial activity and the calculated planarity between the
oxazole and phenyl rings.[13]

A correlation was, however, found between polarity
calculated from hplc retention times and antibacterial
activity. The more polar compounds by this measure were
generally less active, but this failed to distinguish
between electron-poor and electron-rich systems.

Finally, molecular modelling provided an insight
into the connection between the esters and the heterocycles. The electrostatic potential around the ester
(**15**) and oxazole (**16**) was calculated from atomic charges
fitted to reproduce the electrostatic potential calculated from an AM1 waveform.[14] It was found that they
showed a common area of potential which was necessary
for activity, associated with the lone pairs on the
carbonyl oxygen/ring nitrogen.

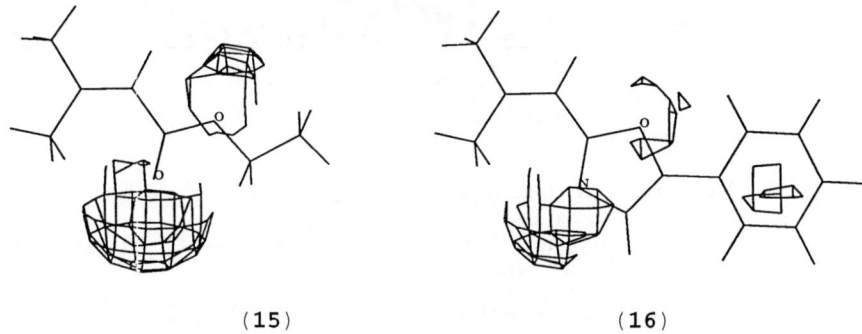

(**15**)   (**16**)

Electrostatic Potential : 10eV contour

Heterocycles not showing electrostatic potential in
this area are inactive. The smaller area of electrostatic potential associated with the oxazole oxygen was
not necessary. To date no functional group in the IRS
enzyme which might interact with this electrostatic
potential has yet been identified.

In conclusion, we have shown how a wide range of
heterocycles, some of which retain antibacterial
activity, may be incorporated into the chemically
fragile pseudomonic acid structure. The most active

series has been identified and methods for the synthesis of compounds in this series have been developed.

References

1.   E.B. Chain and G. Mellows, J. Chem. Soc., Perkin Trans. I, 1977, 294.

2.   M.J. Basker et al in 'Current Chemotherapy and Infectious Diseases' Vol.1, 1980, 471, Eds. J.D. Nelson and C. Grassi, American Society for Microbiology, Washington.

3.   M.J. Crimmin et al, J. Chem. Soc., Perkin Trans. I, 1989, 2047.

4.   I.J. Turchi in 'Chemistry of Heterocyclic Compounds' Vol.45, 1986, 1, Eds. A. Weissberger and E.C. Taylor, Interscience, New York.

5.   K. Clarke, J. Chem. Soc., 1954, 541.

6.   H. Ulrich in 'Chemistry of Imidoyl Halides', 1968, 1, Plenum Press, New York.

7.   H. Bretschneider and N. Karpitschka, Monatsh, 1953, 84, 1021.

8.   B.H. Lipschulz and M.C. Morey, Tet. Lett., 1983, 4583.

9.   M.J. Crimmin et al, J. Chem. Soc., Perkin Trans. I, 1989, 2059.

10.   J.T. Sime, C.R. Pool, and J.W. Tyler, Tet. Lett., 1987, 5169.

11.   J. Tsuji and K. Ohno, Synthesis, 1969, 157.

12.   H.A. Houwing and A.M. van Leusen, J. Het. Chem., 1981, 18, 1127 and 1133.

13.   M.J.S. Dewar, E.G. Zoebisch, E.F. Kealy, and J.J.P. Stewart, J. Amer. Chem. Soc., 1985, 107, 3902.

14.   G.G. Ferenczy, C.A. Reynolds, and W.G. Richards, J. Comp. Chem., 1990, 11, 159.

The Asymmetric Synthesis of (+)-Calyculin A, a Nanomolar Phosphatase 1 and 2A Inhibitor

D. A. Evans,* J. L. Leighton, and J. R. Gage
DEPARTMENT OF CHEMISTRY, HARVARD UNIVERSITY, CAMBRIDGE, MA 02138, USA

Calyculin A was isolated in 1986 by Fusetani and co-workers from the marine sponge *Discodermia calyx* (Figure 1).[1] Its structure (relative configuration) was determined by X-ray analysis and confirmed by the usual complement of spectroscopic tools. Calyculin A has been shown to be a potent tumor promoter, an activity which has been linked to its effective inhibition of protein phosphatases 1 and 2A,[2] two of the four major protein-serine/threonine phosphatases, exhibiting IC_{50} values approximately 1 nM.[3] This activity profile is similar to that of the marine natural product okadaic acid.[4] The activity of both compounds is complementary to and equipotent to the phorbol ester family of protein kinase C activators. Both compounds are of considerable current value in identifying those cellular processes regulated by protein-serine/threonine phosphatases.[5]

Subsequent to the initiation of our synthetic studies, the absolute configuration of calyculin was proposed to be opposite to that illustrated below on the basis of the CD spectrum of the C_{33}-C_{37} γ-amino acid obtained in the degradation of the natural product.[6] More recently, an unambiguous synthesis of this degradation product by Shioiri and co-workers confirmed the Fusetani absolute configuration assignment.[7]

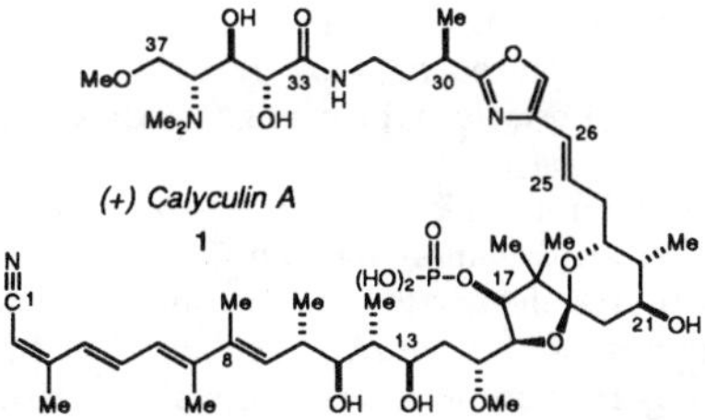

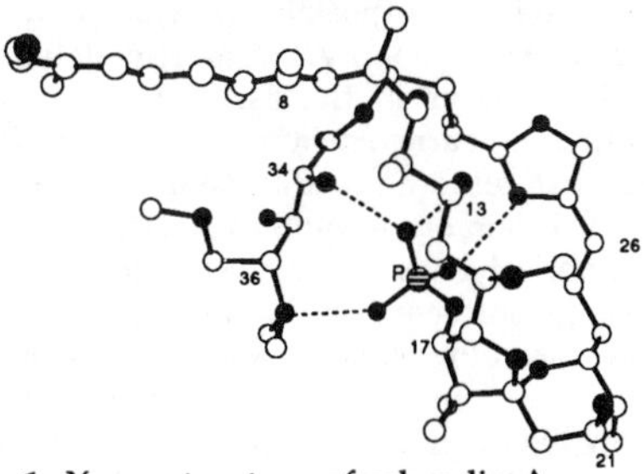

Figure 1. X-ray structure of calyculin A

Since 1986, Fusetani has reported the isolation of seven closely related natural products, calyculins B-H, also from *Discodermia calyx* (Figure 3).[8] Since X-ray crystal structures have not been obtained for any of these analogues, structural assignments rest primarily on NMR experiments and chemical interconversion. Four of the new calyculins, C, D, G, and H, carry an additional methyl group at C_{32}, the configuration of which was deduced from NOE experiments. The only other difference among the eight calyculins lies in the double bond geometries of the conjugated tetraenes, and these have also been assigned based on NOE data. Calyculins A, B, E, and F can be interconverted by exposure to light, as can calyculins C, D, G, and H. Of the eight, calyculin A appears to be the most abundant.[9] The possibility that double bond isomerism is an artifact of the

Calyculin Derivatives Isolated from *Discoderma calyx*

	$\Delta_{6,7}$	R_1	R_2
Calyculin A	E	CN	H
Calyculin B	E	H	CN
Calyculin E	Z	CN	H
Calyculin F	Z	H	CN

Fusetani & Co-workers

	$\Delta_{6,7}$	R_1	R_2
Calyculin C	E	CN	H
Calyculin D	E	H	CN
Calyculin G	Z	CN	H
Calyculin H	Z	H	CN

isolation has not been ruled out. The presence of the additional methyl group at C_{32} in calyculins C, D, G, and H and changes of geometry in the tetraene portion of the molecule have been shown to have a negligible effect on the biological activity.[10]

This lecture describes our successful efforts to develop an asymmetric synthesis of (+)calyculin A, an effort that has culminated in the construction of the enantiomer of the natural product.[11]

The Synthesis Plan

In planning the synthesis of complex organic target structures, it is of vital importance to incorporate the boundary conditions which define the limits of chemical stability of the target into the final stages of the synthesis plan.[12] One might expect this to be true especially in a structure with as much varied functionality as that present in calyculin A. Unfortunately, there was no supply of the natural product available to us in the early stages of the project. To compensate, we relied on information garnered from model systems, literature precedent, and a flexible synthetic plan that could be readily adjusted to circumvent obstacles which might arise during the evolution of the synthesis plan.

Since calyculin A is amenable to standard isolation techniques such as extraction and silica gel chromatography, we expected that the final isolation and purification of synthetic calyculin would be a manageable problem. We also knew from the outset that the phosphate ester moiety is stable to hydrolysis in the natural product. In fact, attempted degradation experiments carried out by Fusetani were unsuccessful in excising the phosphate residue. We speculate that the stability of this group is largely due to the hydrogen bond network which encapsulates the phosphate monoester, a feature which would be lacking in protected intermediates. The information concerning photolability of the conjugated tetraene portion was not published until recently, but it only came as a confirmation of suspicions that we already harbored. We therefore sought to design a scheme in which the conjugated tetraene and phosphate ester could be introduced late in the synthesis.

The value of convergency in the total synthesis of complex targets is well known, and we naturally hoped to make our synthesis as convergent as possible. However, consideration of the demands placed by the functional groups of calyculin A and the paucity of knowledge of their reactivity dictated that flexibility receive equal priority in our

Scheme 1. The Calyculin Subunits

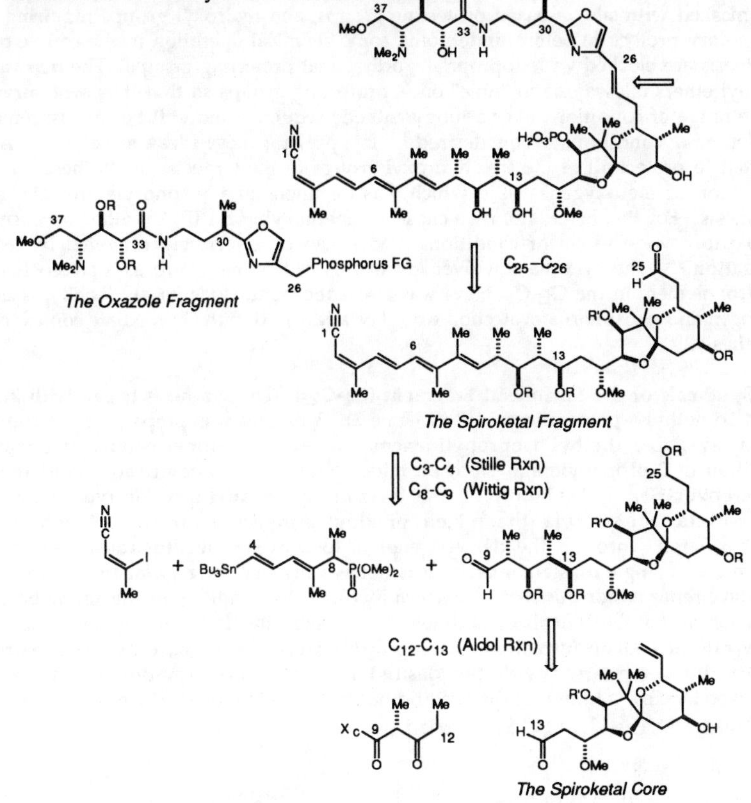

The Oxazole Fragment

The Spiroketal Fragment

C_3-C_4 (Stille Rxn)
C_8-C_9 (Wittig Rxn)

C_{12}-C_{13} (Aldol Rxn)

The Spiroketal Core

thinking. In addition to the conjugated tetraene and the phosphate ester, questions remained about when to install or reveal other potentially troublesome functional groups such as the dimethylamine moiety at C_{36}, the amide, and the *trans* double bond at C_{25}-C_{26}. With these sometimes opposing goals of convergency and flexibility in mind, we set about the task of retrosynthetic analysis hoping to strike an appropriate balance.

We chose to incorporate the spiroketal ring system early in the synthesis in order to simplify the problem of hydroxyl group protection. This decision dictated that major bond disconnections be made on either side of this fragment (Scheme 1). Disconnection of the C_{25}-C_{26} olefin meets this requirement and conveniently segments calyculin A into two relatively large fragments of comparable complexity. A phosphorus-based olefination procedure appeared attractive in this setting due to the relatively mild conditions needed for bond formation. Of the two major fragments produced by disconnection at C_{25}-C_{26} bond, the oxazole-containing fragment, representing C_{26}-C_{37}, appeared to be somewhat simpler, at least in terms of advantageous sites for major bond constructions, if not with regard to reactive functional groups. Disconnection at the amide bond produces two smaller pieces: a stereochemically complex γ-amino acid diol and an amino oxazole which could be prepared from a γ-amino acid.

Disconnection of the C_1-C_{25} spiroketal fragment at the three illustrated sites provided the four smaller fragments which served as our immediate goals for synthesis. Protecting group strategy inevitably becomes a major consideration in a synthetic undertaking of this size. We elected to adopt what could be termed the *cumulative silicon strategy*, an approach successfully demonstrated in our recent total synthesis of cytovaricin.[12b] That is, hydroxyl groups to be protected until the end of the synthesis are masked with silicon-based protecting groups, and hydroxyl groups requiring only temporary protection before undergoing some chemical operation in the course of the synthesis are blocked with appropriate, orthogonal protecting groups. The rich variety of silyl ethers allows one to "tune" one's protecting groups so that they will survive a wide range of conditions over a long synthetic sequence and still be readily removed under mild conditions when desired.[13] In applying these ideas to calyculin A, we elected to protect all of the free hydroxyl groups in the target as silyl ethers with the exception of the oxygen at C_{17} which was destined for phosphorylation late in the synthesis. For this heteroatom we chose a p-methoxybenzyl (PMB) ether because it is stable to a wide range of conditions, and it can be selectively removed by a mild oxidation.[14] The eventual involvement of this protecting group as a stereochemical control element in the C_{15}-C_{16} bond was also anticipated (Vide infra). Finally, issues of nitrogen and phosphorus protection would be addressed in the immediate context of the synthesis.

Synthesis of the Spiroketal Fragment (C_1-C_{25}). The synthesis began with known 3,3,4-trimethyl-4-pentenoic acid[15] (Scheme 2). This acid was prepared from commercially available diethyl isopropylidenemalonate by a copper-catalyzed conjugate addition of 2-propenylmagnesium bromide followed by saponification and thermal decarboxylation. Acylation of the oxazolidinone auxiliary derived from (S)-phenylalanine[16] via the mixed pivaloyl anhydride afforded **1** which was hydroxylated through the derived enolate to provide the illustrated α-hydroxyl imide **2** as a single diastereomer.[17] It is noteworthy that the regular procedure which we have reported for this reaction, which involves slow addition of the amide base to a cold solution of the N-acyloxazolidinone and oxaziridine, failed in this case due to the slow rate of enolate formation, making direct attack of the base on the oxaziridine kinetically competitive with the desired reaction. Preformation of the enolate followed by slow addition of the oxaziridine circumvented this problem.

Scheme 2. The C_{15}-C_{17} Stereotriad

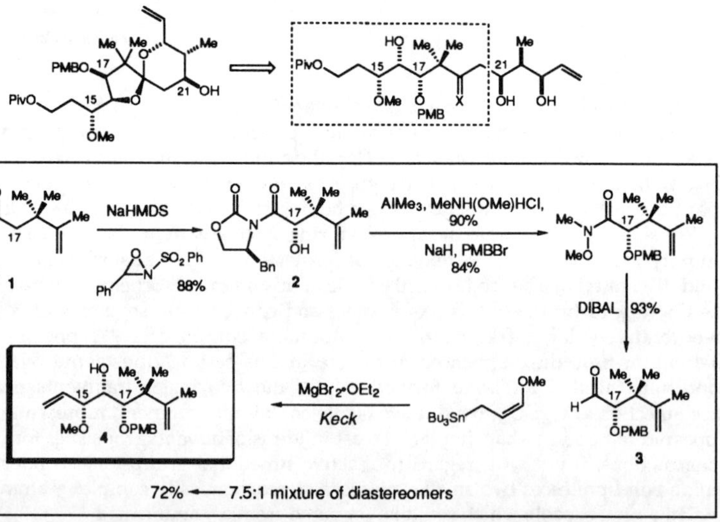

72% ◄── 7.5:1 mixture of diastereomers

Chiral auxiliary removal was effected by transamination under the advertised conditions[18] to provide the derived hydroxamic acid which was transformed into its PMB ether (NaH, p-methoxybenzyl bromide). The presence of the free hydroxyl moiety in the transamination of **2** was found to be critical to the success of the transformation. We speculate that activation of the exocyclic acyl moiety is being achieved by aluminum(III) chelation between the vicinal oxygen heteroatoms.

Chelation also plays a pivotal role in the the stereoselective allylation of 3. Treatment of this aldehyde with the illustrated methoxyallylstannane and magnesium bromide etherate led to the formation of the desired alcohol 4 along with a minor diastereomer in a 7.5:1 ratio according to the precedent of Keck and Koreeda.[19] These diastereomers were readily separated by silica gel chromatography, affording a 71% isolated yield of intermediate 4. During the planning stages of the synthesis the organizational role of the C_{17} oxygen had been anticipated, and for that reason the PMB ether was selected as the protecting group for this heteroatom.

At this stage we were faced with a problem of olefin differentiation. Since we planned to construct the C_{20}-C_{21} prior to the C_{12}-C_{13} bond, we initially attempted to selectively convert the 1,1-disubstituted olefin to the derived methyl ketone. Epoxidation of the hindered, but relatively electron-rich, 1,1-disubstituted olefin proceeded smoothly without affecting the much less-hindered, electron-poor monosubstituted olefin at the other terminus. Unfortunately, all attempts at oxidative scission[20] of this epoxide led to decomposition, either through loss of the PMB ether, or, when the C_{16} alcohol was protected as a methoxyethoxymethyl ether, by participation of this oxygen leading to tetrahydrofuran formation.

Scheme 3. Olefin Differentiation: Catalyzed Hydroboration

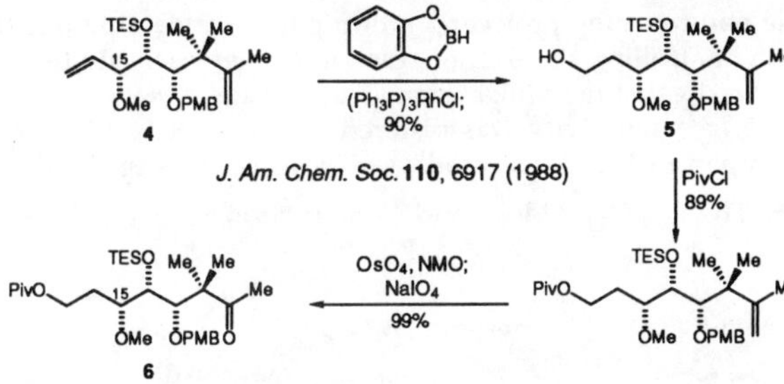

J. Am. Chem. Soc. **110**, 6917 (1988)

A solution to the controlled oxidation of diene 4 was found when it was discovered that the monosubstituted olefin could be hydroborated with complete selectivity under Rh(I) catalysis, a reaction which we have spent considerable energy developing as a synthetic method.[21] In contrast, other hydroborating agents such as 9-borabicyclononane (9-BBN) did not proceed to completion, and, upon oxidation, resulted in a 2:1 mixture of 5 and the regioisomeric alcohol derived from hydroboration at the other

Scheme 4. The C_{21}-C_{25} Synthon

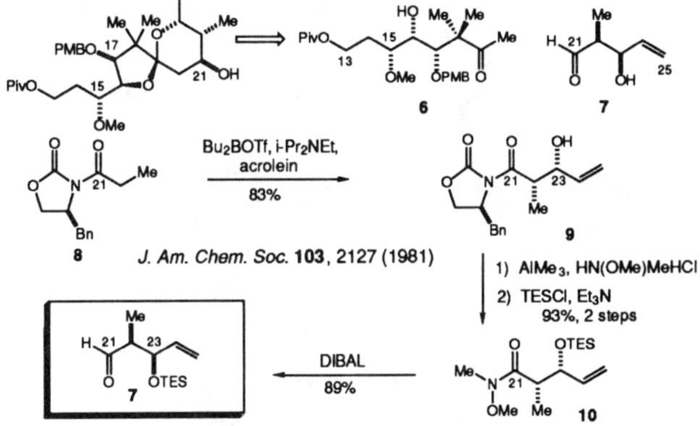

J. Am. Chem. Soc. **103**, 2127 (1981)

olefinic site. The completed C_{13}-C_{20} calyculin fragment **5** was finally obtained after an uneventful olefin osmylation and subsequent periodate cleavage.

The synthesis of the other spiroketal fragment **7** comprising the C_{22}-C_{25} portions of the calyculin skeleton is outlined in Scheme 4. The two stereochemical relationships in this segment where conveniently constructed via the boron enolate aldol reaction developed by us some years ago. The aldol adduct **9** was obtained in excellent yield with complete stereocontrol. The transformation of **9** via the "Weinreb amide" to **10** was then followed by convenient DIBAL-H reduction to give the sensitive aldehyde which was carried on to subsequent transformations without purification.

The incorporation of aldehyde **7** into the spiroketal framework is illustrated below (Scheme 5). The key transformation to be addressed is the C_{20}-C_{21} aldol bond construction. The stereochemical course of this reaction is readily predicted on the basis of the Felkin-Anh hypothesis, and the desired Felkin-Anh diastereomer is that diastereomer required by the synthesis plan. *In the event, union of lithium enolate **11** and aldehyde **7** resulted in the rather unexpected observation that the incorrect (and unanticipated) C_{21} alcohol diastereomer was obtained with good reaction diastereoselectivity.* With this unprecedented observation before us, the effect of enolate and aldehyde structure on reaction diastereoselectivity were evaluated (eq 1-3). In the illustrated model studies, the enolate derived from pinacolone was employed as a suitable surrogate for the actual enolate **11**. As illustrated in these experiments, lithium enolates do indeed exhibit *anti*-Felkin behavior in the addition reactions to this family of chiral aldehydes. Furthermore, the nature of the protecting group β to the aldehyde function plays no significant role in controlling the overall reaction diastereoselectivity. On the other hand, when we investigated the Mukaiyama aldol reaction variant (eq 3),[22] the proper sense of aldehyde face selectivity was restored. This latter result is not altogether unexpected based upon analogous observations made some years ago by Heathcock.[23]

Scheme 5. The C_{20}-C_{21} Aldol Bond Construction

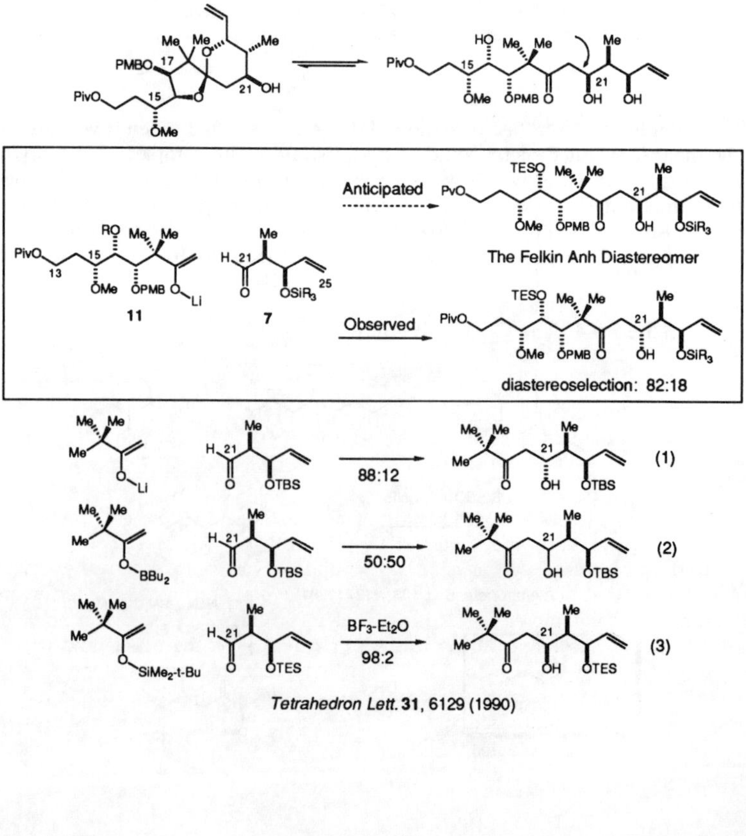

Application of the preceding Mukaiyama aldol reaction to the actual system proceeded with superb diastereoface selectivity (eq 4) wherein the desired desired aldol adduct 12 was obtained as a single diastereomer in 80% isolated yield.

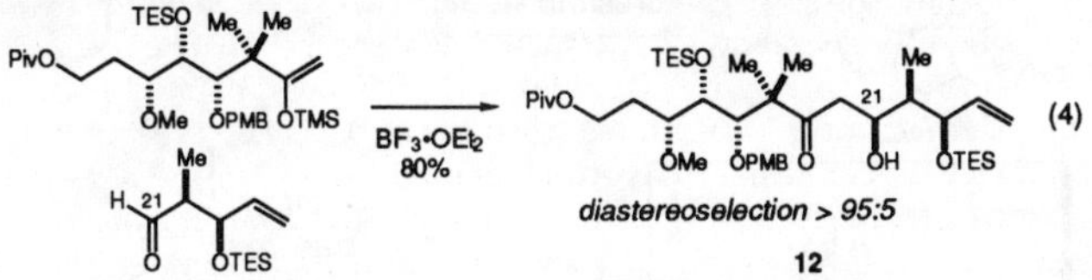

$$BF_3 \cdot OEt_2 \quad 80\%$$

diastereoselection > 95:5

(4)

12

In retrospect, the anti-Felkin behavior observed with hindered methyl ketone enolates with the illustrated aldehydes is unexpected[24] and current models on carbonyl addition do not adequately deal with this observation. One could explain the turnover in face selectivity by invoking β heteroatom chelation on the aldehyde fragment; however, there is absolutely no evidence in the literature that hindered silyloxy substituents of the type found in these systems will participate in such chelate organization.

Upon desilylation of 12 under the usual conditions, two spiroketal products were obtained in 71 % and 14 % yields respectively (Scheme 6). These two diastereomers could be readily separated by silica gel chromatography, and their structures were unequivocally determined by NMR spectroscopy. As anticipated, the major spiroketal diastereomer 13 possessed the requisite stereochemistry needed for the calyculin synthesis. The illustrated mixture of 13 and 14 represents an equilibrium ratio of isomers, a point confirmed by approaching the equilibrium from both diastereomers. *In following the course of this reaction we were surprised to find that the minor spiroketal diastereomer 14 is that isomer which is kinetically favored during spiroketalization.* At the present time we have no compelling explanation for the kinetic selectivity for 14. The analysis of this transformation is obscured by a lack of information on the rate in which silicon protecting groups are being lost from 12 prior to the initiation of spiroketalization. It is also not clear whether loss of the protecting groups is in fact required before spiroketalization is initiated.

Scheme 6. Spiroketalization

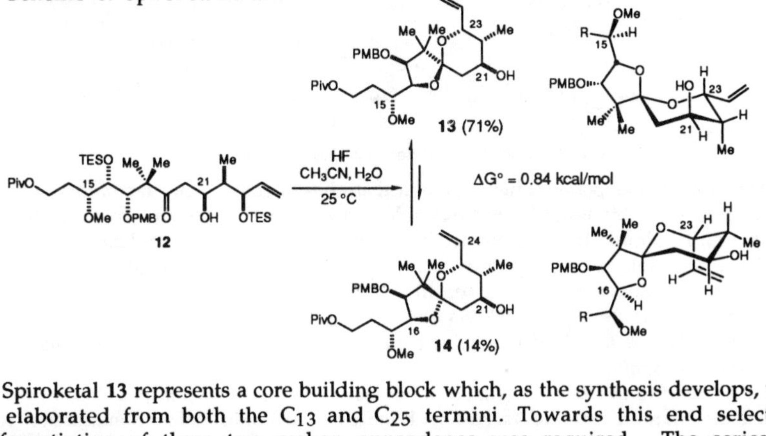

$$\Delta G° = 0.84 \text{ kcal/mol}$$

HF
CH₃CN, H₂O
25 °C

13 (71%)

14 (14%)

Spiroketal 13 represents a core building block which, as the synthesis develops, will be elaborated from both the C_{13} and C_{25} termini. Towards this end selective differentiation of these two carbon appendages was required. The series of transformations that we found to be effective in elaborating 13 to the eventual derivatized spiroketal moiety 17 is illustrated in Scheme 7. One of the noteworthy transformations encountered during this set of reactions is the ultrasound-promoted hydroboration of the C_{24}-C_{25} olefin. We had to resort to this strategy to effect a convenient hydroboration of the substrate after other borane reagents failed. The precedent for the use of ultrasound in this reaction can be found in the work of Brown.[25]

Scheme 7. C_{13} & C_{25} Refunctionalization

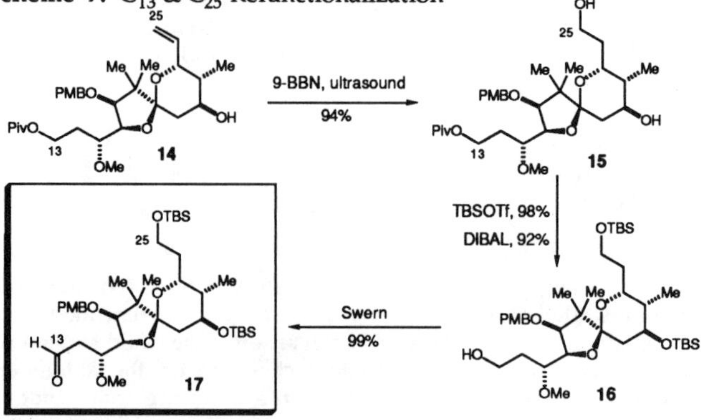

The next stage in the elaboration in the spiroketal fragment was the incorporation of the "polypropionate" region extending from C_9 to C_{13} (Scheme 8). From aldehyde **17**, one in principle could accomplish the construction of these four stereocenters via the sequential application of two *anti*-selective enantioselective aldol reactions. Although the first of these bond constructions should proceed in accord with expectation, the second of the *anti* aldol bond constructions represents a miss-matched double stereodifferentiating aldol bond construction wherein the enolate face selectivity would be required to override the aldehyde face selectivity.[26] The prognosis for such a stereochemical override in the literature is not encouraging.

Scheme 8. The Polypropionate Region

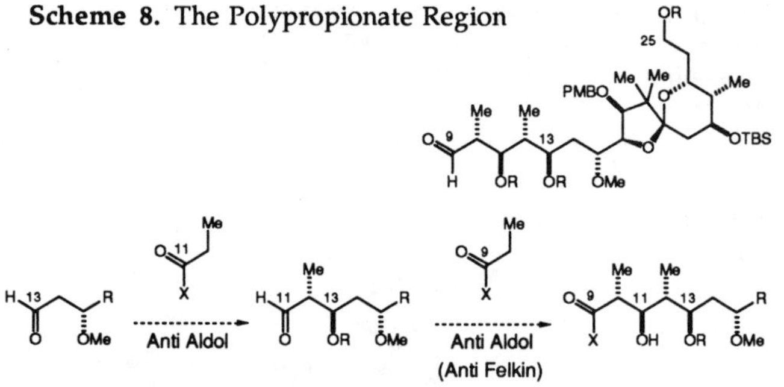

■ The second aldol reaction is a problem: Chirality on reacting partners is unmatched.

Fortunately, independent methodological studies ongoing in our laboratory could be used to address this issue.[27] For example, we have been interested in exploiting the utility of β-ketoimide **18** in stereoselective aldol bond constructions (Scheme 9). To date, we have been able to develop three of the four possible aldol bond constructions from this chiral ethyl ketone derivative. Both face selectivity and aldol stereochemical relationships can be controlled via the selection of the appropriate metal enolate. *Unfortunately, the fourth of the possible stereochemical variants is the one required for the calyculin synthesis!*

In an effort to utilize the available set of aldol bond construction possibilities, we selected the titanium enolate variant with suitable post-aldol refunctionalization reactions to solve the problem (Scheme 10). By inspection, one can see that this reaction establishes the required *syn* stereochemical relationship between the C_{10} and C_{12} methyl groups. This good fortune must be weighed against the requirement that the C_{13} hydroxyl-bearing stereocenter established in this reaction must be inverted in a subsequent step.

Scheme 9. Synthesis of the Propionate Fragments with β-Ketoimides

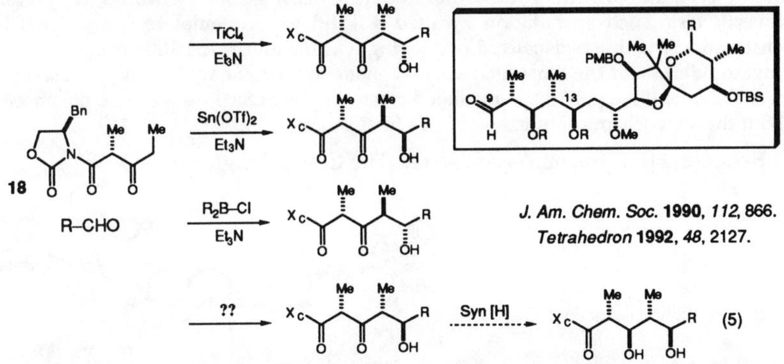

J. Am. Chem. Soc. **1990**, *112*, 866.
Tetrahedron **1992**, *48*, 2127.

When aldehyde **17** was treated with the titanium enolate derived from ethyl ketone **18** the expected aldol adduct **19** was obtained in excellent yield and stereoselectivity. This transformation established the desired methyl-bearing stereocenters at C_{10} and C_{12}, along with the *incorrect* hydroxyl-bearing stereocenter at C_{13}. This stereochemical defect was then adjusted by hydroxyl inversion. Prior to this transformation, we carried out a "directed" borohydride reduction[28] to give the requisite hydroxyl-bearing stereocenter at C_{12}, a transformation which was followed by Mitsunobo inversion[29] at the C_{13} stereocenter. This latter transformation is noteworthy in that the overall reaction was carried out without intermediate protection of the C_{12} hydroxyl group. From our own experience, hydroxyl groups flanked by branching carbon centers on both sides of the hydroxyl moiety are inert to inversion reactions of this type. We were indeed fortunate, that calyculin lacked a methyl-bearing stereocenter at C_{14}. Had this been the case, this portion of the synthesis would have been far more challenging!

Scheme 10. C_{12}–C_{13} Bond Construction: The Solution

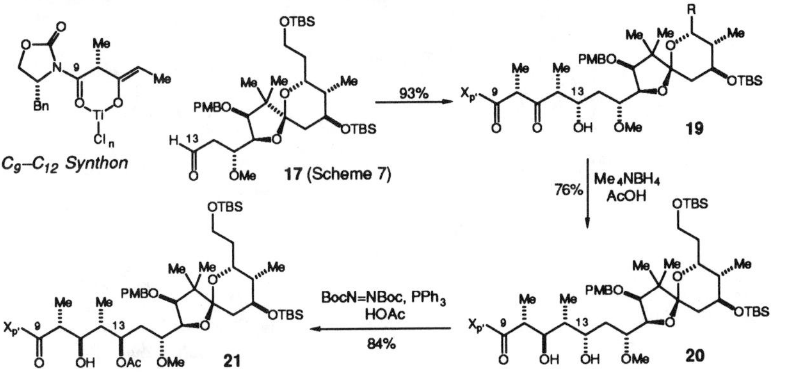

The final task remaining in the synthesis of the C_1-C_{25} calyculin fragment was the appendage of the cyanotetraene portion of the molecule. This subunit is no doubt responsible for the light- and oxygen-sensitivity of this natural product. Accordingly, reactions employed in the generation of this moiety needed to be compatible with the sensitivity of this organic architectural element. The plan which we adopted is shown below (Scheme 11). It was our intention to construct the C_8-C_9 bond through the illustrated Wittig olefination using the vinylstannane **22** as its derived conjugate base. During the initial stages of this investigation we were mildly concerned about the overall (*E*)-stereoselectivity of this bond construction as no adequate stereochemical precedent exists in the literature for this process. However, one might expect, based on thermodynamic considerations, that the (E) olefin geometry would be preferred. Nevertheless, such speculation awaited a valid experimental test. The final bond construction which we visualized employing was the illustrated Stille coupling between vinyl iodide 23 and the illustrated vinylstannane to form the C_3-C_4 bond. Based on the mechanism of these palladium-mediated cross-coupling reactions we were not concerned about the stereochemical outcome of this final process.

Scheme 11. Tetraene Assemblage: The C_1-C_{25} Fragment

■ The Wittig Olefination: C_8-C_9

Δ^8 Stereochemistry problematic

■ The Stille Coupling: C_3-C_4

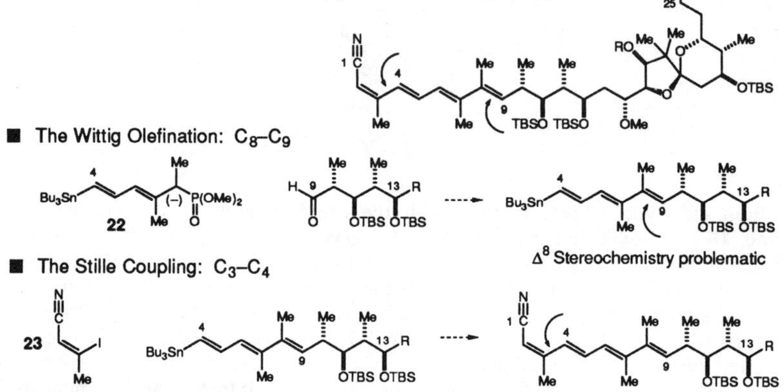

The execution of the tetraene assemblage began with the selective oxidation of diol 27 (Scheme 12). After some experimentation we found that the "Dess-Martin"[30] reagent provided exquisite selectivity for the C_9 alcohol. In this transformation, we speculated that the unprotected C_{17} secondary alcohol, being located in a sterically congested environment, might be oxidized at a slower rate. In fact, this prediction proved to be correct: the desired hydroxyaldehyde 28 was obtained in excellent yield.

Scheme 12. Tetraene Assemblage: The C_1-C_{25} Fragment

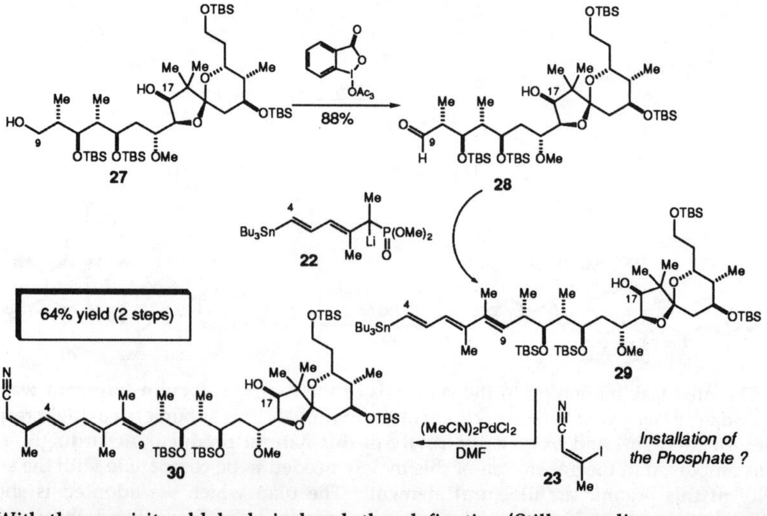

With the requisite aldehyde in hand, the olefination/Stille coupling sequence was attempted. In the first of these reactions, the metallated phosphonate 22 and aldehyde 28 were coupled uneventfully via the Wittig reaction to provide vinylstannane 29 which was carried directly to the Pd-mediated Stille coupling with vinyl iodide 23. As a result of this cascade of carbon-carbon bond constructions we were able to isolate the cyanotetraene 30 in 64% yield over the two-step sequence. The overall stereoselectivity for the two step process averaged between 5 and 7:1, and we strongly speculate that the source of the stereochemical imperfection resides in the Wittig bond construction. Stereochemical assignment at the C_9 olefinic linkage was secured by an unambiguous set of NOE NMR experiments.

Introduction of the Phosphate Ester. In the assemblage of the calyculin skeleton there are several key points at which the phosphate residue might be incorporated. In view of the hindered nature of the C_{17} hydroxyl group, the site of phosphorylation, we elected to introduce the phosphate moiety prior to the Wittig union of the individual subunits comprising the upper and lower portions of the skeleton. This decision brought us to the issue of developing the appropriate phosphorus reagents that might be sufficiently reactive to affect C_{17} hydroxyl phosphorylation without compromising the sensitive cyanotetraene portion of the structure. In order to address these questions we embarked on a series of phosphate ester model studies on the spiroketal fragment 31 (Scheme 13). This model was chosen to reflect the actual steric hindrance that might be encountered during phosphorylation as well as the silicon deprotecting steps that might be required during the terminal stages of the synthesis. After a number of unsuccessful attempts at phosphorylating this hindered secondary alcohol, we found that the expedient use of phosphorus trichloride and pyridine led to the transformation of 31 into the derived dichlorophosphite ester which could be esterified with a variety of alcohols in the same reaction vessel. We further found that an *in situ* oxidation to the corresponding mixed phosphotriester could be achieved with aqueous hydrogen peroxide to give the phosphotriesters 32 carrying representative protecting groups. Each of these phosphotriesters were subjected to not only the conditions of the Wittig reaction, but also to the conditions associated with silicon deprotection. From these experiments we concluded that the PMB moiety would serve our purposes admirably. For example, 32 (R = PMB) survives deprotection of the primary OTBS moiety (HF/pyridine) but is rapidly removed in the presence of aqueous HF in acetonitrile, conditions under which secondary OTBS ethers are normally cleaved.

Scheme 13. Phosphate Model Studies

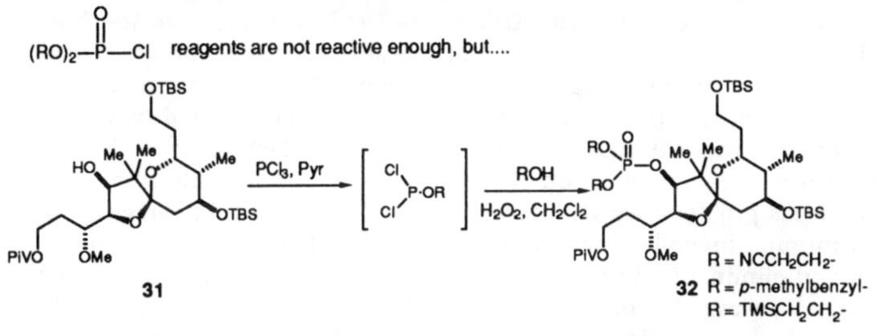

The C26–C37 Calyculin Subunit. The C_{26}-C_{37} calyculin subunit, although lacking the stereochemical complexity of the spiroketal portion moiety, carries an array of synthetic challenges, including a number of sensitive nitrogen heteroatom functional groups. On the other hand, this fragment can be conveniently dissected at the amide bond into two manageable fragments.

The Oxazole Fragment. In turning to the synthesis of the oxazole-containing calyculin subunit, we were faced with selecting an appropriate diastereoselective reaction which might be employed to establish the C_{30} methyl-bearing stereocenter adjacent to the oxazole ring. The two bond constructions which we entertained are illustrated in Scheme 14. The major distinguishing feature in the two transformations lies in the strategy by which the terminal amino substituent is incorporated into the evolving structure. Of the two possibilities we elected to proceed with the illustrated Michael reaction via the titanium enolate.[31] This transformation leading to adduct 33 proceeded in superb diastereoselection and yield to provide the differentiated α-

Scheme 14. Oxazole Fragment: The C_{30} Stereocenter

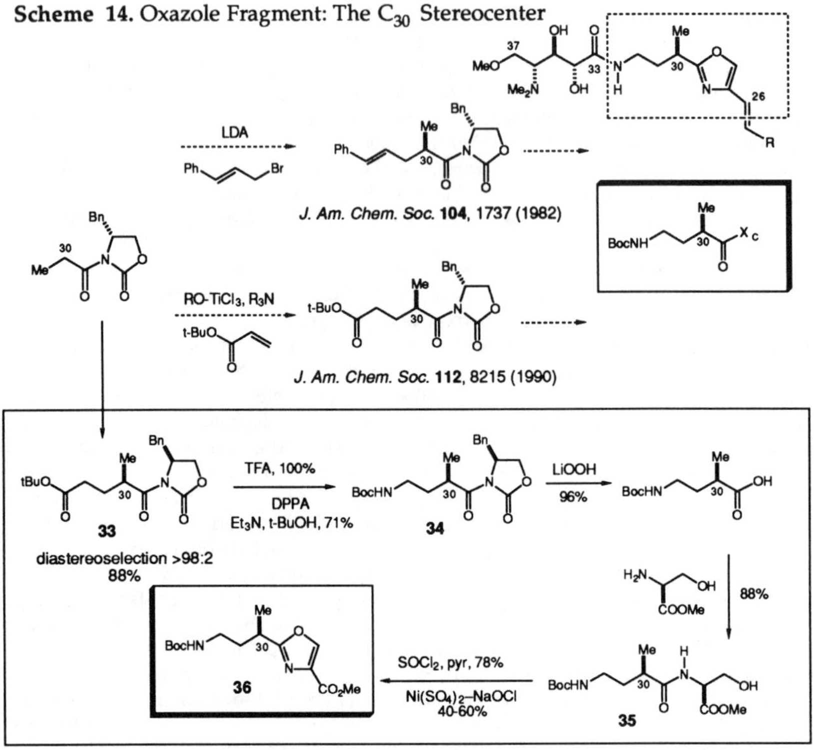

J. Am. Chem. Soc. **104**, 1737 (1982)

J. Am. Chem. Soc. **112**, 8215 (1990)

methylglutaric acid derivative capable of being manipulated from either terminus. Acid treatment of **33** afforded the derived terminal carboxylic acid which was subjected to the Curtius rearrangement with DPPA[32] in *tert*-butanol to give the Boc-protected imide **34** which was readily hydrolyzed to the derived carboxylic acid in an excellent overall yield. The oxazole ring was then appended to this intermediate via amide coupling to serine, closure to the derived oxazoline and a final oxidation[33] of this intermediate to the illustrated oxazole **36**. This last reaction was a cause of considerable difficulty and reaction yields fluctuated unpredictably from 60 to 30%. As the result of this problem, we also developed another method for carrying out this final oxidation through phenylselenation of the oxazoline-derived ester enolate followed by oxidation and elimination to the aromatic system. Although the overall yields of this latter process averaged around 60%, we were able to achieve a level of reliability which allowed us to carry large amounts of material through this reaction sequence.

The γ-amino acid Fragment. The synthesis of this amino acid is outlined in Scheme 15. In planing a rational approach to this highly functionalized fragment, we entertained a number of strategies for controlling the individual stereochemical relationships. In one option, we considered linking the terminal carboxyl and nitrogen moieties so that olefin osmylation might be employed to establish the required relationship between the diol and amino stereogenic centers. Although we did ultimately reject this route in favor of an alternate approach, this strategy for constructing this amino acid has recently appeared. The plan which we finally adopted for the construction of this fragment featured the illustrated aldol reaction shown in Scheme 15.

Scheme 15. γ-Amino Acid Segment: Synthesis Plan

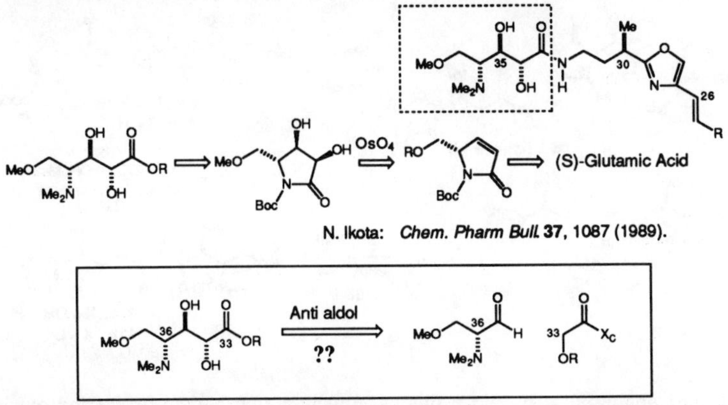

N. Ikota: *Chem. Pharm. Bull.* **37**, 1087 (1989).

The execution of this plan began with the acid-catalyzed methoxymethylation of the titanium enolate derived from **36** (Scheme 16). This reaction proceeded with superb enolate face selectivity to provide the illustrated D-serine derivative which was conveniently reduced to the primary alcohol **37** with methanolic lithium borohydride. In our early experiences with the oxidation of **37** to the derived aldehyde, we experienced extensive racemization under the normal Swern[34] oxidation conditions; however, if the triethylamine in this transformation is replaced with Hunig's base under the otherwise normal oxidation conditions, racemization can be effectively suppressed.

The *anti* aldol reaction illustrated in Scheme 16 was developed from a modest precedent found in the work of Mukaiyama.[35] Although such Sn(II) enolates normally undergo highly syn-selective aldol reactions, if the enolate is pretreated with a chelating diamine such as TMEDA, the overall aldol diastereoselectivity becomes *anti*-selective. When these reaction conditions were applied to the union of the enolate derived from **38** and the aldehyde derived from **37**, the desired aldol adduct **39** was obtained with the desired sense of stereoinduction at both of the newly formed stereogenic centers. At the present time the origin of the *anti* diastereoselectivity in this reaction is a matter of some speculation, and experiments are ongoing to more deeply probe the stereochemical control elements in this reaction.

After some experimentation we found an exceptionally convenient method for uniting the amino acid and oxazole fragments **39** and **36** respectively (Scheme 17). If the aldol adduct **39** is pretreated with trimethylaluminum and amine **36**, the expected coupling product for **40a** is obtained in modest yield along with some of the derived C_{34}-

Scheme 16. Synthesis of the γ-Amino Acid Segment:

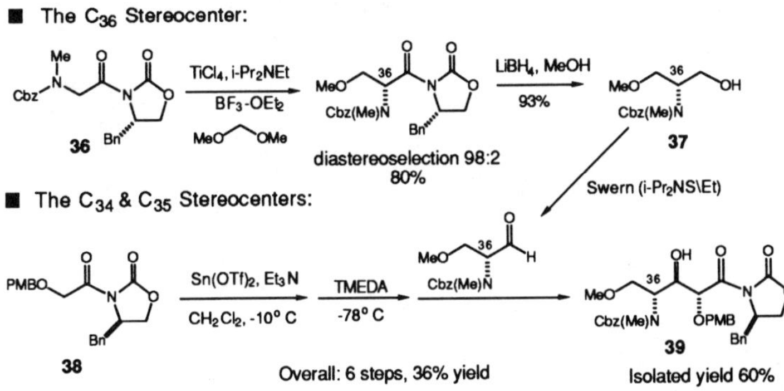

■ The C_{36} Stereocenter:

■ The C_{34} & C_{35} Stereocenters:

Scheme 17. Union of Amino Acid & Oxazole Fragments

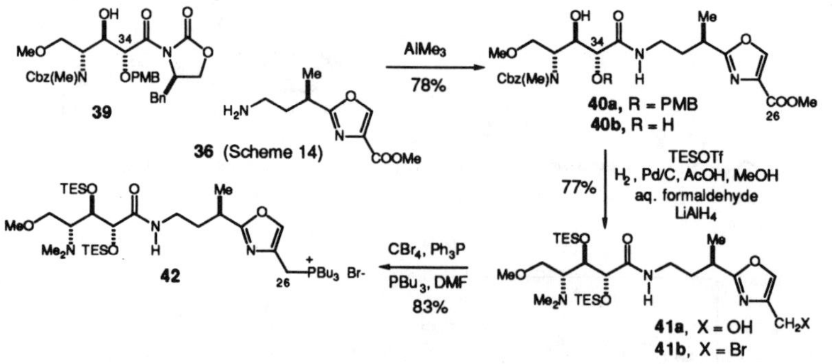

deprotected analogue **40b**. Since this deprotection step had to be carried out in any event, we were able to optimize this two-step transformation to the desired diol amide **40b** in excellent yield. Although we have not evaluated the use of trimethylaluminum as a general reagent for the removal of PMB ethers, the present example serves to illustrate the potential for carrying out such transformations in selected systems. From this intermediate a routine set of refunctionalization reactions provided us with oxazole derivatives **41a** and **41b** through an uneventful set of chemical transformations.

A comment is in order at this point as to our decision to use triethylsilyl (TES) protecting groups for the C_{34} and C_{35} diol functionality. We had anticipated that under the acidic conditions for silicon deprotection the presence of the basic dimethylamino moiety would significantly retard the rate of desilyation of silicon protecting groups in the vicinity of this basic heteroatom. Indeed, relevant control experiments confirmed the fact that TBS protecting groups cannot readily be removed from this diol moiety.

Assemblage of Subunits. All that remained to transform **41b** into a substrate suitable for the Wittig coupling was the illustrated transformation to the tributylphosphonium salt. At this stage we anticipated the successful model studies carried out independently by Professor Armstrong at UCLA and elected to follow the precedent established in his laboratory for this final bond construction.[36]

After considerable experimentation, suitable conditions were developed for the Wittig coupling to form the C_{25}-C_{26} *trans* olefin. The relevant model studies which were instrumental in optimizing reaction conditions are illustrated in Scheme 18. In this sequence, bromomethyloxazole **43** was treated with 2.4 equivalents of

Scheme 18. Model Studies on the Wittig Union

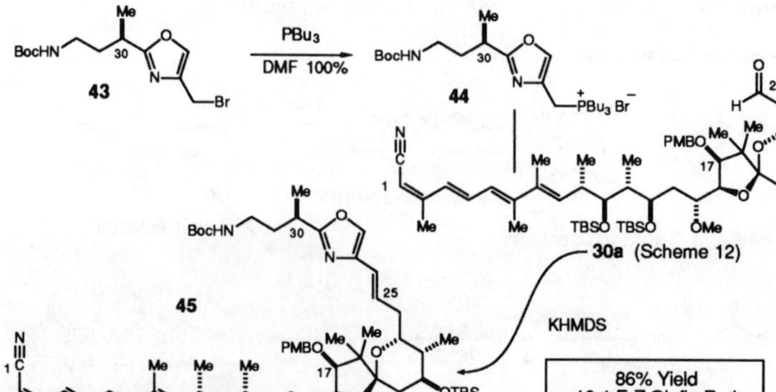

tributylphosphine in DMF at room temperature for several hours. Upon formation of the phosphonium salt **44**, the illustrated aldehyde (1 equiv) was introduced to the reaction, cooled to 0° C and treated with 2.4 equivalents of potassium hexamethyldisilazide (KHMDS) in THF. Under these conditions, an 86% yield of the *trans* olefin adduct **45** was obtained in high stereoselectivity. These reaction conditions, which were used rather extensively by the Kishi laboratory in their synthesis of palytoxin, eliminate the need for stoichiometric quantities of amide base in the formation of the phosphorane. The kinetic selectivity for the deprotonation of phosphonium salt **45** appears to be quite high, a critical issue since aldehyde **30a** is quite base-sensitive.

In preparation for the actual Wittig coupling, we then applied the previously described phosphorylation methodology (Scheme 13) to the phosphorylation of the C_1-C_{25} alcohol **30**. (Scheme 12). After considerable experimentation, the bis-PMB-phosphotriester was selected, and the incorporation of this phosphorus moiety into the fully assembled spiroketal synthon **46** proceeded in 84% yield (Scheme 19). At this juncture we confirmed that the primary C_{25} silicon protecting group could be removed with HF-pyridine without compromising the protected phosphate moiety. On the other hand, deprotection of the secondary silicon protecting groups with aqueous HF/MeCN resulted in complete deprotection of **46** along with removal of the PMB phosphorus protecting groups. With this information in hand, the C_{25} aldehyde was obtained in high yield by successive deprotection and Dess-Martin oxidation.

Scheme 19. The Phosphorylation Step

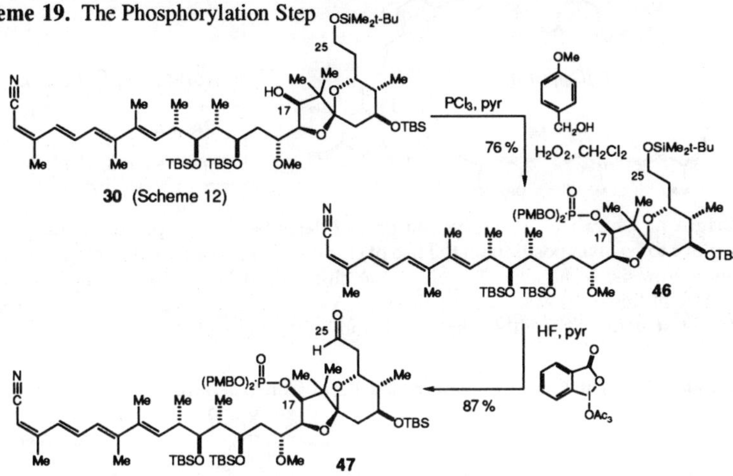

As in the preceding model studies, the Wittig coupling of bromomethyl oxazole **41b**, via its derived phosphonium salt, with aldehyde **47** proceeded in good yield to afford the fully protected calyculin structure, one formal step away from the natural product (Scheme 20). However, after aqueous HF deprotection, a mixture of calyculin and a monosilyl calyculin derivative were obtained after a deprotection time period of approximately one day. Based upon NMR and mass spectroscopic evidence, we speculate that the C_{11} alcohol is the lone heteroatom carrying this offending silicon protecting group. However, when the reaction time was extended to 92 hours, full deprotection to (+)Calyculin A was achieved in 70% yield. The sluggishness with which the last silyl ether was released from the calyculin structure is noteworthy. We speculate that this moiety is rendered inaccessible by external reagents as a result of strong hydrogen bonding between the phosphoric acid residue and the polar functionality resident in both side chains. Once obtained, our synthetic calyculin proved to be identical in all respects with natural calyculin A except for its optical rotation which was equal in magnitude and opposite in sign to that of the natural product. The studies thus confirm the absolute configuration of natural calyculin A to be opposite to that illustrated in the structure illustrated in this discussion.

Scheme 20. The Final Wittig Coupling

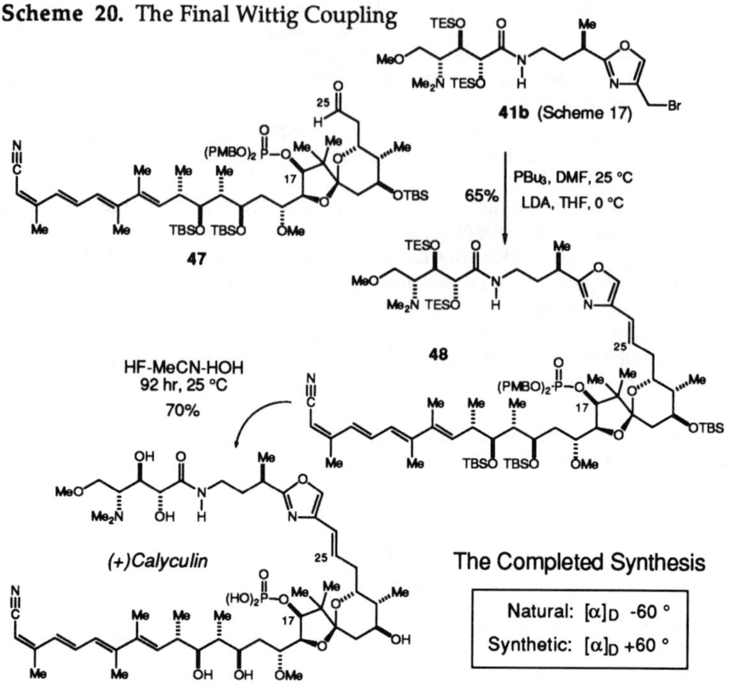

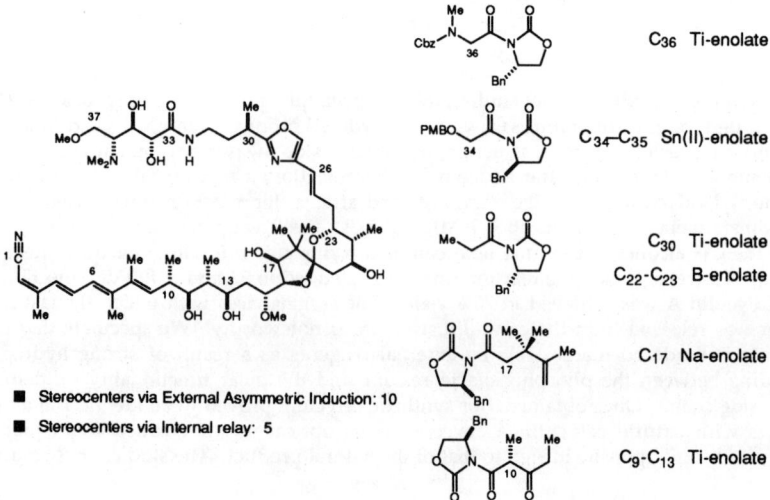

The Completed Synthesis

Natural: $[\alpha]_D$ -60 °	
Synthetic: $[\alpha]_D$ +60 °	

Origin of Chirality. The calyculin project described in the preceding narrative has turned out to be an exposition on the use of chiral imide enolate methodology in stereoselective synthesis. In Scheme 21 below, the chiral imide derivatives employed during the course of this synthesis are summarized. All told, every stereogenic center present in the target structure was either directly or indirectly controlled through this chiral eno-

Scheme 21. The Chiral Enolate-Based Bond Constructions

C_{36} Ti-enolate

C_{34}-C_{35} Sn(II)-enolate

C_{30} Ti-enolate
C_{22}-C_{23} B-enolate

C_{17} Na-enolate

C_9-C_{13} Ti-enolate

■ Stereocenters via External Asymmetric Induction: 10
■ Stereocenters via Internal relay: 5

late methodology. Over the years, we have continued to develop new capabilities for these enolate systems on a need-to-require bases. Challenging total syntheses of this type provide the stimulation for continued research in this area.

Acknowlegements

This project has consumed approximately five and one half man-years of effort in our laboratory. The synthesis was initiated in January of 1989 by graduate student Jim Gage. Approximately two years later, Jim Leighton, a young first-year graduate student, joined the project as an understudy. In the summer of 1991, Gage handed over sole responsibility for completing the project to Leighton, who subsequently finished the synthesis in April of 1992 during the third year of his thesis research. During the course of this study, an able undergraduate, Annette Kim, helped to develop the Sn(II) enolate methodology used during the course of the project. The facility with which this complex synthesis came together is a testament to the diligence of my co-workers and their skill at the bench. Working with people such as this is the ultimate reward that one reaps as an academician.

I would also like to thank both the National Science Foundation, the National Institutes of Health and Merck for their generous support of our research program over the years.

Footnotes and References

1) Kato, Y.; Fusetani, N.; Matsunaga, S.; Hashimoto, K.; Fujita, S.; Furuya, T. *J. Am. Chem. Soc.* **1986,** *108*, 2780-2781.

2) (a) Ishihara, H.; Martin, B. L.; Brautigan, D. L.; Karaki, H.; Ozaki, H.; Kato, Y.; Fusetani, N.; Watabe, S.; Hashimoto, K.; Uemura, K.; Hartshorne, D. J. *Biochem. Biophys. Res. Commun.* **1989,** *159*, 871-877. (b) Suganuma, M.; Fujiki, H.; Furuya-Suguri, H.; Yoshizawa, S.; Yasumoto, S.; Kato, Y.; Fusetani, N.; Sugimura, T. *Cancer Res.* **1990,** *50*, 3521-3525.

3) Ishihara, H.; Martin, B. L.; Brautigan, D. L.; Karaki, H.; Ozaki, H.; Kato, Y.; Fusetani, N.; Watabe, S.; Hashimoto, K.; Uemura, D.; Hartshorne, D. J. *Biochem. Biophys. Res. Comm.* **1989,** *159*, 871-877.

4) Tachibana, K.; Scheuer, P. J.; Tsukitani, Y.; Kikuchi, H.; Van Engen, D.; Clardy, J.; Gopichand, Y.; Schmitz, F. J. *J. Am. Chem. Soc.* **1981,** *103*, 2469-2471.

5) For a review on the biology of okadaic acid and calyculin A, see: Cohen, P.; Holmes, C. F. B.; Tsukitani, Y. *Trends Biochem. Sci.* **1990,** *15*, 98-102.

6) Matsunaga, S.; Fusetani, N. *Tetrahedron Lett.* **1991,** *32*, 5605-5606.

7) Hamada, Y.; Tanada, Y.; Yokokawa, F.; Shioiri, T. *Tetrahedron Lett.* **1991,** *32*, 5983-5986.

8) (a) Kato, Y.; Fusetani, N.; Matsunaga, S.; Hashimoto, K.; Koseki, K. *J. Org. Chem.* **1988,** *53*, 3930-3932. (b) Matsunaga, S.; Fujiki, H.; Sakata, D.; Fusetani, N. *Tetrahedron* **1991,** *47*, 2999-3006.

9) In one isolation run, the following amounts were obtained from 800 g of the sponge: A, 123 mg; B, 72 mg; C, 31 mg; D, 13 mg; E, 14 mg; F, 10 mg; G, 7 mg; H, 7mg. Data is from reference 3b.

10) IC_{50} values against protein phosphatase 2A activity are as follows: A, 0.9 nM; B, 6.0 nM; C, 1.0 nM; D, 5.2 nM; E, 2.7 nM; F, 3.2 nM; G, 5.6 nM; H, 6.0 nM. Data is from reference 3b.

11) For preliminary communications from this laboratory on this subject, see: (a) Evans, D. A.; Gage, J. R. *Tetrahedron Lett.* **1990,** *31*, 6129-6132. (b) Evans, D. A.; Gage, J. R. *J. Org. Chem.* **1992,** *57*, 1958-1961. (c) Evans, D. A.; Gage, J. R.; Leighton, J. L.; Kim, A. S. *J. Org. Chem.* **1992,** *57*, 1961-1963. (d) Evans, D. A.; Gage, J. R.; Leighton, J. L. *J. Org. Chem.* **1992,** *57*, 1964-1966.

12) See, for example: (a) Kishi, Y. *J. Nat. Prod.* **1979**, *42*, 549-568. (b) Evans, D. A.; Kaldor, S. W.; Jones, T. K.; Clardy, J.; Stout, T. J. *J. Am. Chem. Soc.* **1990**, *112*, 7001-7031.

13) Greene, T. W. *Protective Groups in Organic Synthesis*; John Wiley & Sons: New York, 1981.

14) Oikawa, Y.; Yoshioka, T.; Yonemitsu, O. *Tetrahedron Lett.* **1982**, *23*, 885-888.

15) Andersen, N. H.; Hadley, S. W.; Kelly, J. D.; Bacon, E. R. *J. Org. Chem.* **1985**, *50*, 4144-4151.

16) Gage, J. R.; Evans, D. A. *Org. Synth.* **1989**, *68*, 77-91.

17) (a) Evans, D. A.; Morrissey, M. M.; Dorow, R. L. *J. Am. Chem. Soc.* **1985**, *107*, 4346-4348. (b) Morrissey, M. M. Ph.D. Thesis, Harvard University, 1986.

18) (a) Nahm, S.; Weinreb, S. M. *Tetrahedron Lett.* **1981**, *22*, 3815-3816. (b) Levin, J. I.; Turos, E.; Weinreb, S. M. *Syn. Commun.* **1982**, *12*, 989-993.

19) (a) Keck, G. E.; Abbott, D. E.; Wiley, M. R. *Tetrahedron Lett.* **1987**, *28*, 139-142. (b) Koreeda, M.; Tanaka, Y. *Tetrahedron Lett.* **1987**, *28*, 143-146.

20) For a typical procedure for oxidative cleavage of an epoxide to a ketone, see: Nagarkatti, J. P.; Ashley, K. R. *Tetrahedron Lett.* **1973**, 4599-4600.

21) (a) Fu, G. C. Ph.D. Dissertation, Harvard University, 1991. (b) Evans, D. A.; Fu, G. C.; Hoveyda, A. H. *J. Am. Chem. Soc.* **1988**, *110*, 6917-6918.

22) Mukaiyama, T.; Banno, K.; Narasaka, K. *J. Am. Chem. Soc.* **1974** *96*, 7503-7509.

23) Heathcock, C. H.; Flippin, L. A. *J. Am. Chem. Soc.* **1983**, *105*, 1667-1668.

24) Evans, D. A.; Gage, J. R. *Tetrahedron Lett.* **1990**, *31*, 6129-6132.

25) (a) Brown, H. C.; Racherla, U. S. *Tetrahedron Lett.* **1985**, *26*, 2187-2190. (b) Crimmins, M. T.; O'Mahony, R. *Tetrahedron Lett.* **1989**, *30*, 5993-5996.

26) Masamune, S.; Choy, W.; Peterson, J. S.; Sita, L. R. *Angew. Chem. Int. Ed. Eng.* **1985**, *24*, 1-76.

27) (a) Evans, D. A.; Clark, J. S.; Metternich, R.; Novack, V. J.; Sheppard, G. S. *J. Am. Chem. Soc.* **1990**, *112*, 866-868. (b) Evans, D. A.; Ng, H. P.; Clark, J. S.; Rieger, D. L. *Tetrahedron* **1992**, *48*, 2127-2142.

28) Evans, D. A.; Chapman, K. T.; Carreira, E. M. *J. Am. Chem. Soc.* **1988**, *110*, 3560-3578.

29) Mitsunobu, O. *Synthesis* **1981**, 1-28. For a recent relevant study see: Martin, S. F.; Dodge, J. A. *Tetrahedron Lett.* **1991**, *32*, 3017-3020.

30) Dess, D. B.; Martin, J. C. *J. Org. Chem.* **1983**, *48*, 4156-4158.

31) (a) Evans, D. A.; Urpí, F.; Somers, T. C.; Clark, J. S.; Bilodeau, M. T. *J. Am. Chem. Soc.* **1990**, *112*, 8215-8216; (b) Evans, D. A.; Bilodeau, M. T.; Somers, T. C.; Clardy, J.; Cherry, D.; Kato, Y. *J. Org. Chem.* **1991**, *56*, 5750-5752.

32) Ninomiya, K.; Shiori, T.; Yamada, S. *Tetrahedron* **1974**, *30*, 2151-2157.

33) Evans, D. L.; Minster, D. K.; Jordis, U.; Hecht, S. M.; Mazzu, Jr., A. L.; Meyers, A. I. *J. Org. Chem.* **1979**, *44*, 497-501.

34) Mancuso, A. J.; Swern, D. *Synthesis* **1981**, 165-185.

35) Mukaiyama, T.; Iwasawa, N. *Chem. Lett.* **1984**, 753-756.

36) Zhao, Z.; Scarlato, G. R.; Armstrong, R. W. *Tetrahedron Lett.* **1991**, *32*, 1609-1612.

10
Iron Transport-mediated Drug Delivery

M. J. Miller,* F. Malouin,[1] E. K. Dolence, C. M. Gasparski,
M. Ghosh, P. R. Guzzo, B. T. Lotz, J. A. McKee,
A. A. Minnick, and M. Teng
DEPARTMENT OF CHEMISTRY AND BIOCHEMISTRY, UNIVERSITY OF NOTRE
DAME, NOTRE DAME, IN 46556, USA
[1]LABORATOIRE ET SERVICE D'INFECTIOLOGIE, CENTRE HOSPITALIER DE
L'UNIVERSITÉ LAVAL, QUEBEC, GIV 4G2, CANADA

Rational design of therapeutic agents requires detailed knowledge of the biological target, the skill to prepare the drug, and the ability to deliver it to its effective site. Significant advances in X-ray crystallographic and spectroscopic techniques, especially multidimensional NMR and electrospray mass spectroscopy, coupled with concurrent advances in chemistry, biochemistry, molecular biology and biology, promote the rapid isolation, purification and characterization of enzymes, related biological targets, and substrates or inhibitors. Molecular modelling facilitates the design of unnatural inhibitors of biological reactivity. Recent advances in organic synthesis, especially asymmetric synthesis, now make it conceptually possible, but not yet always practical, to synthesize essentially any molecule of biological, industrial or theoretical interest. After a potential drug has been identified and synthesized, much less is known about the design of an effective mode for delivering a drug to its target. This paper summarizes our efforts to utilize microbial iron assimilation processes for active transport of drugs to, and possibly into, pathogenic microbes.

Microbial iron assimilation is a beautifully detailed example of molecular recognition. Microbes synthesize and excrete relatively low molecular weight (<1500 Da) iron-selective chelators (siderophores) to facilitate active iron transport.[1] Most siderophores contain three bidentate ligands, usually hydroxamic acids, catechols, α-hydroxy acids or combinations thereof, to provide iron-selective complexing agents with association constants generally ranging from 10^{30} to 10^{52}! The resulting iron complex is then recognized by an outer-membrane cellular protein and actively transported into the cell where the iron is released by reduction, siderophore hydrolysis, or siderophore modification processes. The structures in figures 1-3 represent the various types of the more than 200 known siderophores.[2]

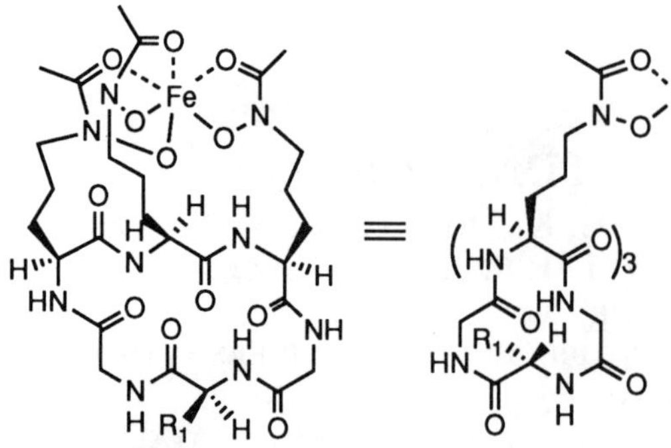

1, ferrichromes
a, ferrichrome, $R_1 = H$
b, ferricrocin, $R_1 = CH_2OH$

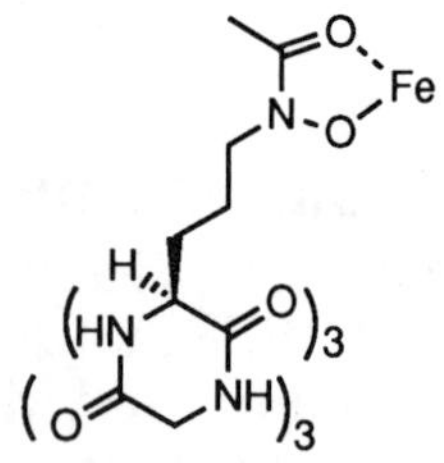

1a, ferrichrome

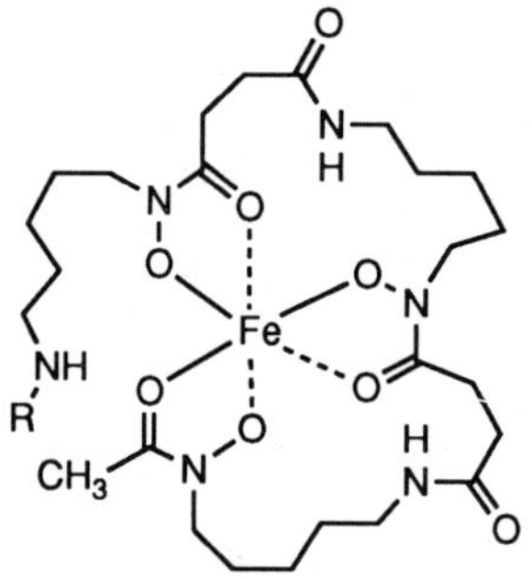

2a, ferrioxamine B, R = H
2b, ferrimycin, R =

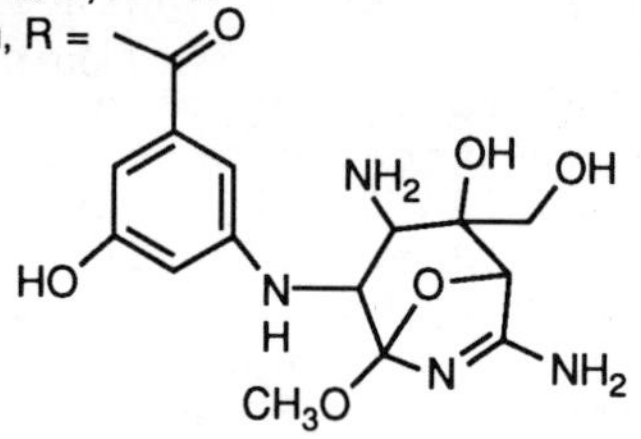

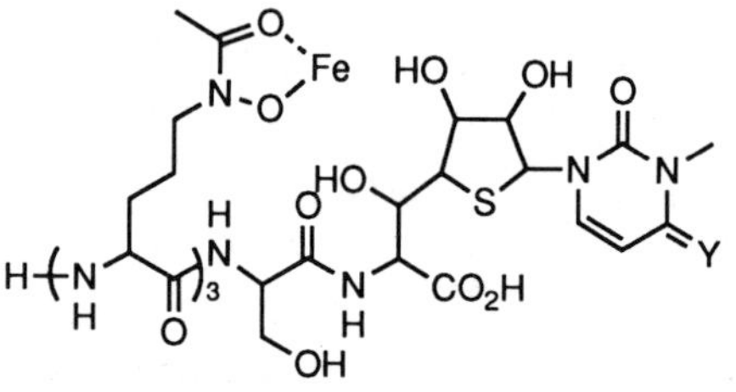

3a, albomycin δ_1, Y = O
3b, albomycin δ_2, Y = NCONH$_2$
3c, albomycin ϵ, Y = NH

Figure 1. Hydroxamic acid-based siderophores:

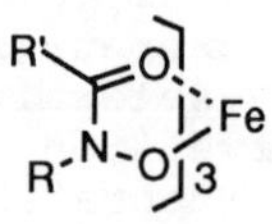

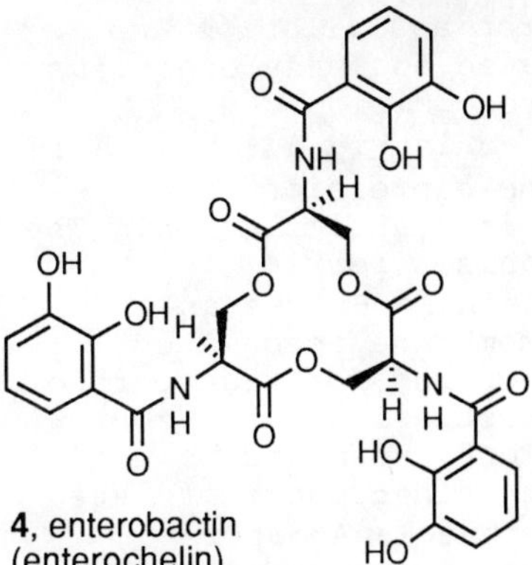

4, enterobactin
(enterochelin),
(shown without iron)

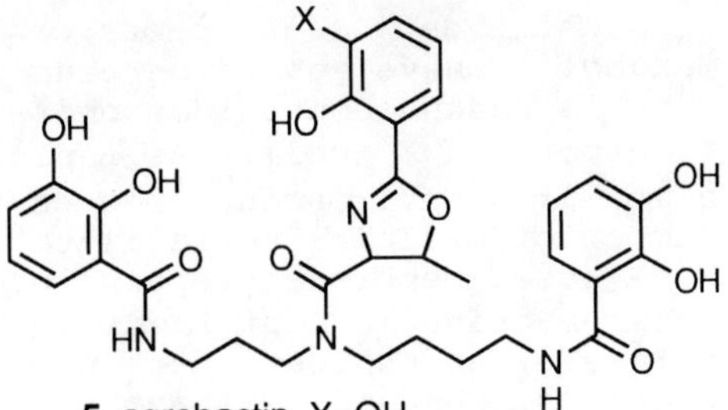

5, agrobactin, X=OH
6, parabactin, X=H (shown without iron)

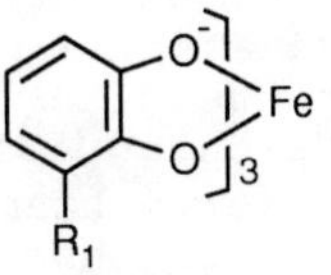

catecholate

Figure 2: Catechol containing siderophores

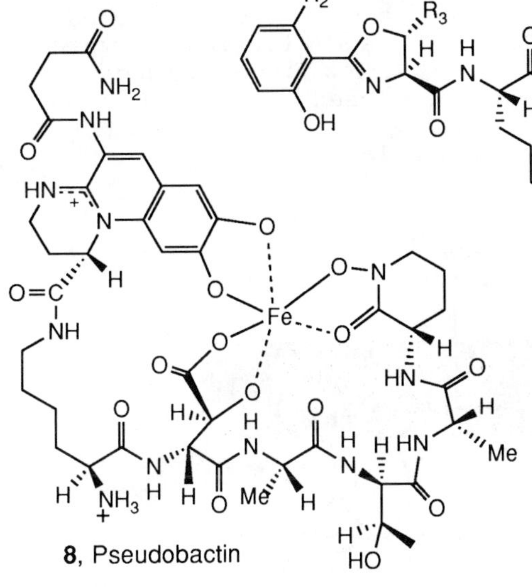

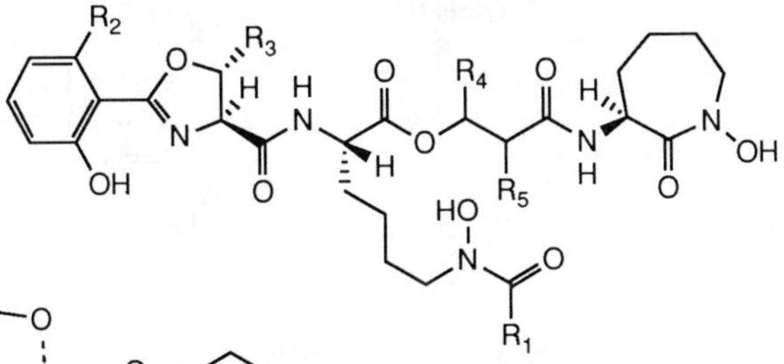

7, mycobactins
(shown without iron)

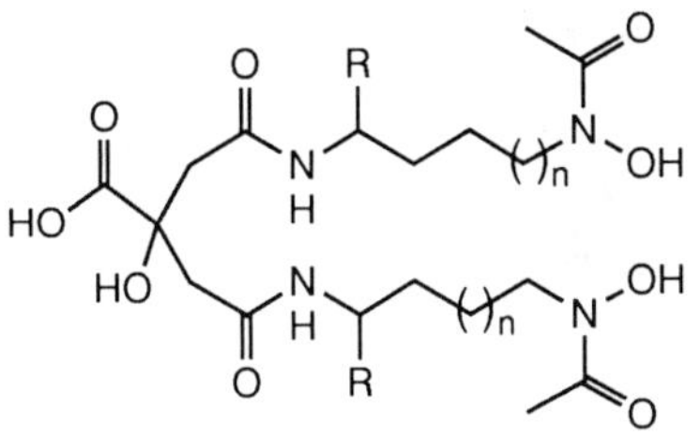

8, Pseudobactin

9, citrate-based siderophores
a, awaitin A   R = COOH, n = 1
b, awaitin B   R = H, n = 1
c, awaitin C   R = COOH, n = 0
d, aerobactin   R = COOH, n = 2
e, schizokinen   R = H, n = 0
f, arthrobactin   R = H, n = 2
(all shown without iron)

Fig 3. Mixed Ligand Siderophores

While detailed aspects of iron assimilation and
metabolism have yet to be elucidated for many organisms,
much is known about siderophore recognition and
transport in *E. coli*. As indicated in Fig 4,[3] growth in
a low iron environment induces the expression of a
number of proteins to facilitate iron assimilation.   The
FepA outer membrane protein recognizes ferric
enterobactin.[4]   Periplasmic protein FepB acts as a
permease, and three other inner-membrane associated
proteins (FepC, D, G) also are required for transport of
ferric enterobactin. TonB, in association with ExbB
that influences the function of TonB, seems to be
essential for all energy-requiring siderophore-mediated
iron assimilation.[5]   *E. coli* uses the FhuA outer
membrane protein to recognize ferrichrome and other
hydroxamate-based siderophores, even though it does not
synthesize ferrichrome.[6]   FhuA is also the receptor
used by bacteriophages T1, T5, and ø80. FhuB, C, and D,
associated with the cytoplasmic membrane, are involved
in the transport of ferrichrome, ferric complexes of
aerobactin, coprogen, rhodotorulic acid and related
siderophores.[7]   In addition, two iron-regulated outer
membrane proteins (IROMP), Cir and Fiu, are inducible
under low iron conditions.[8] Although their function is
uncertain, these proteins have been implicated in a
TonB-dependent transport of ferri-monocatechols. Cir is
also the receptor for colicin Ia.

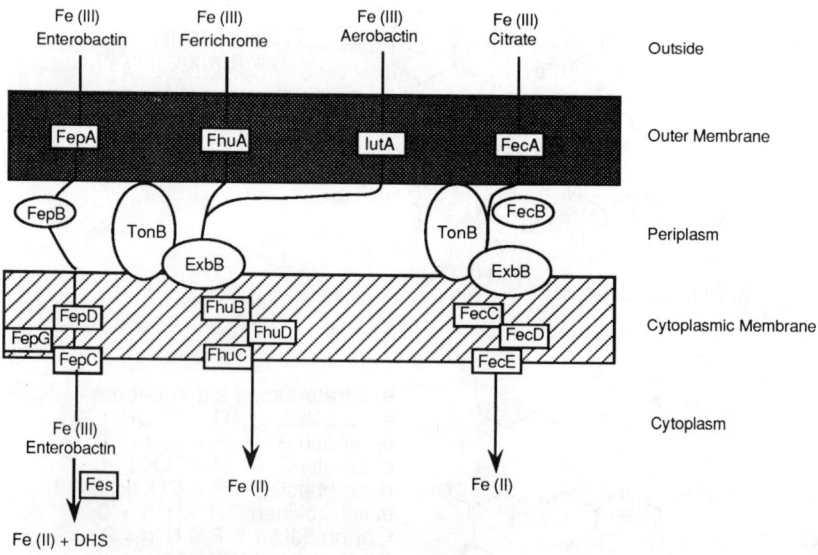

Figure 4.   Iron Transport in *E. coli*

Conceptual iron transport mediated drug delivery

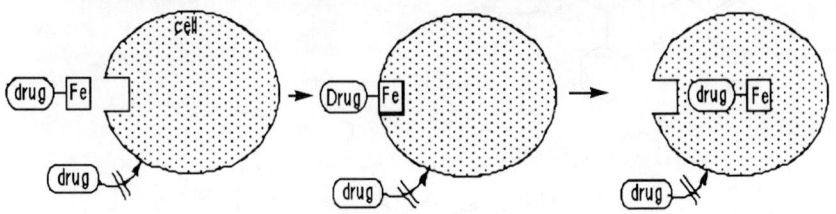

As shown in the cartoon, drugs that may be effective against isolated biochemical targets, but which are therapeutically ineffective because of their inability to permeate a cell, might be smuggled into cells by attachment to siderophores. Of course, consideration must be given to the type of siderophore and drug utilized, as well as the type of linkage used in the conjugate. The drug and linkage must not interfere with siderophore recognition and transport. In some instances the intact conjugate may appropriately react with the biochemical target and in others the drug may need to be released by chemical or enzymatic processes within the cell. The design of conjugates obviously depends on the availability and structure of the siderophores. While many siderophores are available in quantity by fermentation processes, some of the more common ones have no peripheral functionality suitable for conjugation with antimicrobial agents. For example, while ferrichrome and enterobactin appear to be highly functionalized, they really are simple repetitive units of amino acids, and except for key esters (enterobactin) or peptide bonds (ferrichrome) which hold the subunits together, all of the labile functionality is required for iron chelation. However, Ferricrocin (**1b**), a natural ferrichrome analog, contains a serine in place of one of the glycine residues in the cyclic hexapeptide. Early attempts to test the concept of siderophore-mediated drug delivery involved reaction of ferricrocin's serine hydroxyl group with arylsulfonyl isocyanates to give sulfa drug conjugates.[9] General antimicrobial testing indicated that the conjugates were inactive. More recently, a number of ß-lactam antibiotics (**10-16**) have been shown to have improved efficacy when even a single bidentate iron chelating group is attached to the antibiotic.[10] (Details of the biological activity of many of these compounds were discussed at the 1984 meeting on Recent Advances in the Chemistry of ß-Lactam Antibiotics.)

Natural examples of siderophore-drug conjugates have been reported. The albomycins (**3**), like the ferrichromes (**1**), utilize δ-N-hydroxy-δ-N-acetyl-L-ornithine residues for iron chelation; but rather than incorporating them in a cyclic hexapeptide, they are attached through a linear peptide to a toxic thioribosyl derivative.[11] In effect, a number of organisms recognize albomycin as being similar

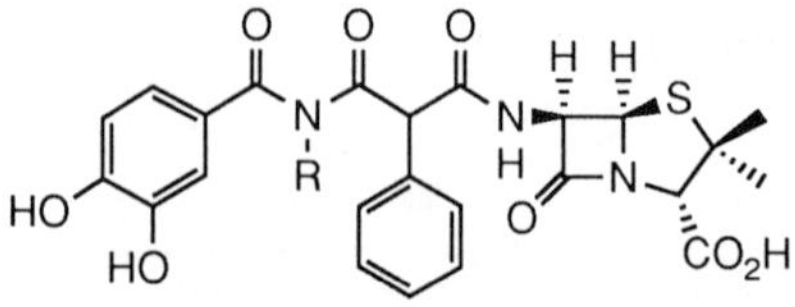

10a, R = CH₃
10b, R = (CH₂)₃OH

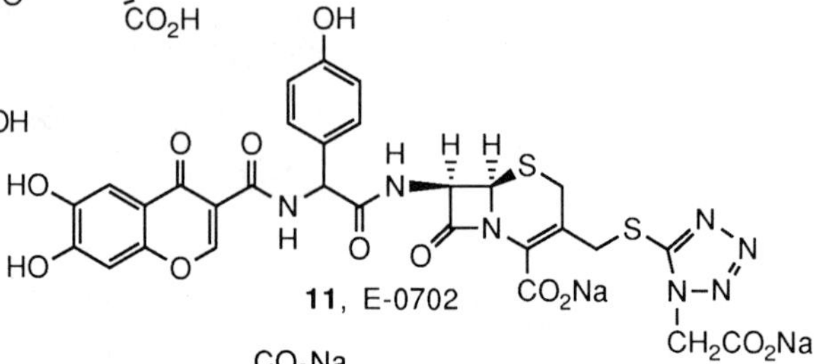

11, E-0702

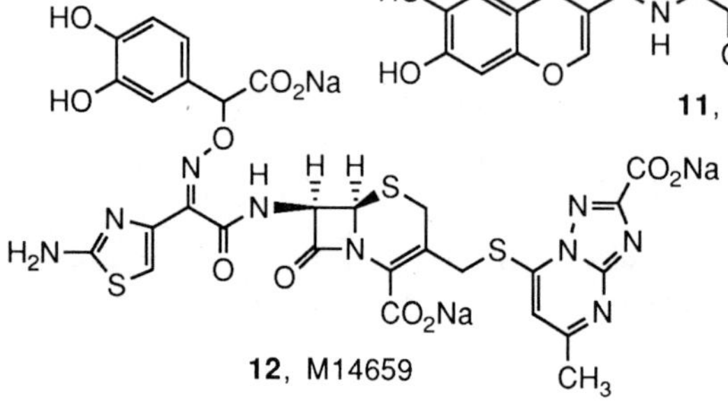

12, M14659

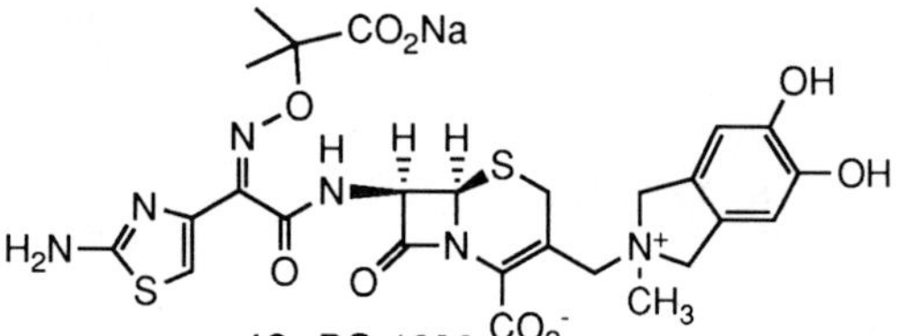

13, BO-1236

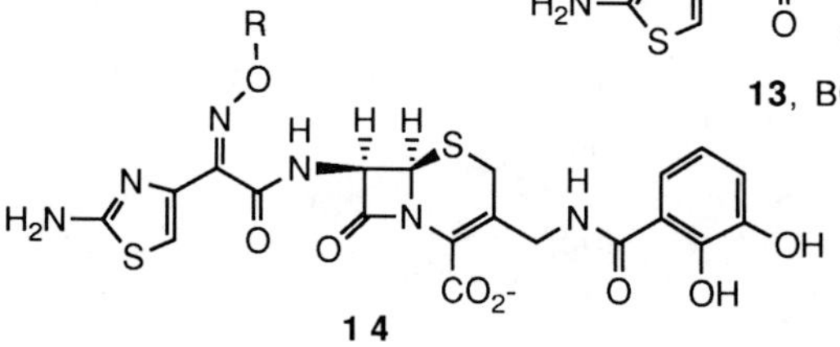

1 4

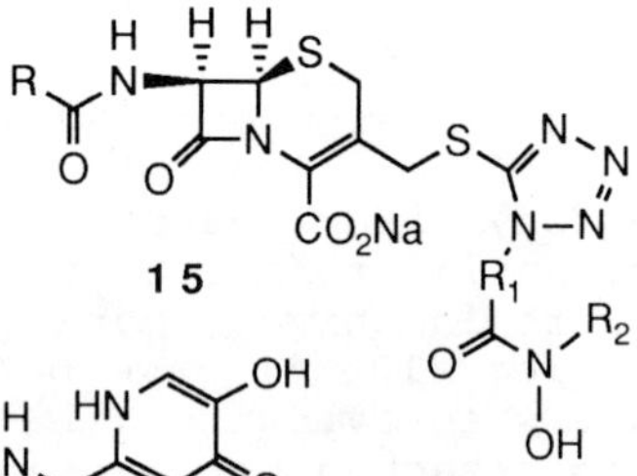

1 5

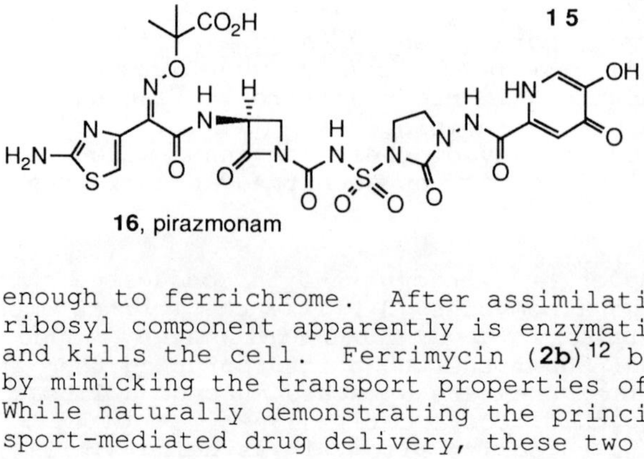

16, pirazmonam

enough to ferrichrome. After assimilation, the thioribosyl component apparently is enzymatically released
and kills the cell. Ferrimycin (**2b**)[12] behaves similarly
by mimicking the transport properties of ferrioxamine B.
While naturally demonstrating the principle of iron transport-mediated drug delivery, these two compounds have

not found extensive therapeutic use. Neither is readily available. The exact mode of action of the toxic components is not known and indications are that organisms rapidly develop resistance to these natural conjugates. In order to systematically explore the therapeutic potential of iron transport-mediated drug delivery, our goals were to synthesize siderophore components, determine the minimal structural requirements for iron binding, recognition, and transport, and then covalently attach antimicrobial agents with known modes of action.

Because it is used clinically for the treatment of transfusional-induced iron overload,[13] desferrioxamine B [DFO, the iron-free form of ferrioxamine B (2a)] is available in quantity. Conceptually, a number of derivatives of DFO can be made by simple modification of its terminal primary amine. In fact, standard carboxyl activation (DCC/N-hydroxysuccinimide) of nalidixic acid and subsequent coupling to DFO in a mixed THF/water solvent gave the simple conjugate 17. Nalidixic acid is a quinolone antibiotic with known activity as a DNA gyrase inhibitor. Although 17 is insoluble in water, it was subjected to a broad antimicrobial screen. The results were disappointing in that the conjugate was apparently devoid of activity. Reasons for the inactivity may include conjugate insolubility, decreased siderophore recognition and transport, improper location of the conjugate if it did enter the cell, or a mechanism may be needed to release the antibiotic from the siderophore once in the cell since quinolone antibiotics usually require free carboxylic acid groups. Alternatively, as discussed later, standard broad-screen testing methods may not be adequate for determination of the activity of siderophore-drug conjugates.

Since ferrichrome (1a) appears to be a more widely used siderophore and it contains the same iron-chelating tri-δ-N-hydroxy-δ-N-acetyl-L-ornithine segment as the albomycins (3), attention was turned to the synthesis and study of related antibiotic conjugates. The prerequisites for this effort included accessing appropriate

amino acid-based iron chelating components, determining
the minimal structural requirements for effective iron
chelation, molecular recognition and transport, and
finally assessing the nature of the drug and its linkage
to the siderophore. Eventually, useful quantities of
siderophores and their components may be obtained
biosynthetically. Meanwhile, chemical synthesis of the
relevant siderophore components has been effective. Much
of this effort has been reviewed[14] and only highlights
will be given here. Our focus has been on the
incorporation of key iron binding hydroxamate components
by direct alkylation of the nitrogen of preformed O-
protected hydroxamic acids. Alternatively, many of the
same hydroxamates can be made by condensation of O-
protected hydroxylamines with aldehydes and subsequent
reductive acylation (Scheme 1).[13a]

Scheme 1: Generalized syntheses of hydroxamate components

$Y-(CH_2)_n-X$ +

(pK=7-12)

$H-\underset{OR_1}{\overset{N}{}}$

$Y = PNH, RO_2C, halo, RO_2C-CH-$
                                    NHP

$X = halo, OH$

$()_n-\underset{Y}{\overset{N}{}}-OR_1$

$NaCNBH_3$
$(RCO)_2O$

or: $Y-(CH_2)_{n-1}-CH=O$ + H_2NOR_1 ⟶ $Y-(CH_2)_{n-1}-CH=N-OR_1$

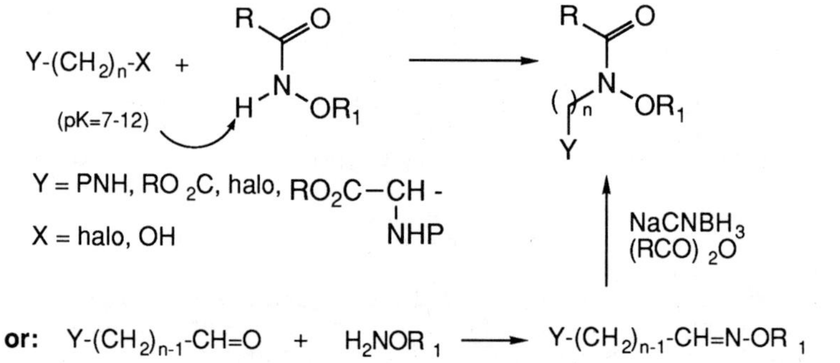

The utility of the hydroxamate alkylation procedure
was first demonstrated by the synthesis of several of the
citrate-based siderophores (**9**, Figure 3). Most
challenging of these was the synthesis of aerobactin (**9d**)
since the constituent hydroxamates are derived from the
unusual amino acid, ε-N-hydroxy-ε-N-acetyl-L-lysine. The
same amino acid appears in both linear and cyclic forms
in another class of siderophores, the mycobactins (**7**).
The syntheses of both forms of this key amino acid
facilitated the first total syntheses of aerobactin[15] and
mycobactin S2 (Scheme 2).[16] Antibiotic conjugates of the
citrate-based siderophores have not yet been reported.

While the synthesis of the N-hydroxy lysine derivatives of aerobactin and mycobactin required an enzymatic
resolution, the synthesis of the δ-N-hydroxy-δ-N-acetyl-L-
ornithine constituent of ferrichrome and the albomycins
utilized L-glutamic acid (**27**) as the chiral starting
material (Scheme 3). The key hydroxamate N-alkylation
reaction (**28** to **29**) was performed using the versatile
Mitsunobu process.[17] Subsequent elaboration led to the
total syntheses of rhodotorulic acid (**31**),[18] the diketo-

piperazine form of δ-N-hydroxy-δ-N-acetyl-L-ornithine and foroxymithine, a siderophore with angiotensin converting enzyme (ACE) inhibition properties.[19]

Scheme 2

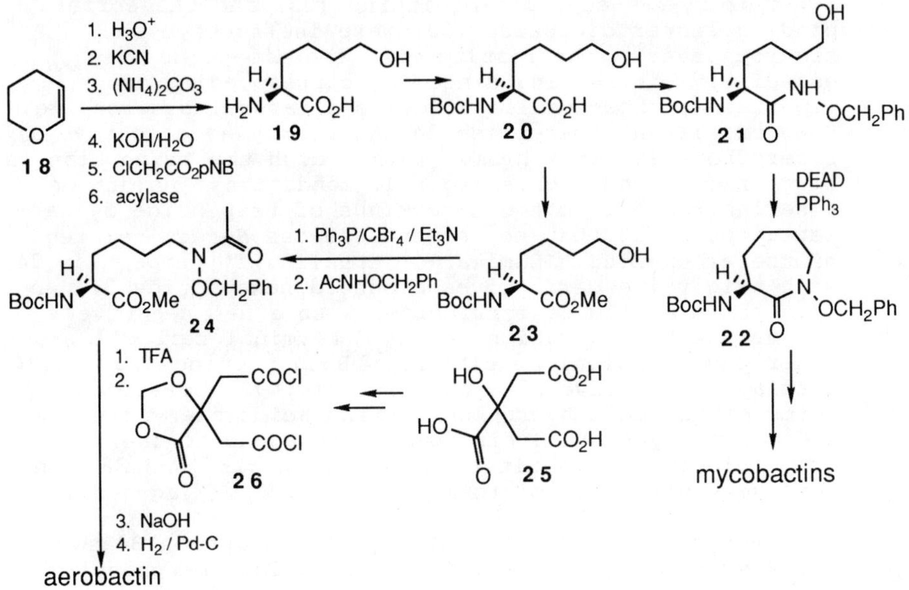

Scheme 3

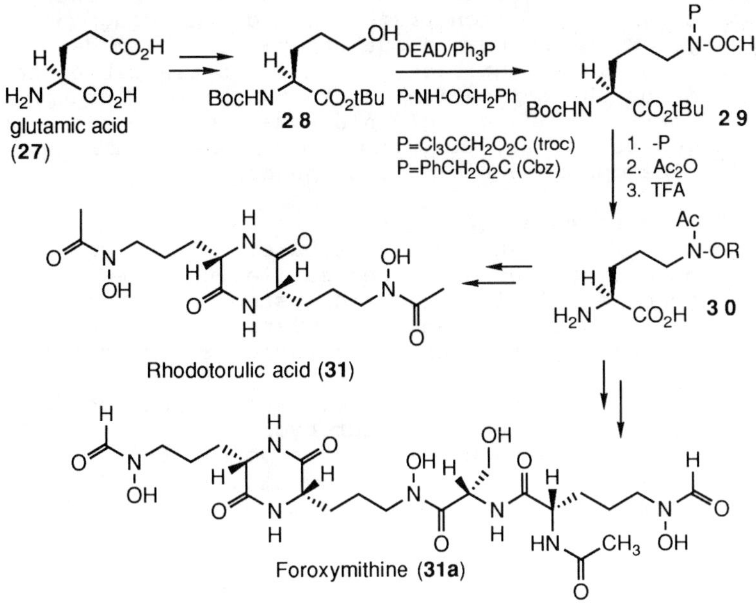

Rhodotorulic acid (**31**)

Foroxymithine (**31a**)

With a reasonable synthesis of δ-N-hydroxy-δ-N-acetyl-L-ornithine derivatives available, attention turned to the determination of the minimal structural requirements for siderophore activity of related peptides. In siderophore growth-promotion assays, monomeric δ-N-hydroxy-δ-N-acetyl-L-ornithine (**32**) and the corresponding linear dipeptide (**33**) were ineffective as siderophores.[20] The results with the dipeptide were especially interesting since the corresponding diketopiperazine, rhodotorulic acid, is a natural siderophore. However, linear tripeptide **34** was nearly as effective a siderophore as ferrichrome, even though the tripeptide is zwitterionic under physiological conditions. Addition of a serine residue to the C-terminus of tripeptide **34** gave tetrapeptide **35** that corresponds to the N-terminal region of the albomycins. Similarly, coupling of tripeptide **34** with triglycine derivatives produced hexapeptide **36**, an open-chain analog of ferrichrome with a net negative charge due to ionization of the C-terminal carboxyl group under physiological conditions. Both peptides **35** and **36** displayed effective siderophore activity. Less common D-forms of the more hydrophobic amino acids phenylglycine and p-hydroxyphenylglycine were also added to the tripeptide. The resulting tetrapeptides,**37** and **38**, again were very effective microbial iron transport agents.

Besides testing the ability of tripeptide **34** to carry hydrophobic amino acids into cells, the phenylglycine residues were chosen because they are effective side chains for Lorabid,[21] a potent new carbacephalosporin being developed at Eli Lilly and Co. Siderophore-lorabid conjugates were considered ideal candidates to test the proposed concept of iron transport-mediated drug delivery. Much is known about the biological mode of action of ß-lactam antibiotics and it is plausible that promotion of the permeation of the antibiotic through the outer membrane of target organisms may improve antibiotic effectiveness. Thus, compounds **39** and **40** became the target siderophore-antibiotic conjugates.

The carbacephalosporin class of antibiotics offered another chemically attractive feature. All of the siderophore syntheses described earlier utilized hydrogenolytic debenzylation as a final step to minimize handling of the free ligands that tend to chelate trace metals in solvents and render characterization difficult. Most ß-lactam antibiotics, including penicillins, cephalosporins and penems, contain sulfur that poisons the usual catalysts employed for hydrogenolyses reactions. The carbacephalosporins do not contain sulfur, although they do have a relatively sensitive vinyl chloride in the bicyclic ring.

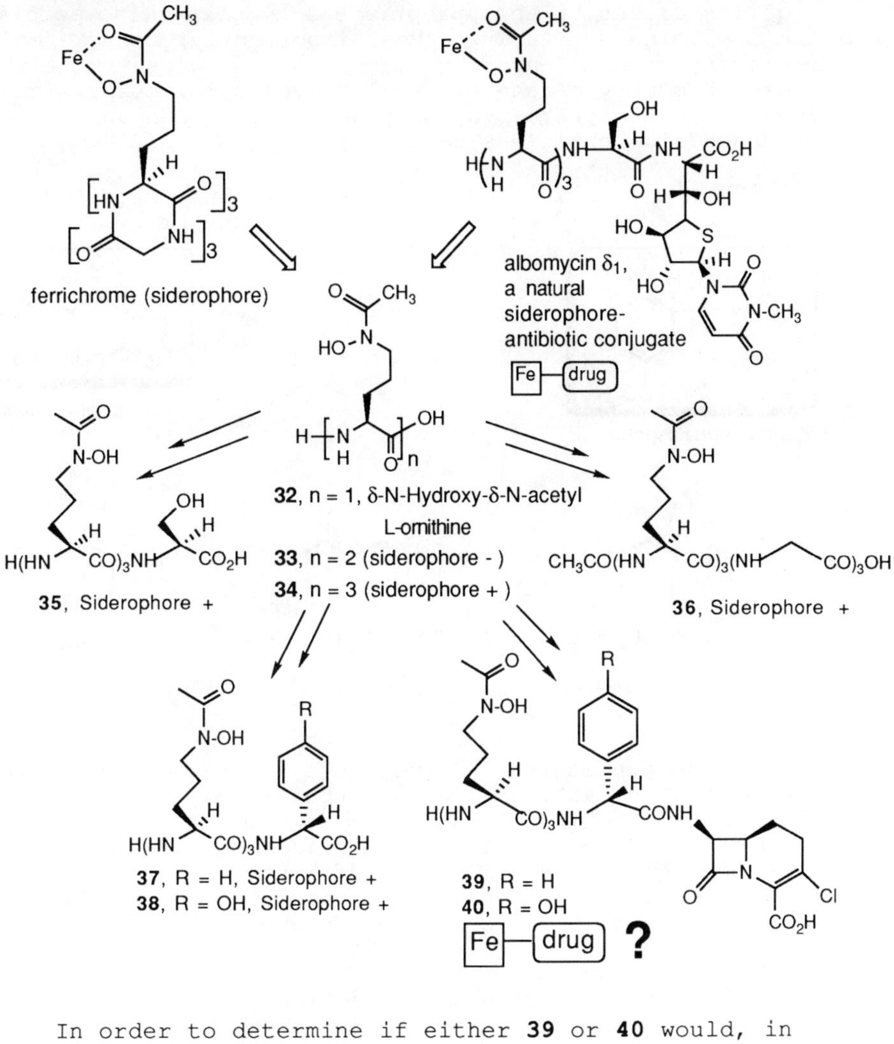

ferrichrome (siderophore)

albomycin δ₁, a natural siderophore-antibiotic conjugate

$$Fe\!-\!drug$$

32, n = 1, δ-N-Hydroxy-δ-N-acetyl L-ornithine
33, n = 2 (siderophore -)
34, n = 3 (siderophore +)

35, Siderophore +

CH₃CO(HN CO)₃(NH CO)₃OH

36, Siderophore +

37, R = H, Siderophore +
38, R = OH, Siderophore +

39, R = H
40, R = OH

$$Fe\!-\!drug \quad ?$$

In order to determine if either **39** or **40** would, in effect, be "the magic bullet," syntheses and biological studies were required. Appropriately protected forms of siderophore component **34** were available from our earlier work and the carbacephalosporin eventually utilized for the conjugation was kindly provided by Eli Lilly and Company. However, our general interest in the synthesis of ß-lactams[22] prompted us to consider new approaches to carbacephalosporins.

One retrosynthetic analysis of the carbacephalo-sporin nucleus (Scheme 4) reveals that this important antibiotic may be constructed from an appropriate dipeptide **44**. The key step would be formation of the ß-lactam by a direct cyclization of the peptide (**44** to **43**), as we have described for other amino acid derived ß-lactams.[22] Construction of the prerequisite peptide requires access to an appropriate ß-hydroxy amino acid **45** and a glycine derivative. Most importantly, the stereochemistry of the ß-hydroxy amino acid would determine the stereochemistry of the eventual bicyclic ß-lactam. Thus, ideally an L-*erythro*-ß-hydroxy amino acid with an extended side chain was needed.

Scheme 4

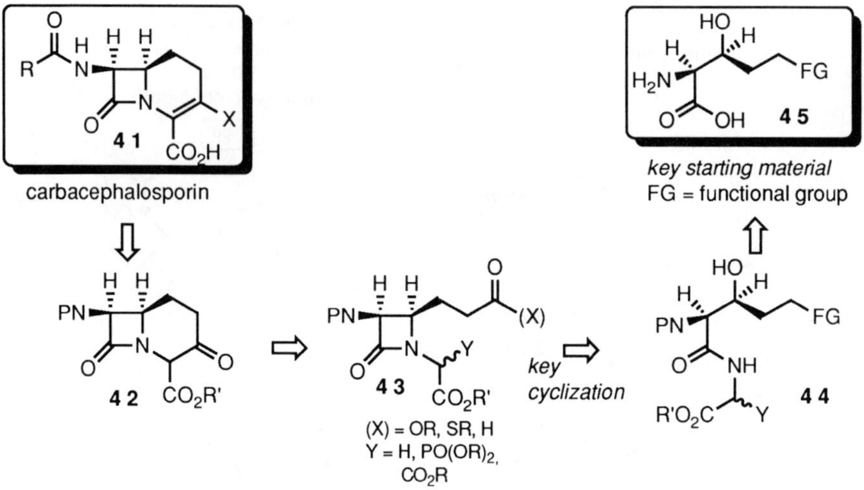

carbacephalosporin

key starting material
FG = functional group

(X) = OR, SR, H
Y = H, PO(OR)$_2$,
CO$_2$R

While considering asymmetric syntheses of this type of amino acid (**45**), a racemic route was also explored (Scheme 5)[23] to test the validity of the retrosynthetic analysis. Interestingly, aldol condensation of an Ox-protected glycine allyl ester (**46**) with various aldehydes, including pentenal, gave the desired protected *erythro* ß-hydroxy amino acid **47**. Removal of the allyl group, followed by coupling with protected aminophosphonoacetate (**49**) gave peptide **50** with the correct number of carbons and relative stereochemistry for construction of the carbacephalosporin nucleus. Much to our delight, the amazing Mitsunobu reaction promoted direct cyclization of **50** to ß-lactam **51** in 76% isolated yield! Ozonolysis of **51** converted the Ox protecting group to a simple benzamide side chain and the terminal alkene to the corresponding aldehyde, which without isolation, was treated *in situ* with triethylamine to produce the carbacephalosporin nucleus **52**. While not yet done, alternative oxidation of the terminal alkene to an acid followed by subsequent conversion to an active ester and base initiated formation of the bicyclic ring would provide the desired oxidation state in the six-membered ring. Overall, this represents a short and efficient, but racemic synthesis of carbacephalosporins.

Scheme 5

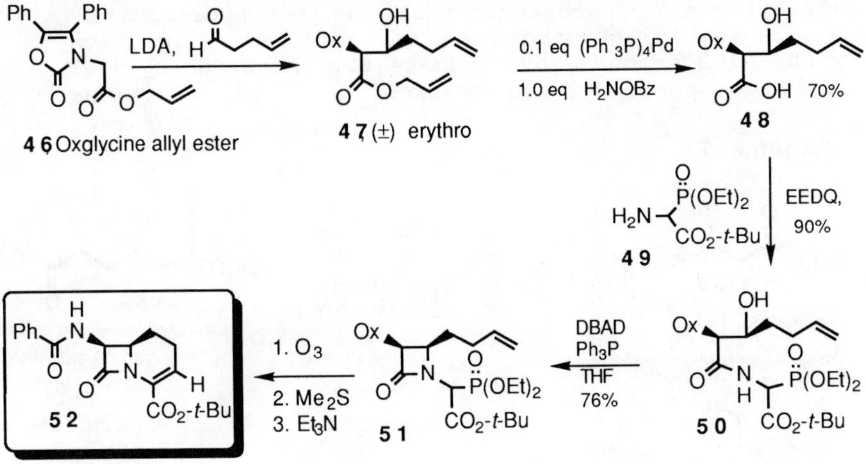

46 Oxglycine allyl ester

47 (±) erythro

48 70%

49

50

51

52

To effect an asymmetric synthesis of carbacephalo-sporins by the amino acid / peptide route, the only requirement is preparation of the starting ß-hydroxy amino acid in optically pure form. Preliminary indica-tions are that this can be accomplished enzymatically.[24] Serinehydroxymethyltransferase (SHMT)[25] is an aldolase that naturally degrades L-serine and L-*allo*-threonine to glycine and formaldehyde or acetaldehyde, respectively. We demonstrated that the reaction can be run in the aldol direction with a number of aldehydes to give the desired L-*erythro (anti)*-ß-hydroxy-α-amino acids (Scheme 6). Extension of this methodology to the asymmetric synthesis of ß-lactams, including carbacephems is in progress.

Scheme 6

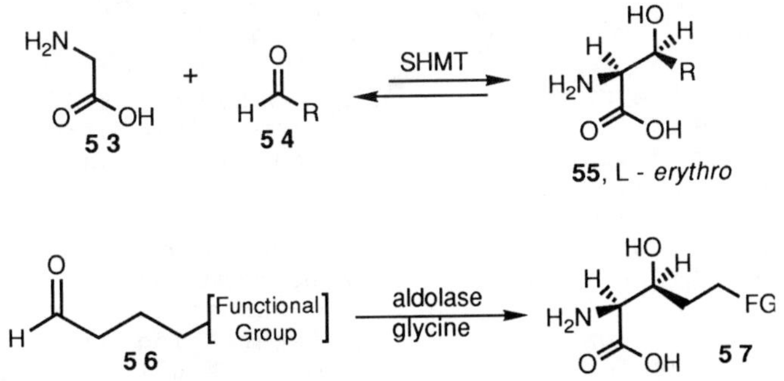

53

54

55, L - *erythro*

56

57

We also recognized that an alternative to the amino acid approach to carbacephalosporins might involve use of relatively simple ß-hydroxy acids. Many such acids can be prepared by a variety of asymmetric reductions. We found the Taber modification[26] of the Noyori ruthenium-BINAP reduction[27] to be especially suitable for asymmetric

reduction of appropriate ß-keto esters. For instance,
catalytic asymmetric reduction of ß-oxoadipate esters
(58, Scheme 7)[28] proceeded well. Direct hydroxaminolysis
of the resulting ß-hydroxy ester followed by reaction
with CbzCl produced hydroxamate **60,** which was cyclized to
the corresponding ß-lactam under the usual conditions.[22]

Scheme 7

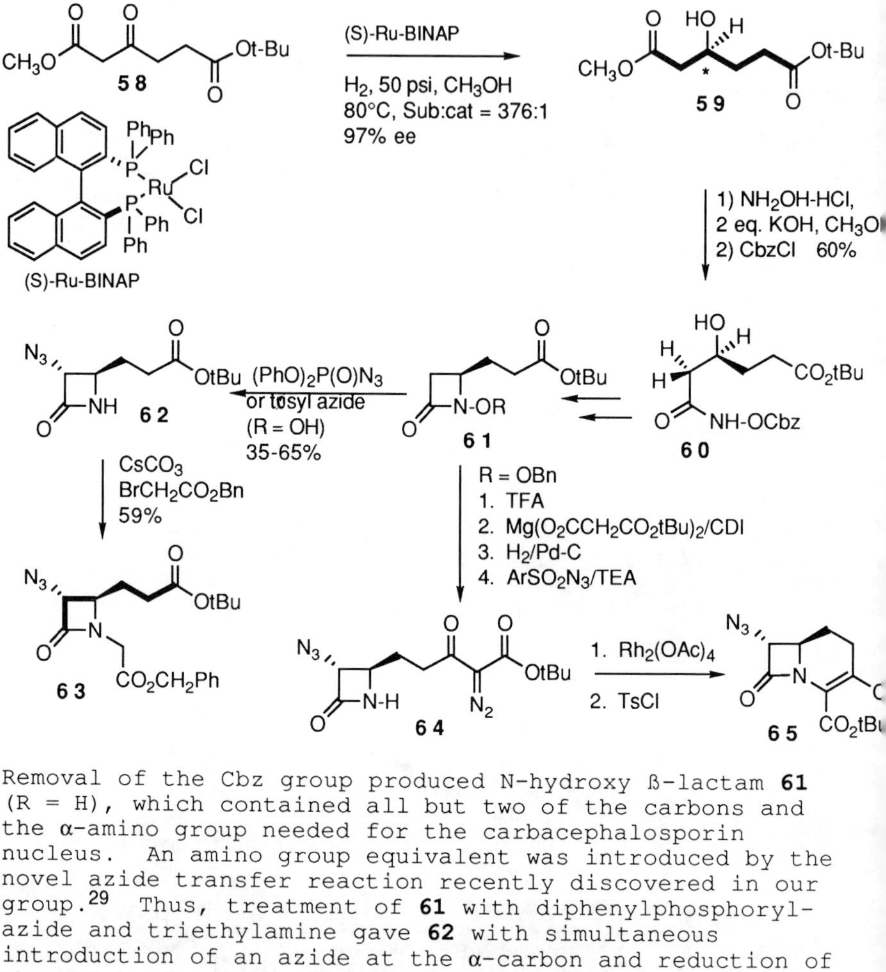

Removal of the Cbz group produced N-hydroxy ß-lactam **61**
(R = H), which contained all but two of the carbons and
the α-amino group needed for the carbacephalosporin
nucleus. An amino group equivalent was introduced by the
novel azide transfer reaction recently discovered in our
group.[29] Thus, treatment of **61** with diphenylphosphoryl-
azide and triethylamine gave **62** with simultaneous
introduction of an azide at the α-carbon and reduction of
the N-O bond! Extension of the interesting azide
transfer reaction to other nucleophiles will be reported
elsewhere.[30] Subsequent reaction of **62** with a bromo-
acetate gave **63,** which contains all of the carbons and
functionality required for the carbacephalosporin
synthesis. Closure to the bicyclic ring has precedent.
The remaining requirement is epimerization of the *trans*
azido group, or related amine derivative, to the desired
cis stereochemistry. In addition, carbon extension of

the side chain of the initially formed ß-lactam, followed
by deprotection of the N-hydroxy group and treatment with
an arylsulfonylazide or phosphorylazide effected three
desired transformations in one reaction to produce N-O
reduced diazo azide **64**. Treatment of **64** with rhodium
acetate and subsequent tosylation of the resulting
bicyclic ß-keto ester produced bicyclic ß-lactam **65**,
which again contained the entire carbacephalosporin
framework, but with the undesired *trans* stereochemistry.[31]

As indicated earlier, while developing alternative
syntheses of carbacephalosporins, we did obtain a sample
from Lilly for conjugation to the siderophore component
34. First, the phenylglycyl and p-hydroxy phenylglcycyl
side chains were separately attached to the carbacephalosporin to give protected antibiotics **67a,b**. An N-
hydroxysuccinimide active ester (**68**) of the N-Cbz
tripeptide was formed and coupled to both **67a** and **67b**.
Initial hydrogenolytic deprotection not only removed the
Cbz, benzyl and p-nitrobenzyl protecting groups, but also
reduced the vinyl chloride of the carbacephems to the 3-
unsubstituted carbacephams. Subsequent biological assay
of the over reduced compounds indicated that they had
siderophore activity (i.e., promoted microbial growth as
iron transport agents) rather than antibiotic activity.
However, this suggested that the iron binding tripeptide
could carry a ß-lactam into bacterial cells. Extensive
study provided hydrogenation conditions compatible with
production of the desired carbacephalosporin conjugates
(**39** and **40**).[32] Finally, we were ready to test the
postulated iron transport-mediated drug delivery.

Scheme 8

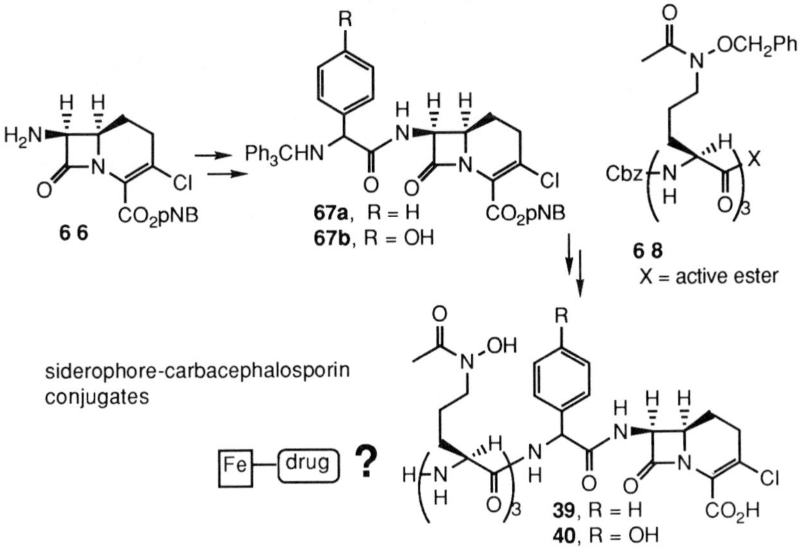

siderophore-carbacephalosporin
conjugates

Fe—$drug$ **?**

Much to our dismay, neither conjugate **39** nor **40** displayed significant activity in the same broad standard antimicrobial screen used to demonstrate the tremendous potency of Lorabid! However, detailed studies with *E. coli* X580, a representative organism from the broad screen testing, revealed that while growth did occur over extended time (up to 36 h), significant delay of the onset of growth was reproducibly noted in cultures containing the siderophore-carbacephalosporin conjugates. Further studies revealed that reincubation of the organism that did eventually grow in the same culture medium with the same conjugate resulted in no delay of growth. This suggested that the organism that grew was a mutant selected from the initial *E. coli* and that the mutant lacked the outer membrane receptor or transport proteins for hydroxamate-based siderophores. Consistent with this, attempted promotion of growth of the new organism by ferrichrome, or related tripeptide **34**, under iron-deficient conditions failed. Apparently the mutant relied on one of the other microbial iron assimilation pathways (Fig 4).

A separate study determined that enterobactin (**4**) and various other synthetic catechol siderophores stimulated growth of the hydroxamate conjugate-resistant strain (HCRS) under iron deficient conditions. Thus, related antibiotic conjugates were anticipated to be effective inhibitors of the HCRS. Neither enterobactin nor the natural spermidine-based catechol siderophores agrobactin (**5**) and parabactin (**6**) have suitable functionality for drug conjugation. However, the syntheses of the spermidine-based siderophores and analogs reported by Bergeron's group[33] and our group[34] were compatible with preparation of drug conjugates. Simple modification of the synthesis of spermexatol, a synthetic siderophore analog with effective microbial growth promotion ability,[34] gave the first spermidine-based catechol carbacephalosporin conjugate **71** (Scheme 9).[35] Conjugate **71** contains only two bidentate catechol units rather than the usual three. Still, incubation of **71** with the previously described hydroxamate conjugate resistant strain of *E. coli* (HCRS) resulted in inhibition of growth. Moreover, similar incubation with the original *E. coli* X580 also resulted in delay of growth with apparent selection of new mutants that are resistant to the new conjugate and related catechol-based siderophores.

Simultaneous incubation of the parent strain of *E. coli* X580 with both the catechol (**71**) and hydroxamate (**39**) conjugates was anticipated to further inhibit growth.[36] In fact, using half the concentration (0.5 mM) of each conjugate, so that the total concentration of ß-lactam remained the same, did result in further delay of growth. Apparently, the organism that did eventually grow is a double mutant lacking both hydroxamate- and catechol-siderophore receptors.

Scheme 9

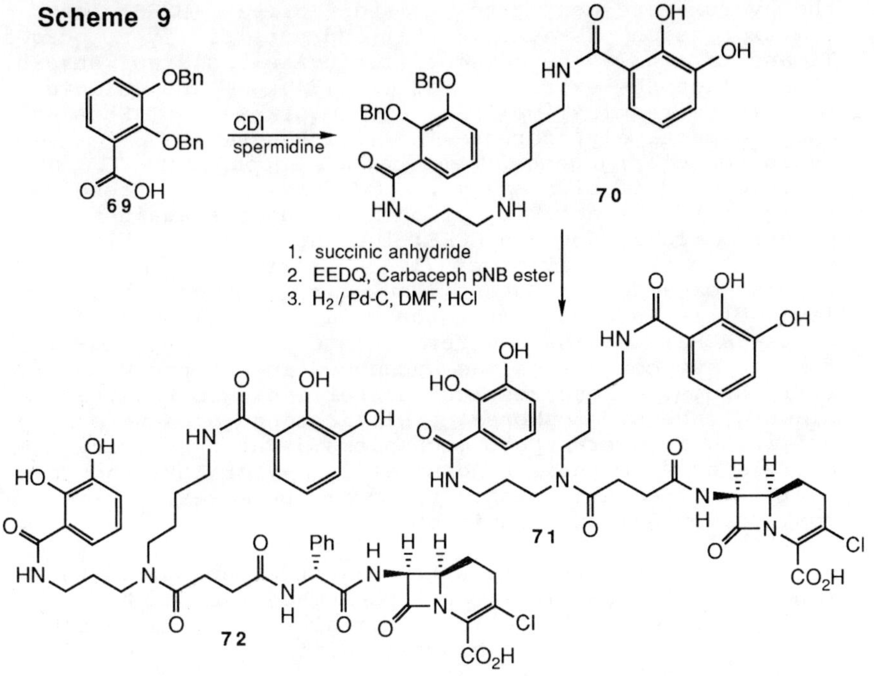

Thus, it was apparent that *the antibiotics were being actively carried into the E. coli by the sidero-phore components*. These results indicate that routine MIC (minimum inhibitory concentration) studies, which involve evaluation of inhibition of growth only after a given period of time, may miss compounds active against parent microbial strains, and thus, many potential lead compounds may not be detected. While the ultimate fate of the antibiotic conjugates is still not known, detailed studies of the mode of action of the conjugates have been informative.[37]

In separate agar-diffusion tests, both conjugates generated large zones of inhibition against *E. coli* X580 The resistant strains selected from individual exposure of the *E. coli* to conjugates **39** and **71** displayed no cross resistance. A competitive assay with ^{125}I labelled penicillin V determined the separate affinity of conjugates **39** and **71** for isolated inner membrane penicillin-binding proteins (PBPs).[38] Thus, hydroxamate conjugate **39** appeared to target PBPs 1A/B and 3 while catechol conjugate **71** apparently interacted with PBPs 1A/B and 5/6. Introduction of a ß-lactamase-encoding plasmid into *E. coli* X580 resulted in the loss of activity of each of the conjugates. The hydroxamate- and catechol-resistant strains were separately grown on agar plates under iron- deficient conditions and the outer membrane protein (Omps) profiles were analyzed by SDS-PAGE. While OmpC and Ompf porins and OmpA were normal,

the hydroxamate conjugate resistant strain (HCRS) lost
the expression of FhuA (78 kDa) and sensitivity to phages
T1 and T5, while the catechol conjugate-resistant strain
(CCRS) lost the expression of Cir (74 kDa) and colicin
Ia. These results implicate the requirement of FhuA and
Cir, respectively, for the inhibitory activity of
hydroxamate conjugate **39** and catechol conjugate **71**. In
antibiotic diffusion assays, 1 mM ferrichrome strongly
antagonized the activity of both conjugates against the
parent *E. coli* X580 and CCRS; indicating a possible role
of FhuA in the activity of **71**. Consistent with this,
studies of a double mutant (FhuA-Cir-) showed a higher
level of resistance to conjugate **71**. The susceptibility
of the HCRS, lacking the ferrichrome receptor (FhuA-),
for the two conjugates was unchanged in the presence of
ferrichrome. These results further indicate that, as
planned, the siderophore-antibiotic conjugates were
utilizing iron transport mechanisms. Their activity
correlates with their binding affinity for PBPs and the
activity of each is sensitive to ß-lactamases, as are
most common ß-lactam antibiotics.

While these results were very encouraging, it was
important to determine if the conjugate resistant strains
of *E. coli* were pathogenic to mammals. To test this, *E.
coli* X580 and HCRS were grown in diffusion chambers
implanted in the peritoneal cavity of rats to simulate
iron-restricted *in vivo* growth conditions. Compared to
the parent strain, the growth of HCRS was impaired!
Furthermore, the post-incubation bacteria recovered from
rats was found to be susceptible to hydroxamate conjugate
39, suggesting that the resistant organism was not
viable, and hence less virulent *in vivo*. Apparently,
only reversion to the parent strain, containing a full
set of iron assimilation mechanisms, allowed growth.
Such a reversion might not be detrimental since the
revertants containing a hydroxamate transport system
would be susceptible to the hydroxamate conjugate.
Overall, these biological studies clearly demonstrated
that the concept of iron transport-mediated drug delivery
has considerable potential for the rational design of new
therapeutic agents. Furthermore, the conjugates may
serve as valuable biological tools for the elucidation of
important aspects of microbial iron metabolism by the
selection of mutants lacking various iron assimilation
machineries.

These results encouraged the preparation and study
of catechol conjugate **72** which incorporated the phenylglycyl side chain of the carbacephalosporin.[39] At 10mM of
conjugate **72**, complete inhibition of the growth of *E.
coli* X580 was observed. As before, at 1mM selection of
resistant strains was noted. Further study is needed to
confirm that the mode of action of this conjugate is the
same as the original catechol conjugate (**71**).

Another bis-catechol conjugate, **73b**, was designed after a natural bis-catechol siderophore bis(2,3-dihydroxybenzoyl)-L-lysine (**73a**), isolated from *Azotobacter vinelandi*.[40,41] Carbacephalosporin conjugate **73b** was synthesized[51] and initial growth inhibition studies with *E. coli* X580 gave results nearly identical to the related spermidine-based conjugate **71**. Details of the mode of action of the lysine-based conjugate have not yet been determined.

Next, the related triscatechol **74a** and the corresponding conjugate **74b** were selected as appropriate targets for synthesis and study, since a triscatechol siderophore would presumably bind iron better than a biscatechol. The synthesis[42] of acid **74a** and conjugate **74b** paralleled those of previous conjugates **71** and **73b**. Interestingly, preliminary studies of the incubation of tricatecholate conjugate **74b** with *E. coli* X580 indicated no growth inhibition. Instead, conjugate **74b** and the triscatechol **74a** were both found to be effective *growth stimulants* of *E. coli* X580. More study is required to determine if the decreased solubility or the increased bulk of **74b** is a factor for its lack of antibiotic activity or if a drug release mechanism is required.

The determination of the mode of action of carba-cephalosporin conjugates **39** and **71** suggests that detailed biological studies of the sulfa drug conjugates of ferri-crocin and the quinolone antibiotic (nalidixic acid) conjugate of desferrioxamine B (**17**) should be performed. Synthesis and study of other antibiotic-siderophore conjugates should also be considered. The possible combinations are almost limitless. Our program related to the total synthesis of pseudobactin (**8**)[43] is also intended to facilitate the synthesis of conjugates of pseudobactin and its components to determine if a selective, antipseudomonal drug can be prepared. Meanwhile, we have conjugated other drugs to tripeptide **34** to begin structure-activity relationship studies of the conjugates.

Erythromyclamine, a macrolide antibiotic with ribosomal RNA inhibitory activity similar to erythromycin,[44] was coupled to protected tripeptide active ester **68**. Hydrogenolytic deprotection provided the desired conjugate (**75**) cleanly.[45] Again, preliminary broad screen assays indicate that the conjugate has no activity against normally erythromycin-sensitive organisms. While detailed growth studies must be performed to confirm the lack of activity, it might be surprising if the conjugate was indeed effective since the drug is larger than the delivery agent and no easily cleaved linkage was incorporated.

Scheme 10

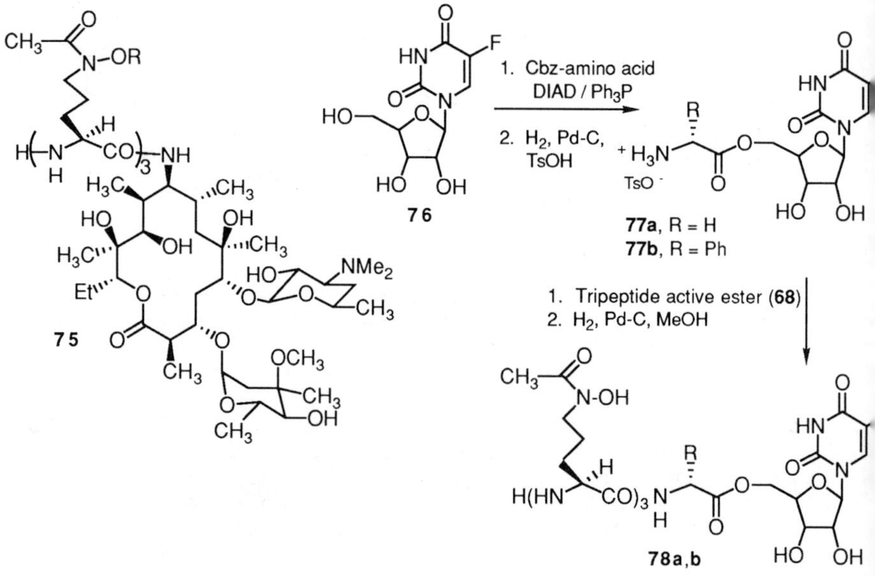

The studies described above have all concerned the development of new antibacterial agents. The use of iron transport-mediated drug delivery might also facilitate the design of new antifungal agents. Fungal infections are a significant concern, especially in immuno-compromised patients. In fact, most AIDS patients succumb to opportunistic infections from strains of *Candida*, *Cryptococcus*, *Pneumocystis*, *Histoplasma*, *Toxoplasma*, *Aspergillus* and related organisms. The development of microbially-selective antifungal agents would be a significant advance in prolonging the lives of AIDS patients. First, however, it is necessary to determine which siderophores can be used by the target organisms.

Our initial focus has been on *Candida albicans*. Catechols and hydroxamates are reportedly generated and utilized by *Candida* for siderophore activity.[46] However, no structural details are known and no assay for siderophore recognition and transport in *Candida* was available.

Utilizing the variety of natural and unnatural siderophores and components available from our antibacterial
conjugate studies, we recently developed a bioassay for
siderophore utilization by *Candida albicans*.[47] Neither
enterobactin nor the other catechol siderophores and
components described throughout this paper were
effective. However, ferrichrome, the constituent tri-δ-N-
hydroxy-δ-N-acetyl-L-ornithine (**34**), and several of the
extended peptides served as candidal siderophores. We
are in the process of learning how to attach a number of
antifungal agents to these peptides. Enough progress has
been made in one case to allow preliminary discussion.
5-Fluorocytosine, which displays both fungicidal and
fungistatic activity against *Candida albicans*,[48] was
chosen as the first antifungal agent for conjugation to
34. To facilitate the synthesis, and with the intent of
providing more labile linkages to the siderophore, 5-
fluorouridine was first esterified with N-protected
glycine and D-phenylglycine (Scheme 10). The N-terminal
Cbz groups were removed and each product was coupled
directly to protected tripeptide active ester **68**.
Deprotection gave desired conjugates **78a,b**. Studies of
the stability of the conjugates under potential microbial
assay conditions in Lee's media were readily accomplished
by following the hydrolytic release of free 5-fluoro-
uridine by ^{19}F NMR. In this manner, conjugate **78a** with
the glycine spacer was determined to have a half life of
13 h, whereas the D-phenylglycyl-containing conjugate's
half life was reduced to 9.5 h. Still, preliminary
antibacterial and anticandidal assays have been
performed. Both conjugates were two to four times more
active than 5-fluorocytosine (5-FC) itself against
various strains of *Staphylococcus aureus* (MICs = 1-4 μg/mL
vs. 2-8 for 5-FC) and somewhat less active against
various strains of *Streptococcus* (4-16 μg/mL vs. 0.25-
4.0). Both conjugates and 5-FC were apparently inactive
against selected strains of *E. coli*, *Salmonella*, and
Pseudomonas. The activity of the conjugates against
Candida albicans and *C. parapsilosis* CP18 depended
significantly upon the media, with MIC data ranging from
0.625 to 80 μg/mL. The presence of competitive exogenous
siderophores in some of the media or other hydrolytic
processes may drastically alter the assays. Related
studies and detailed growth-curve generation and analysis
must be done to confirm the effectiveness of these
conjugates, or less hydrolytically labile conjugates,
against *Candida*, as well as other yeasts and fungi.

In summary, iron transport-mediated drug delivery
has been demonstrated to be a feasible process, at least
for ß-lactam conjugates. More study is required to
determine the potential utility of this concept.

Acknowledgements

MJM gratefully acknowledges the NIH for support of this research. Significant thanks are due to Eli Lilly and Company for providing the carbacephalosporin, erythromycylamine, and general antimicrobial assays. The assistance of A. Brochu and N. Brochu with the biological evaluation of the conjugates is sincerely appreciated. Helpful discussions with Dr. Thalia Nicas, Jan Turner, Robert Gordee, Bill Turner (Lilly) and Professor J. B. Neilands (Berkeley) are appreciated. Thanks are due to Dr. Andy Sheppard of Ferring Research Institute, and Drs. Peter Crocker and Ihab Darwish for assistance in preparation of the manuscript. Desferal was a gift from the late Professor Tom Emery.

References

1. G. Winkelman, Ed. Handbook of Microbial Iron Chelates, CRC: Boca Raton, FL, 1991.
2. a) R. C. Hider, Structure Bonding (Berlin), 1984, 58, 25. b) G. Winkelman, D. van der Helm and J.B. Neilands, Eds. Iron Transport in Microbes, Plants, and Animals, VCH: New York, 1987.
3. Taken in part from the Ph.D. Thesis of J.A. McKee, University of Notre Dame, Notre Dame, IN, 1991.
4. a) R. Wayne, K. Frick and J.B. Neilands, J. Bacteriol., 1976, 126, 7. b) R.R. Wayne, and J.B. Neilands, Fed. Proc. Fed. Am. Soc. Exp. Biol., 1976, 35, 1453.
5. a) J.A. Crosa, Microbiol. Rev., 1989, 53, 517. b) K. Hankte and V. Braun, J. Bacteriol., 1978, 135, 190. c) C.C. Wang and A. Newton, J. Biol. Chem., 1971, 245, 2147. d) S. Silver and M. Walderhaug, Microbiol Rev., 1992, 56, 195.
6. V. Braun, K. Hankte, K. Eick-Helmerich, W. Köster, U. Preßler, M. Sauer, S. Schäffer, H. Schöffler, H. Staudenmaier and L. Zimmermann, in Iron Tranport in Microbes, Plants and Animals, G. Winkelmann, D. van der Helm and J.B. Neilands, Eds.; VCH: Weinheim, FRG, New York, 1987, Chapter 2.
7. a) K. Hankte, Mol. Gen. Genet., 1983, 191, 301. b) Interestingly, the intact ferrichrome complex apparently is not taken into the cytoplasm. Instead the iron presumably is reductively removed in the membrane and a free hydroxamate of the deferriferrichrome is acetylated to reduce its affinity for ferric ion and initiate its recycle: A. Hartman and V.J. Braun, Bacteriol., 1980, 143, 246.
8. H. Nikaido and E.Y. Rosenburg, J. Bacteriol., 1990, 172, 1361.
9. H. Zahner, H. Diddens, W. Keller-Schierlein and H.-U. Nageli, J. Antibiotics, 1977, 30, S-201.

10. a) N. Ohi, B. Aoki, T. Shinozaki, K. Moro, T.
     Noto, T. Nehashi, H. Okazaki and I. Matsunaga, J.
     Antibiotics, 1986, 39, 230. b) N. Ohi, B. Aoki, K.
     Moro, T. Kuroki, N. Sugimura, T. Noto, T. Nehashi,
     M. Matsumoto, H. Okazaki and I. Matsunaga, J.
     Antibiotics, 1986, 39, 242. c) N. Ohi, B. Aoki, T.
     Shinozaki, K. Moro, T. Kuroki, T. Noto. T. Nehashi,
     M. Matsumoto, H. Okazaki and I. Matsunaga, Chem.
     Pharm. Bull., 1987, 35, 1903. d) H. Mochizuki, H.
     Yamada, Y. Oikawa, K. Murakami, J. Ishiguro, H.
     Kosuzume, N. Aizawa and E. Mochida, Antimicrob.
     Agents Chemother., 1988, 32, 1648. e) N-A.
     Watanabe, T. Nagasu, K. Katsu and K. Kitoh,
     Antimicrob. Agents Chemother., 1987, 31, 497. f)
     K. Katsu, K. Kitoh, M. Inoue and S. Mitsuhashi,
     Antimicrob. Agents Chemother., 1982, 22, 181. g) S.
     Nakagawa, M. Sanada, K. Matsuda, N. Hazumi and N.
     Tanaka, Antimicrob. Agents Chemother., 1987, 31,
     1100. h) N.A.C. Curtis, R.L. Eisenstadt, S.J.
     East, R.J. Cornford, L.A. Walker and A.J. White,
     Antimicrob. Agents Chemother., 1988, 32, 1879. i)
     T. Harada, A. Yoshisato, Y. Imai, Y. Takano, Y.
     Ichikawa and Y. Suzudi, JP 63,192,781 (Chem.
     Abstr., 1988, 109, 230641h). j) ICI Pharma, Europ.
     Pat. Appl., 87309767. k) W. Nagata, T. Aoki and Y.
     Nishitani, EP 0117143 A2. l) C.E. Newall, in
     Recent Advances in the Chemistry of ß-Lactam
     Antibiotics, P.H. Bentley and R. Southgate, Eds.
     Royal Society of Chemistry, Special Publication No.
     70 1989, Chap. 24.
11. a) G. Benz, T. Schroder, J. Kurz, C. Wunsche, W.
     Karl, G. Steffens, J. Pfitzner and D. Schmidt,
     Angew Chem. Suppl., 1982, 1322. b) G. Benz, L.
     Born, M. Brieden, R. Grosser, J. Kurz, H. Paulsen,
     V. Sinnwell and B. Weber, Liebigs Ann. Chem., 1984,
     1408.
12. H. Bickel, E. Gaumann, G. Nussberger, P. Reusser,
     E. Vischer, W. Vosser, A. Wettstein and H. Zahner,
     Helv. Chim. Acta, 1960, 43, 2105.
13. A.E. Martell, Ed., Inorganic Chemistry in Biology
     and Medicine, American Chemical Society,
     Washington, D. C., 1973.
14. a) M.J. Miller, Chem. Rev., 1989, 89, 1563. b)
     R.J. Bergeron and J.S. McManis, in Handbook of
     Microbial Iron Chelates, G. Winkelmann, Ed. CRC
     Press, Boca Raton, FL, 1991, p. 271.
15. P.J. Maurer and M.J. Miller, J. Am. Chem. Soc.,
     1982, 104, 3096. For related syntheses of
     arthrobactin and schizokinen see: B.H. Lee and
     M.J. Miller, J. Org. Chem., 1983, 48, 24.
16. a) P.J. Maurer and M.J. Miller, J. Am. Chem. Soc.,
     1983, 105, 240. b) P.J. Maurer and M.J. Miller, J.
     Org. Chem., 1983, 46, 2835.
17. O. Mitsunobu, Synthesis, 1981, 1.

18.   B.H. Lee, G.J. Gerfen and M.J. Miller, J. Org.
      Chem., 1984, 49, 2418.

19.   E.K. Dolence and M.J. Miller, J. Org. Chem., 1991,
      56, 492.

20.   E.K. Dolence, C.-E. Lin, M.J. Miller and S.M.
      Payne, J. Med. Chem., 1991, 34, 956.

21.   a) K. Mochida, T. Ogasa, J. Shimada, T. Hirata, K.
      Sato, and R. Okachi, J. Antibiot., 1988, 42, 283.
      b) R.N. Jones and A.L.J. Barry, Antimicrob. Ther.,
      1988, 22, 315. c) J.H. Jorgensen, J.S. Redding,
      L.A. Maher, Antimicrob. Agents Chemother., 1988,
      32, 1477. d) A.J. Howard and K.T. Dunkin, J.
      Antimicrob. Ther., 1988, 22, 445. e) C.C. Bodurow,
      B.D. Boyer, J. Brennan, C.A. Bunnell, J.E. Burkes,
      M.A. Carr, C.W. Doecke, T.M. Eckrich, J.W. Fisher,
      J.P. Gardner, B.J. Graves, P. Hines, R.C.Hoying, B.
      G. Jackson, M.D. Kinnick, C.D. Kochert, J.S. Lewis,
      W.D. Luke, L.L. Moore, J.M. Morin, Jr., R.L. Nist,
      D.E. Prather, D.L. Sparks and W.C. Vladuchick, Tet.
      Lett., 1989, 30, 2321. f) T. Ogasa, H. Saito, Y.
      Hashimoto and K. Sato, Chem. Pharm. Bull., 1989,
      37, 315. g) H. Saito, H. Matsushima, C. Shiraki
      and T. Hirata, Chem. Pharm. Bull., 1989, 37, 275.
      h) For the first asymmetric synthesis of
      carbacephalosporins see: D.A. Evans and E.B.
      Sogren, Tet. Lett., 1985, 26, 3787.

22.   M. J. Miller, Accts. Chem. Res., 1986, 19, 49.

23.   B.T. Lotz, Ph.D. Thesis, University of Notre Dame,
      Notre Dame, IN, 1991.

24.   B.T. Lotz, C.M. Gasparski, K. Peterson and M.J.
      Miller, J. Chem. Soc. Chem. Commun., 1990, 1107.

25.   L. Schirch, Adv. Enzymol. Relat. Areas Mol. Biol.,
      1982, 53, 83.

26.   D.F. Taber and L.J. Silverberg, Tet. Lett., 1991,
      34, 4227.

27.   R. Noyori and H. Takaya, Accts. Chem. Res., 1990,
      23, 345.

28.   P.R. Guzzo and M.J. Miller, to be submitted.

29.   C.M. Gasparski, M. Teng and M.J. Miller, J. Am.
      Chem. Soc., 1992, 114, 2741.

30.   M. Teng, and M.J. Miller, J. Am. Chem. Soc., in
      press.

31.   C.M. Gasparski, Ph.D. Thesis, University of Notre
      Dame, Notre Dame, IN, 1991.

32.   E.K. Dolence, A.A. Minnick, C.E. Lin and M.J.
      Miller, J. Med. Chem., 1991, 34 969.

33.   a) R.J. Bergeron, K.A. McGovern, M.A. Channing and
      P.S. Burton, J. Org. Chem.,1980, 45, 1589. b) R.J.
      Bergeron, P.S. Burton, S.J. Kline and K.A.
      McGovern, J. Org. Chem., 1981, 46, 3712. c) R.J.
      Bergeron, S.J. Kline, N.J. Stolowich, K.A. McGovern
      and P.S.Burton, J. Org. Chem., 1981, 46, 4524. d)
      R.J. Bergeron and S.J. Kline, J. Am. Chem. Soc.,
      1982, 104, 4489. e) R. J. Bergeron, J. S. McManis,

J.B. Dionis and J.R. Garlich, <u>J. Org. Chem.</u>, 1985, <u>50,</u> 2780.

34. S.K. Sharma, M.J. Miller and S.M. Payne, <u>J. Med. Chem.</u>, 1989, <u>32</u>, 357.

35. J.A. McKee, S.K. Sharma and M.J. Miller, <u>Bioconj. Chem.</u>, 1991, <u>2</u>, 281.

36. A.A. Minnick, J.A. McKee, E.K. Dolence and M.J. Miller, <u>Antimicrob. Agents Chemother.</u>, 1992, <u>36</u>, 840.

37. A. Brochu, N. Brochu, T.I. Nicas, T.R. Parr, Jr., A.A. Minnick, E.K. Dolence, J.A. McKee, M.J. Miller and F. Malouin, <u>Antimicrob. Agents Chemother.</u>, 1992, in press.

38. F. Malouin, S. Chamberland, N. Brochu and T.R. Parr, Jr. <u>Antimicrob. Agents Chemother.</u>, 1991, <u>35</u>, 477.

39. M.J. Miller, J.A. McKee, A.A. Minnick and E.K. Dolence, <u>Biol. Metals,</u> 1991, <u>4</u>, 62.

40. J.L. Corbin and W.A. Bulen, <u>Biochemistry</u>, 1969, <u>8</u>, 757.

41. A. Chimiak and J.B. Neilands, in <u>Structure and Bonding</u>; Springer-Verlag: New York, 1984, 54, 89.

42. J.A. McKee and M.J. Miller, <u>Bioorg. Med. Chem. Lett.</u>, 1991, <u>1</u>, 513.

43. a) T. Kolasa and M.J. Miller, <u>J. Org. Chem.</u>, 1990, <u>55</u>, 1711. b) T. Kolasa and M.J. Miller, <u>J. Org. Chem.</u>, 1990, <u>55</u>, 4246.

44. M. Cannon, in <u>Comprehensive Medicinal Chemistry</u>, C. Hansch, P.G. Sammes and J.B. Taylor, Eds, Pergamon Press, Oxford, 1990, Vol 2, Chap. 10.5

45. J.A. McKee, Ph.D. Thesis, University of Notre Dame, Notre Dame, IN, 1991.

46. a) A. Ismail, G.W. Bedell and D.M. Lupan, <u>Biochem. Biophys. Res. Comm.</u>, 1985, <u>130</u>, 885. b) A. Ismail, G.W. Bedell, D.M. Lupan, <u>Biochem. Biophys. Res. Comm.</u>, 1985, <u>132</u>, 1160. c) A. Ismail and D.M. Lupan, <u>Mycopathologia</u>, 1986, <u>96</u>, 109.

47. A.A. Minnick, L.E. Eizember, J.A. McKee, E.K. Dolence and M.J. Miller, <u>Anal. Biochem.</u>, 1991, <u>194</u>, 223.

48. M. Vederame, Ed. "Antifungal Agents," in <u>CRC Handbook of Chemotherapeutic Agents</u>, Vol 1, CRC Press, Inc.: Boca Raton, FL, 1986, pp. 219-260.

Chemistry of Antifungals

11
The Need for New Antifungals – A Clinical Appraisal

R. J. Hay
ST JOHN'S INSTITUTE OF DERMATOLOGY, GUY'S HOSPITAL, LONDON
SE1 9RT, UK

Summary

As our ability to cope with bacterial infection in even the sickest patient improves it is inevitable that other disease targets should become more prominent. Viral, parasitic and fungal diseases are increasingly recognised as complications which have to be controlled. While the past twenty years have seen a major expansion in the number of antifungal drugs available there are still weaknesses in both the range and scope of antifungal chemotherapy. New developments in this field have included the refinement of existing drug molecules in order to eliminate toxicity and improve the spectrum of activity, for instance with the development of the triazoles from the imidazoles, or the use of new less toxic formulations including the amphotericin B lipid complexes. The development of new groups of drugs with different modes of action such as the morpholines and the echinocandins has also been possible. Even in the superficial mycoses where treatment is often regarded as being highly effective it has been possible to reduce the durations of therapy and attack diseases such as onychomycosis where, previously, the success rate of treatment was low. The development of new drugs to confront new infections or infections regarded as being untreatable has had to be matched with a challenge to clinicians – the need to construct methods of assessment which provide an objective measure of efficacy. All this is

occurring against the familiar background of reduced
health budgets and the introduction of different
schemes for the "efficient" management of limited
financial resources. This has resulted in a dilemma in
that good treatments which are not cost effective are
a major source of concern to health authorities.

Introduction

The fungal infections or mycoses range from the
commonplace (thrush, athletes foot) to the rare and
arcane, diseases such as systemic penicillinosis now
being seen with infrequent but increasing regularity
in AIDS patients. The commonest of these diseases are
those affecting the skin or mucous membranes, the
superficial mycoses. These are a problem in all parts
of the world. They include infections of the otherwise
healthy patient such as dermatophytosis or ringworm as
well as those which principally affect the compromised
host. Candidosis affecting the integument is an
example of the latter; in AIDS patients oropharyngeal
candidosis is the single commonest infectious
complication of HIV infection. Yet in the case of
vaginal *Candida* infection most affected women have no
evidence of detectable underlying abnormality. Other
superficial infections include diseases such as tinea
nigra, a tropical infection caused by a pigmented
fungus, *Phaeoanellomyces werneckii* and black and white
piedra where the invasion of the skin is restricted to
the hair shaft. The subcutaneous mycoses are generally
uncommon disorders, occurring principally in the
tropics and subtropics. More aptly named mycoses of
implantation they generally arise through traumatic
injury where environmental fungi gain entry to the
skin or subcutaneous tissue and survive long enough to
cause indolent infections. They are notoriously
difficult to treat and even with seemingly effective
drugs the process of recovery is slow. The best known
of these diseases are sporotrichosis,
chromoblastomycosis and mycetoma, an infection which

can be caused by either actinomycetes, filamentous bacteria, or fungi. The systemic mycoses are often divided into two groups, the endemic respiratory infections and the opportunistic infections. This division, which depends on the effect of host resistance on the clinical manifestations of disease, is somewhat artificial as host immune status will affect the clinical expression of most disease processes. However the use of these groupings does have some advantage. The endemic mycoses are confined to well defined endemic areas; none, for instance, is endemic in the UK. They all gain entry via the respiratory tract and can disseminate to other sites from the lung. The majority of those exposed to infection develop no clinical signs of disease and remain well. The only evidence of exposure is the development of a positive skin test. This is similar to the pathogenesis of tuberculosis. The host condition will nonetheless affect the response to infection. Those with chronic lung disease often appear to be susceptible to indolent pulmonary infections, those with severe immunodeficiency to widespread disseminated disease. AIDS patients come into this latter category. By contrast the opportunistic infections almost always occur against a background of host defects. Their clinical manifestations are much less predictable because the appearance of the infection is determined by a number of variables such as route of entry as well as the host response. These opportunistic infections are more correctly thought to represent different disease syndromes. The manifestations of systemic candidosis, for instance, vary greatly depending on whether it occurs in the neutropenic cancer patient, the intravenous drug abuser or the premature neonate. A different spectrum of disease is seen with *Aspergillus* infections. The similarities between the two types of systemic mycoses, the endemic and the opportunistic, are seen in cryptococcosis. Here respiratory infection

may be followed in some patients by dissemination to
other sites particularly the meninges. Cryptococcal
meningitis may occur in an otherwise healthy patient
or in an immunocompromised patient particularly the
AIDS patient. There are two varieties of *Cryptococcus
neoformans* known as *C. neoformans* var *neoformans* and
C. neoformans var *gattii*. The former is usually, but
not exclusively, associated with disease in the
compromised host; the latter with disease in the
otherwise healthy individual. This shows that small
variations in virulence may also play a part in
determining the pathogenesis of infection. In the case
of *Cryptococcus* differences in antigen structure,
melanin biosynthesis and capsule production have all
been thought to be potential virulence factors.

Antifungal drugs

Compared with the large number of antibacterial
drugs, there are far fewer antifungals. There are
three main families of antifungal drugs: the polyenes,
the azoles and the allylamines. There is also a
miscellaneous group of compounds such as flucytosine
and griseofulvin which do not belong to a single
family of drugs. This is not a static picture and
there are new groups of antifungals under development
all the time. These include the morpholine antifungals
such as amorolfine, a new fungal sterol biosynthesis
inhibitor,[1] and cell wall antagonists such as the
echinocandins, eg cilofungin,[2] or the nikkomycins.

Polyene antifungals and their use

The polyene antifungals comprise a large family
of drugs which are derived from *Streptomycete* species[3]
but only three, amphotericin B, nystatin and
natamycin, are used for human disease. The activity of
the polyene antifungals depends on the inhibition of
the formation of the fungal cell membrane. These drugs
bind avidly at this site and there is a close
relationship between cell membrane binding and

inhibitory activities. The consequences are, for instance, increased cell permeability and, eventually, death. However the lethal effects of the polyenes can be separated from the increase in cell permeability which is reversible at low concentrations.[4] Similar binding to mammalian cell membranes is believed to form the basis for their toxicity in man and is probably dependent on the affinity between the polyene and sterol membrane constituents such as cholesterol.

The polyenes are generally poorly soluble in aqueous solvents and poorly absorbed after oral administration, but are topically active. As topical therapies they are principally used for superficial candidosis. They are less effective in established AIDS cases where orally absorbed drugs for oropharyngeal candidosis are necessary. The only drug of this series which can be given parenterally is amphotericin B. It disappears rapidly from serum, only about 10% of the administered dose being present, bound to plasma proteins, 12 hours after dosage, and is thought to be bound thereafter to tissue cell membranes.[5] Penetration into CSF or urine is poor.

Amphotericin B has a broad range of antifungal activity including the major pathogens involved in systemic mycoses apart from the zygomycete fungi. Failures certainly occur, particularly in the severely neutropenic patient. There is some evidence that isolates from patients failing to respond to amphotericin B are less sensitive to the drug than would be expected;[6] also the bioavailability of the compound in infected tissue has been reported to be lower than necessary for adequate responses.[7] Amphotericin B is given intravenously in 5% dextrose, but is commonly toxic.[8] The side effects of amphotericin B include immediate reactions of hyperpyrexia, severe malaise and hypotension.[9] A progressive rise in creatinine and hypokalaemia, as well as chronic anaemia may also occur. In the longer term chronic usage, above a total dose of 4g is associated with a high risk of irreversible renal tubular damage. Various ingenious ways have been devised to circumvent some of these side effects from the use of antihistamines, heparin or pethidine to concurrent induction of an osmotic diuresis and the gradual increase of dosage to reach a threshold at which side effects appear. All these manoeuvres are only partially successful and none can prevent the long term renal tubular damage.

Other approaches to this problem have been the development of amphotericin B analogues, notably ester derivatives and reformulation of the drug in a different vehicle. The most extensively investigated of the analogues was the methyl ester of amphotericin B which, although clinically active and less nephrotoxic, caused cerebral lipid deposition and proved difficult to purify.

An attempt has been made to avoid the side effects of amphotericin B by incorporating the drug into liposomes [11,12] or lipid complexes. One unilamellar liposomal preparation of amphotericin B (Ambisome) is now available in some countries. It has been used to treat patients with invasive candidosis and aspergillosis, hepatosplenic candidosis and mycetoma. The daily dose can be rapidly increased without any adverse effects appearing. The limited human toxicology available suggests that it does not cause renal impairment despite doses of up to 4mg/kg daily or total doses over 4-6g. The initial clinical trial indications required that all patients had either renal failure, amphotericin B related renal impairment or had failed to respond to amphotericin B. [12] In no patient was there further deterioration of renal function and none of the other expected side effects, such as hepatotoxicity or anaemia, were manifest. These patients received between 2 and 3 mg/kg daily, some two and a half to three times the normal dosage.

Seventy five percent of the patients with systemic candidosis were reported to be in clinical and mycological remission. Other lipid complexes under investigation include a lipid complex (ABLC) and a colloidal dispersion (ABCD). One potential problem with some of these formulations is the high cost of manufacture.

The azole antifungals and their use

The azole antifungals are a rapidly expanding family of drugs.[13] The first group to be developed, the imidazoles, contained a large number of compounds primarily aimed at topical use. The only members available for parenteral or systemic usage are miconazole, which can be administered intravenously, or ketoconazole which can be given orally. Clotrimazole is well absorbed following oral administration but it is a potent enzyme inducer and serum levels are greatly reduced after two weeks of therapy. It is still used occasionally in the USA in the troche form for oropharyngeal infections.

The principal mode of action of this series is the inhibition of cytochrome P 450 dependent C-14 demethylation in the formation of ergosterol in the fungal cell membrane. One of the potential disadvantages of this group is that in many cases there is some interference with human cytochrome P450[14] as well as affecting some human metabolic processes. Ketoconazole, for instance, is a potent blocker of adrenal androgen biosynthesis and also affects prostacycline/leukotriene biosynthesis. The imidazoles vary considerably in their absorption after oral administration and the doses of two, ketoconazole and itraconazole, have to be increased in patients with AIDS and bone marrow transplant recipents.

The newer triazole series (fluconazole, itraconazole, terconazole) act by the same mechanism as the imidazoles - via inhibition of cytochrome P450. The triazoles in current use are said to have little affinity with mammalian cytochrome P450, although there may be variation between individual triazoles in this respect. Adverse effects because of blockade of human cytochrome P450 have not been reported with these drugs.[14,15] Fluconazole, being very water soluble, has excellent absorption and is widely distributed throughout the body.[15,16] It is excreted unchanged in the urine and reduced creatinine clearance prolongs its half life.[17] If this is below 20 ml/min, the elimination half life is prolonged to 98 hours. Reduction in dosage is, therefore, advised in patients with renal impairment. An intravenous preparation of this azole is also available. Although itraconazole is absorbed in low concentrations tissue levels in liver, brain and skin[18] are considerably higher and there is evidence that in many of these sites there is prolonged excretion. The most dramatic example of this is the nail, where levels of drug can be detected 3 months after the end of therapy. Little itraconazole is present in urine and CSF.

The imidazole antifungals affect most of the common superficial fungal pathogens. For superficial infection they are available as 1% creams, ointments or powders, athough some special formulations are available, eg ketoconazole shampoo (seborrhoeic dermatitis) and econazole powder spray (tinea pedis). Ketoconazole, in addition is active via oral route in doses of 200-400mg but its use in long term therapy, for instance for onychomycosis, is limited because of the risk of hepatoxicity estimated to occur in about 1:10000 cases. It is however effective after a single dose in vaginal candidosis and pityriasis versicolor. A single dose of 150mg fluconazole rapidly induces clinical and mycological remission in over 90% of patients with vaginal candidosis,[19] although there will be some relapses five to six weeks after treatment. This rapidity of response is seen with other *Candida* infections. In oropharyngeal candidosis in AIDS patients, 74% had a significant reduction in

Candida infection to between 0 and 15 colonies by day five of treatment.[20] Similar mycological results are to be expected with other azoles but the response is usually not as rapid, although a comparative study with ketoconazole in AIDS patients, using a lower than normal dose of the latter was reported to show superior efficacy with fluconazole.[21] A single dose of 150mg has been shown to produce the same clearance rate and to be as quick as daily dosing in one pilot study of AIDS patients.[22] Similar rapid responses of oral candidosis have been seen in neutropenic cancer patients.[23] Remission in CMC patients has occurred within five days.[24]

Itraconazole has been most fully assessed in the superficial mycoses,[25] where it is effective in dermatophytosis, pityriasis versicolor and candidosis. The usual courses given in dermatophytosis are 14 days (100mg daily) for tinea corporis/ cruris and 30 days for tinea pedis. In pityriasis versicolor 800-1000mg are necessary to produce responses; in vaginal candidosis 600mg can be given over a single day as an effective therapy. Comparative studies indicate that it has some advantages over griseofulvin particularly in tinea corporis and preliminary data on nails show that it useful where there has been no response to griseofulvin.[26] Given that it is retained in nail it should be possible to use treatments of limited duration in onychomycosis; this has now been established using 3 months therapy for onychomycosis of the toe nails.[27] Itraconazole is also active in certain subcutaneous infections such as sporotrichosis and chromomycosis,[28] but it has little effect in mycetoma.

In systemic infections the uses of the azoles are less well established. Few are active, however, against aspergilli and zygomycetes. The azoles appear to be more selective in their activity than the polyenes. Amphotericin is effective against most of the major fungal pathogens whereas the imidazoles usually have at least one potential weakness in the extent of their antifungal cover. Ketoconazole and fluconazole,[16] for example, have little in vitro effect on *Aspergillus*, although it may respond at higher dosages.[29] Itraconazole is active against a broader spectrum of fungi with the possible exception of the zygomycetes. Generally it has proved difficult to evaluate drugs such as intravenous miconazole.[30] The same is true of ketoconazole particularly in cancer patients.[31] although it has proved possible to carry out studies in the endemic systemic mycoses.[32]

Cryptococcal meningitis in AIDS patients and those with soft tissue cryptococcosis responds to fluconazole both as initial and as suppressive therapy[33] The initiating dose is still not clear, although a large multicentre study in the USA using 400mg showed equal efficacy when used as primary therapy compared with amphotericin B even though there were more early treatment failures amongst those receiving fluconazole.[34] The ability to give the drug orally to prevent the disease recurring is a major advantage and as it allows early discharge from hospital has cost advantages as well.[35] In addition to cryptococcosis fluconazole has been used in the treatment of a variety of different syndromes due to *Candida*. These have included urinary tract infections, peritoneal candidosis associated with continuous ambulatory dialysis as well candidaemias.[36,37,38] The results have been variable but adequately documented remissions have been recorded in reasonable numbers of cases.

In systemic mycoses itraconazole has been assessed more fully in the systemic pathogen infections such as histoplasmosis and paracoccidioiodomycosis where it has been found to induce clinical and mycological remissions more rapidly than ketoconazole.[39,40]. Assessments of the drug in the management of invasive aspergillosis have shown that in certain cases itraconazole can produce documented remissions even in neutropenic or organ transplant patients.[41,42] As itraconazole penetrates the CSF only very poorly its use in cryptococcal meningitis has been in question on theoretical grounds. However in practice at high dosage, particularly if combined with flucytosine, it can induce remissions. [42] When used as a long term suppressant, some 30% of patients have developed negative cryptococcal antigen titres, suggesting complete clearance.

The final area for use of azoles, in this case mainly fluconazole, is in oral antifungal prophylaxis.[15] Studies confirm that this may protect against oropharyngeal fungal infection. [43] There is a possible disadvantage - the poor in vitro spectrum of activity against *Aspergilli*. It has also been shown that other potentially pathogenic *Candida* species such as *C.krusei* may emerge either in superficial or gastrointestinal carriage or causing true infections; *Candida krusei* and *C. glabrata*, for instance are often less sensitive in vitro to the drug, suggesting that some form of selection has taken place.[44] In view of its anti-*Aspergillus* activity the possibility of using itraconazole as a prophylactic agent affecting both candidosis and aspergillosis is currently being explored. A clinical trial, comparing the prophylactic effects of ketoconazole and itraconazole for *Aspergillus* infections in neutropenic patients, showed itraconazole to be more effective than ketoconazole, but as the drugs were given sequentially this would have to be confirmed in a randomized study. [45] A new oral formulation of itraconazole in cyclodextrin has now been developed which is better absorbed in the neutropenic patient.

The allylamines and their clinical use

There are two allylamine antifungals in current

use, naftifine and terbinafine.[46] The allylamines appear to have a similar mode of action as the tolcyclate antifungals, such as tolnaftate, - squalene epoxidase. This step is, therefore, not affected by blockade of cytochrome P 450. While naftifine has both antifungal and antinflammatory activity, the fungicidal concentrations (MFC's) of terbinafine are very similar to those required for mere inhibition of growth (MIC's) in vitro.[47] Its chief clinical use is in dermatophyte infections but in vitro it has activity against *Aspergillus* and some of the dimorphic pathogens, such as *Histoplasma* and *Blastomyces*.[47] Terbinafine is effective against dermatophytes when given orally and against a slightly broader range of superficial pathogens including *Pityrosporum* yeasts when applied topically. In one comparative study of terbinafine 250mg bd versus griseofulvin 500mg bd in chronic dry type tinea pedis caused by *Trichophyton rubrum*, only 45% of the griseofulvin treated patients achieved remission 12 weeks after the end of therapy compared to 100% in the terbinafine group.[48] Other similar studies confirm this high recovery rate in infections of the skin. Onychomycosis also responds well to therapy with terbinafine [49] with a very low relapse rate. The duration of therapy has now been established as 6 weeks for finger nail and 12 weeks for toe nail infections.[50] In both instances treatment is stopped before clinical recovery. The therapeutic range may be greater than this and, for instance, activity in *Candida parapsilosis* paronychia and sporotrichosis have been recorded.

Other antifungal agents

There is also a large and miscellaneous group of antifungal agents, many of which are only available for topical use. These include compounds such as tolnaftate, cyclopiroxolamine and haloprogin, all of which are effective treatments for superficial mycoses although in the case of the first, activity is confined to the dermatophytes. Griseofulvin, an oral agent which acts by the inhibition of intracellular microtubule formation, is widely used for dermatophytosis and is the treatment of choice for scalp infections. While in finger nail infection griseofulvin produces over 70% remission rates, it has been less useful in toe nail disease. For instance one long term study found a 30% remission rate in patients receiving up to 2 years of therapy.

Flucytosine is still widely used in systemic antifungal chemotherapy. It interferes with the formation of RNA and DNA, but is actively taken up by fungi via a permease system and converted to fluorouracil within the cell. However the frequency of resistance and other side effects such as bone marrow depression at high serum levels has made this a less desirable choice for treatment. It does have one major advantage - good absorption after oral administration coupled with good penetration into urine and cerebrospinal fluid. The main use of flucytosine is as part of a combination therapy together with amphotericin B - an effective approach to chemotherapy of cryptococcal meningitis in the non-AIDS patient.[51] The use of this approach in the AIDS patient is still debated. The combination is also used, with less supportive evidence of efficacy in both *Candida* and *Aspergillus* infections. The basis for its use in combination is the synergy seen in vitro when the two drugs are tested together. This means that a lower dose of amphotericin B is possible and in addition with the superior penetration of flucytosine the treatment of infections affecting the CSF or peritoneum is facilitated.

At present there is no clinical data on the value of the inhibitors of cell wall synthesis such as the echinocandins and nikkomycin derivatives.

Amorolfine is a morpholine antifungal drug which inhibits at least two stages in cell membrane biosynthesis. In vitro it also shows fungicidal activity at concentrations close to inhibitory concentrations. In vitro it is active against all the main superficial fungal pathogens but clinical trial work has focussed on the use of a 5% amorolfine nail lacquer which is used twice weekly. Recovery rates as high as 54% have been reported although it is clearly more active in early nail disease without extensive involvement of the nail bed.[52]

Discussion

The current range of antifungal drugs, although extensive, is by no means ideal. There are, for instance, several mycoses which are refractory to existing therapy. Examples include common conditions such as chronic tinea pedis due to *Trichophyton rubrum*, and onychomycosis, although some of the newer additions such as short term terbinafine and itraconazole may provide an answer. The topical approach to the treatment of nail disease has been more disappointing despite evidence that by using penetration enhancers it is possible to obtain active concentrations in vitro at the site of fungal invasion; the new 5% nail lacquer of amorolfine has been reported to produce better responses. There are still some untreatable mycoses such as those caused by *Scytalidium dimidiatum* (*Hendersonula toruloidea*) and *S hyalinum*, common infections in immigrants from the tropics. Likewise there are problems in the management of life threatening infections such as invasive mycoses in the severely immunocompromised patient. The response of systemic *Candida* and *Aspergillus* infections in the neutropenic patient to any form of antifungal therapy is poor.[53,54] While in the case of some fungi such as the zygomycetes (mucormycosis), *Fusarium* and *Trichosporon* infections, this appears to follow primary resistance to the available drugs, there are other reasons such as lack of bioavailability of drugs in tissue [7] and failure to destroy organisms in the presence of totally defective host immunity.

Drug resistance to antifungal agents is rare,[55] apart from flucytosine where primary and secondary resistant yeasts are recognised frequently in clinical practice. Polyene antifungal resistance is extremely rare but has been recorded. Azole resistance has been reported in patients receiving long term ketoconazole therapy. It has been described mainly in patients with chronic mucocutaneous candidosis (CMC) and in AIDS. While in some instances the "resistance" cannot be substantiated, as it has not proved possible to compare pre- and post-therapy isolates, there is little doubt that resistance can appear in the CMC group, at least. These resistant organisms may show cross resistance to other azoles.[56] Patients infected by yeasts in which this has occurred may still respond clinically to therapy with ketoconazole or other azoles although not as well and there is no evidence that the organism recovers its original sensitivity. Recently fluconazole resistance has been reported in strains of *Candida glabrata* [57] as well as other *Candida* species.

In addition, since the introduction of antifungal drugs, as with other antibiotics, there has been a regular need to revise and update their use in clinical practice. There are a number of reasons for this. Firstly the pattern of fungal disease is constantly changing.[58] This follows, for instance, the introduction of new therapeutic practices affecting the choice of different immunosuppressive regimens or the extension of organ transplantation to different diseases.[59] While new diseases occur less frequently, the emergence of the AIDS has also led to changes in the prevalence of different fungal infections and their clinical expression.[60] Oropharyngeal candidosis is one of the commonest secondary infections in HIV positive patients and cryptococcosis occurs in between

3-11 %, depending on their country of origin.[61] All these changes have required appropriate modifications in the application of antifungal chemotherapy, most notably the use of long term suppression of infection as cure appears to be well nigh impossible to achieve in AIDS patients.

Whatever the problems there remains a need for further antifungal drugs which can fill some of these gaps in our armamentarium. There is equally a need for physicians to group together to carry out larger clinical trials as without the numbers the value of studies, particularly in the systemic mycoses is likely to be meaningless.

References

1. A.Polak. In Antifungal Drugs. Ed.St Georgiev V. Annals N York Acad.Sci. 1988 544, 221.

2. M.Debono, BJ Abbott, JR Turner et al. In 'Antifungal Drugs'.Ed. St Georgiev V. Annls.N.York Acad.Sci.1988 544 141.

3. G.Medoff and GA Kobayashi. 'In 'Antifungal Chemotherapy'. Ed.DCE Speller. John Wiley and Sons, Chichester 1980 p 3.

4. JK Brajtburg, G Medoff, GS Kobayashi and D Schlessinger. Biochem.Pharmacol. 1976, 26 705.

5. DD Bindschadler and JE Bennett. J.Infect.Dis. 1969, 120 427.

6. WG Powderly, GS Kobayashi, GP Herzig and G Medoff Am.J.Med.,1988 84, 826.

7. KJ Christiansen, EM Bernard, JWM Gold and D Armstrong. J.Infect.Dis., 1985 152 1037.

8. R Miller and JH Bates. Annals Intern.Med. 1969, 71, 1090.

9. WT Butler, JE Bennett, D Alling and PT Wertlake. Annals.Intern.Med., 1964, 61, 175.

10. G Lopez-Berestein, V Fainstein, R Hopfer et al. J.Infect.Dis.1985, 151, 704.

11. J-P Sculier, A Coune, F Meunier et al.

Eur.J.Cancer Clin.Oncol.1988 24, 527.

12. O Ringden, F Meunier, J Tollemar, P Ricci et al. J.Antimicrob.Chemother 1991. 28 (Supplement B), 73.

13. RA Fromtling. Clin.Microbiol.Revs 1988, 1 187.

14. H Van Cauteren, A Lampo, J Vandenberghe, Ph Vanparys, W Coussement, R de Coster and R Marsboom. Mycoses 1989, 32 (Suppl 1), 60.

15. SM Grant and SP Clissold. Drugs 1990, 39 877.

16. MS Marriott and K Richardson. In 'Recent trends in the Discovery, Development and Evaluation of Antifungal agents'. Ed. RA Fromtling. JR Prous Publishers, Barcelona 1987 p 81.

17. MJ Humphrey, S Jevons and MH Tarbit. Antimicrob.Ag.Chemother. 1985 28, 648.

18. J Heykants, A Van Peer, V Van de Velde, P Van Rooy, W Meuldermans, K Lavrijsen, R Woestenborghs, J Van Cutsem and G Cauwenbergh. Mycoses 1989, 32, (Suppl 1) 67.

19. KW Brammer and LJ Lees. In 'Recent trends in the Discovery, Development and Evaluation of Antifungal agents'. Ed. RA Fromtling. JR Prous Publishers, Barcelona 1987, p 151.

20. B Dupont and E. Drouhet. J.Med.Vet.Mycol. 1988 26 67.

21. S de Wit, H Gloosens, D Weerts and N Clumeck. Lancet, 1990, i, 746.

22. JP Chave, A Cajot, J Bille and MP Glauser. J. Infect.Dis., 1989 159 806.

23. F Meunier, J Gerain, R Snoeck, F Libotte,C Lambert and AM Ceuppens. In 'Recent trends in the Discovery, Development and Evaluation of antifungal agents'. Ed. RA Fromtling. JR Prous Publishers, Barcelona 1987 p 169.

24. RJ Hay, YM Clayton, MK Moore and G Midgley. Br.J.Dermatol. 1988, 119, 683

25. H Degreef, K Marien, H de Veylder, K Duprez, A Borhys and L Verhoeve. Revs.Infect.Dis. 1987 9 (Suppl 1), 104.

26. RJ Hay, YM Clayton, MK Moore and G Midgley.

Br.J.Dermatol. 1988 119, 359.

27. M Willemsen, P de Doncker, J Willems, R Woestenborghs et al. J.Am.Acad.Dermatol. 1992, 26, 731.

28. A Restrepo, J Robledo, I Gomez et al. Arch.Dermatol.1986 122, 413.

29. TF Patterson, P Miniter and VT Andriole. Revs Infect.Dis.1990 12 (Supplement 3), s281.

30. JE Bennett Annls.Intern.Med. 1981, 94, 708.

31. V Fainstein, GP Bodey , L Elting et al. Antimicrob.Ag.Chemother 1987, 31 11.

32. National Institutes of Allergy and Infectious Diseases Mycoses Study Group. Annls Intern.Med., 1985 103 861.

33. SA Bozzette, RA Larsen, J Chiu et al. N.Eng.J.Med. 1991. 324, 580.

34. MS Saag, WG Powderly, GA Cloud, P Robinson et al. N.Eng.J.Med. 1992 326, 83.

35. AM Sugar and C Saunders. Am.J.Med. 1988 85, 481.

36. JW Van't Wout, H Mattie and R Van Furth. J.Antimicrob.Chemother.1988 21, 665.

37. J Cohen. J.Antimicrob.Chemother., 1989 23 294.

38. J Levine, DB Bernard, BA Idelson, H Farnham, C Saunders and AM Sugar. Am.J.Med. 1989 86, 825

39. R Negroni, O Palmieri, K Koren et al. Revs Infect.Dis. 1987 9 s47.

40. JR Graybill. J.Infect.Dis. 1988, 1578, 623.

41. DW Denning, RM Tucker, LH Hanson and DA Stevens. Am.J.Med. 1989 86 791.

42. MA Viviani, AM Tortorano, M Langer et al. J.Infection 1989, 18, 151.

43.M Rozenberg-Arska, AW Dekker, J Branger, and J Verhoef. J.Antimicrob.Chemother. 1991 27, 369.

44. JR Wingard, WG Merz, MG Rinaldi, IR Johnson et al. N.Eng.J.Med 1992 325, 12.

45. GE Tricot, MA Joosten, MA Boogaerts, J Van de Pitte and G Cauwenbergh. Revs.Infect.Dis. 1987 9 s94

46. G Petranyi, JG Meingassner and H Mieth. Antimicrob.Ag.Chemother.1987 31, 1365.

47. S Shadomy, A Espinel Ingroff and RJ Gebhart. Sabouraudia 1985 <u>23</u>, 125.

48. RJ Hay, R Logan, YM Clayton, G Midgley and MK Moore. <u>J.Am.Acad.Dermatol</u>. 1991, <u>24</u>, 233.

49. MJD Goodfield, NR Rowell, RA Forster, EGV Evans and A Raven. <u>Br.J.Dermatol</u>. 1989, <u>121</u>, 753.

50. JG Van der Schroeff, PKS Cirkel, MB Crijns, TJA Van Dijk et al. <u>Br.J.Dermatol</u>. 1992 <u>126</u>, (Supplement) <u>39</u>, 36.

51. JE Bennett, WE Dismukes, RJ Duma et al. <u>N. Eng.J.Med</u>. 1979, <u>301</u> 121.

52. M Zaug and M Bergstrasser. <u>Clin.Exp.Dermatol</u>, 1992, <u>17</u> (Supplement A), 55

53. J Aisner, SC Schimpff and PH Wiernik <u>Annals Intern.Med.</u> 1977. <u>90</u>, 539.

54. AW Maksymiuk, S Thonprasert and R Hopfer. <u>Am.J.Med</u>. 1984, <u>77</u>, 20.

55. D Kerridge, M Fasoli and FJ Wayman. In Antifungal Drugs. Ed V St.Georgiev. <u>Annals.N York Acad.Sci</u>. 1988 <u>544</u>. 245.

56. JF Ryley, RG Wilson and K Barrett-Bee. <u>J.Med.Vet.Mycol</u>.1984, <u>22</u>, 53.

57. DW Warnock, J Burke, NJ Cope, EM Johnson, NA Von Fraunhofer and EW Williams. <u>Lancet</u> 1988 <u>ii</u> 1310.

58. Eds DW Warnock and MD Richardson. 'Fungal infection in the immunocompromisd patient'. John Wiley and Sons Ltd, Chichester UK. 1982

59. DJ Weber and WA Rutala. In 'Diagnosis and therapy of Systemic Fungal Infections'. Eds K Holmberg, R Meyer. Raven Press. New York 1989 p 1.

60. RM Welik, ET Starcher and JW Curran. <u>AIDS</u> 1987 <u>1</u>, 175.

61. WE Dismukes. <u>J.Infect.Dis</u>. 1988 <u>157</u> 624.

12
Recent Developments in Azole Antifungal Agents

K. Richardson
DISCOVERY CHEMISTRY, PFIZER CENTRAL RESEARCH, SANDWICH, KENT CT13 9NJ, UK

Introduction

Early azole antifungal agents for the treatment of infections in man were imidazole derivatives such as miconazole (Janssen), clotrimazole (Bayer) and tioconazole from Pfizer. These agents showed potent in vitro activity against a wide range of fungi which are pathogenic to man[1], due to a selective inhibition of fungal sterol biosynthesis[2]. When administered topically, they show good efficacy against vaginal candidosis and dermatomycoses. However, their metabolic vulnerability results in poor oral bioavailability and low systemic exposure preventing these agents from showing useful antifungal efficacy following oral dosing. Research workers at many of the major pharmaceutical companies set out to solve these problems in the knowledge that, if successful, the result could be a breakthrough in the treatment of systemic fungal infections.

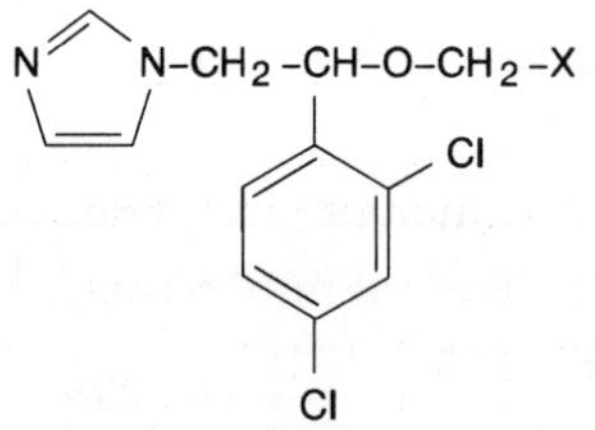

Miconazole, X =

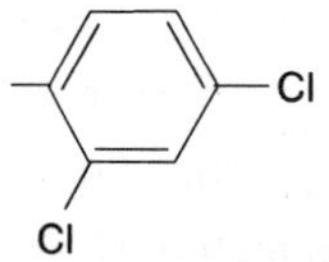

Tioconazole, X =

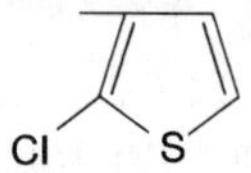

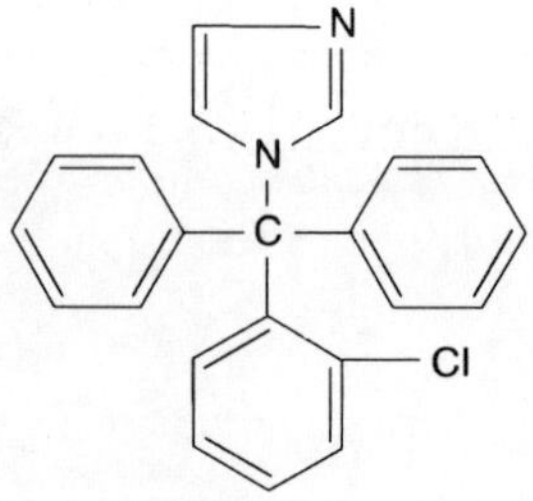

Clotrimazole

There was clearly a need for such a drug. The incidence of systemic, fungal infections was increasing at an alarming rate[3] as the numbers of immune-suppressed patients grew. Patients being treated for cancer or leukemia, and those undergoing organ transplantation are all immune-suppressed to some extent and therefore have reduced ability to fight off infections, including those caused by fungi. The race was on to see who could solve the problem first!

Oral Bioavailability

The early leader was clearly Janssen with ketoconazole[4]. Ketoconazole showed good oral bioavailability in animals and efficacy, by the oral route, in a wide range of fungal infection models[5]. Progression to studies in man showed that, although ketoconazole was a major step forward due to its activity against a number of fungal infections, it was certainly not the breakthrough drug that everyone was seeking. Although more metabolically-stable than earlier imidazole derivatives, ketoconazole was still susceptible to metabolic inactivation, with less than 1% being excreted unchanged in the urine. It was also proving to be less selective than had been hoped for, since there were reports of effects on testosterone levels leading to impotence and gynaecomastia. In addition, ketoconazole could cause severe hepatoxicity, even after only a few doses[6].

Ketoconazole

At Pfizer, our approach was to use tioconazole as a starting point and to seek a derivative which would be as metabolically-stable as possible. This was expected to result in good oral bioavailability and sustained blood levels of unchanged drug. In addition, we wanted to reduce the overall lipophilicity of the molecule as an approach to producing high blood levels of unbound drug (ketoconazole, tioconazole, etc were ~99% bound to plasma proteins resulting in very low levels of free drug). Our work resulted in a large number of different structural types including dioxolanes, dithiolanes, tetrahydrofurans and tertiary hydroxyl derivatives (Fig 1) and we frequently produced compounds with ketoconazole-like pharmacokinetic and antifungal properties. However, after $2^1/_2$ years research, and despite many variations in structure and the synthesis of over 400 derivatives, we were unable to obtain any compounds which were significantly better *in vivo* than ketoconazole.

Although we had made many changes to the structures of our compounds,

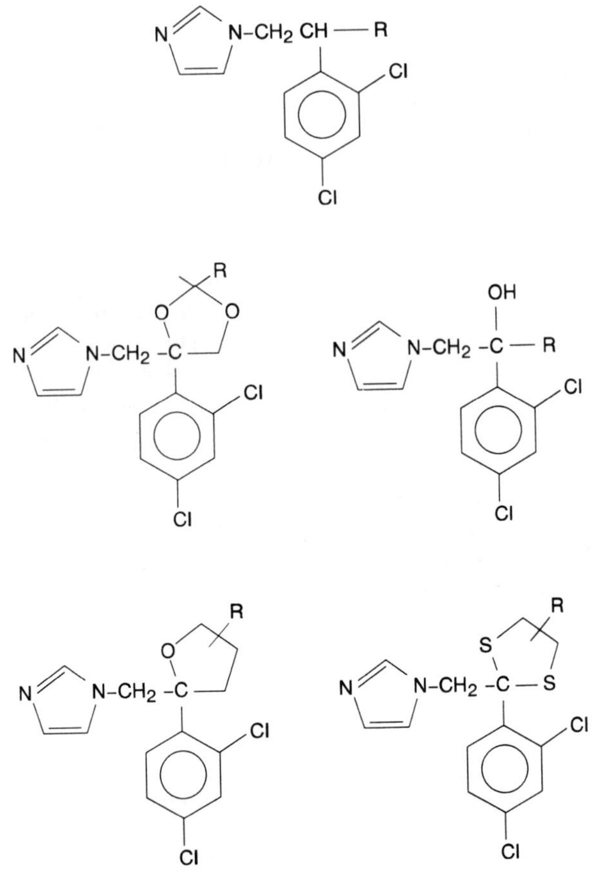

Figure 1   Structures of five series of imidazole derivatives prepared at Pfizer

they had all been imidazole derivatives and we realised that this moiety could be metabolically vulnerable. We therefore decided to seek a metabolically-stable imidazole replacement. We prepared 20 derivatives (Fig 2), and 19 were completely inactive. However, the analogue containing a 1,2,4-triazol-1-yl group UK-46,245 was very encouraging because, although it was 4 times less active than the corresponding imidazole derivative in vitro, it was 3 times more active in vivo. We believed that we had blocked one site of metabolism and therefore added a second triazole in place of the metabolically-vulnerable hexyl group.

The resulting derivative, UK-47,265, was a remarkable compound. It showed an unprecedented level of antifungal activity being up to 100 times more

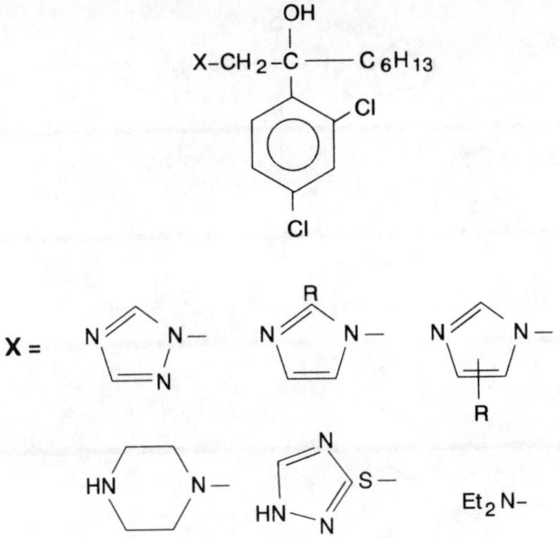

Figure 2   A selection of the compounds prepared seeking a more stable replacement for the usual imidazole moiety

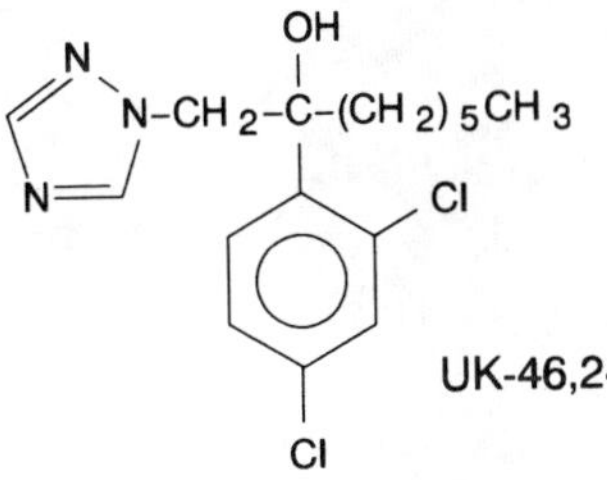

potent than ketoconazole in our various fungal infection models. It also had excellent oral bioavailability and a long plasma half-life in mouse, rat, rabbit and dog. We had clearly made the breakthrough that we, and others, had been seeking and so UK-47,265 was progressed into development.

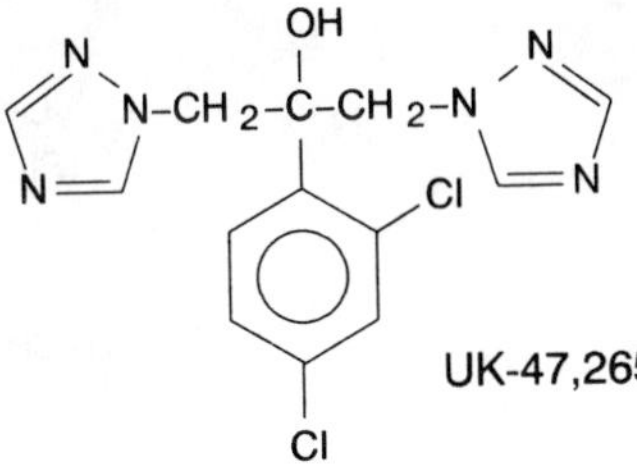

The results were extremely disappointing since it proved to be hepatotoxic in rats and dogs, and teratogenic in rats. It clearly could not be developed. By this time, we had prepared and studied over 100 bis-triazole derivatives and we had a number of compounds with excellent anti-fungal and pharmacokinetic properties (Fig 3). However, almost all of these derivatives proved to have unacceptable toxicity. The exception was the difluorophenyl derivative, UK-49,858. This compound combined the remarkable antifungal and pharmacokinetic properties of the earlier derivatives with an excellent safety profile in animals. In addition, it was water-soluble which would greatly assist formulation for i.v. administration. It was therefore given the generic name of fluconazole[7] and progressed to evaluation in man.

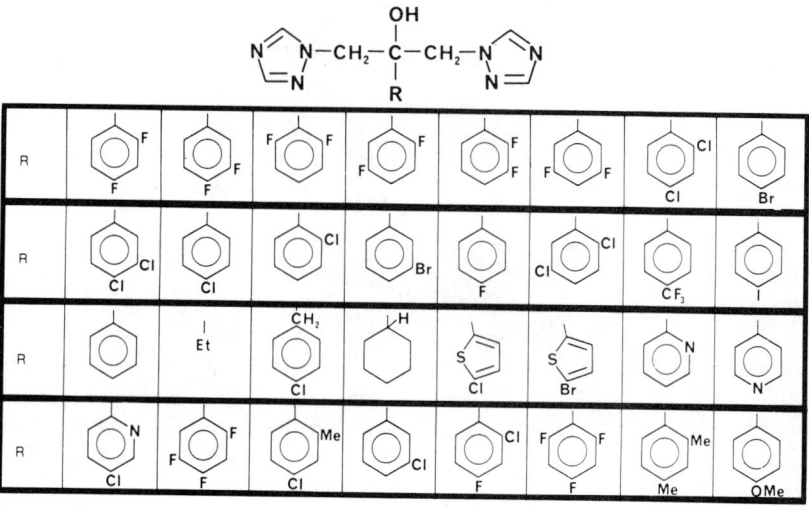

Figure 3   Structures of 32 substituted-1,3-bis-triazolyl propan-2-ol derivatives

UK-49,858
(Fluconazole)

Time does not allow more than a brief summary of the clinical results. However, it is true to say that <u>fluconazole is even better than we had hoped to achieve.</u> It has excellent oral bioavailability, a half-life of ~30 hours and >80% of drug is excreted, unchanged, in the urine. It has also proved to be remarkably safe. In summary, fluconazole shows excellent efficacy as :-

1.   Treatment of systemic infections in cancer, leukemia, transplant and AIDS patients.

2.   Daily dosing to prevent infection, or re-infection, in AIDS and cancer patients.

3.   A single dose for vaginal candidosis.

4.   Daily, or even weekly, treatment of nail and skin infections.

It is clear that fluconazole[8]. which is sold as DIFLUCAN, is a major breakthrough for the treatment and prevention of fungal infections. However, the competition have not given up!

Janssen has developed the triazole derivative itraconazole which is a structural relative of ketoconazole. In vitro, itraconazole[9] has similar activity to fluconazole with improved activity against Aspergillus . However, itraconazole is a very lipophilic compound which results in very low solubility leading to variable oral bioavailability and the lack of an intravenous dosage form. It also has to be given as a 2x100 mg or 2x200 mg dose with toxicity occurring if the dose exceeds 400mg a day. However, despite these problems, itraconazole has proven to have some clinical utility.

Saperconazole   R=

Itraconazole   R=

Workers at Janssen tried to overcome some of these difficulties by development of the slightly less lipophilic analogue, saperconazole[10], but this has been reported to cause ovarian cancer in animals, leading to its demise. Bayer followed up their early work on clotrimazole by the preparation of the triazole derivative vibunazole[11] but this proved to induce its own metabolism in mice and dogs following multiple dosing, and it was dropped from development. They followed this with Bay-3783, which was reported to have anti-fungal activity similar to fluconazole. However, Bay-R3783 was not developed since it is metabolised[12] to the corresponding nitrile Bay-U3625 which has a very long half-life leading to accumulation on multiple dosing.

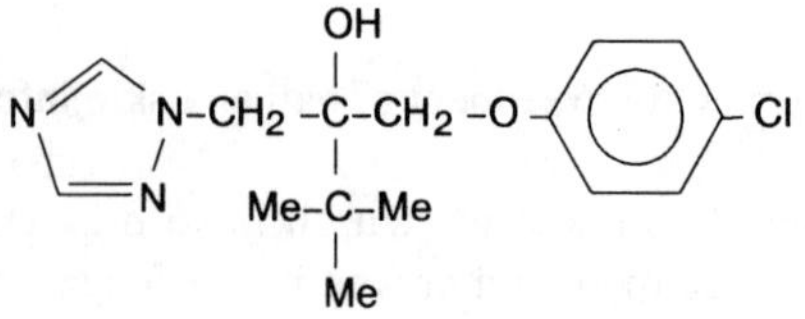

Vibunazole

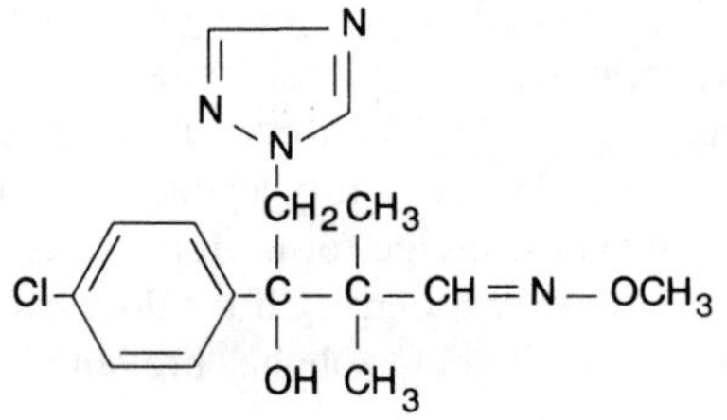

Bay R3783

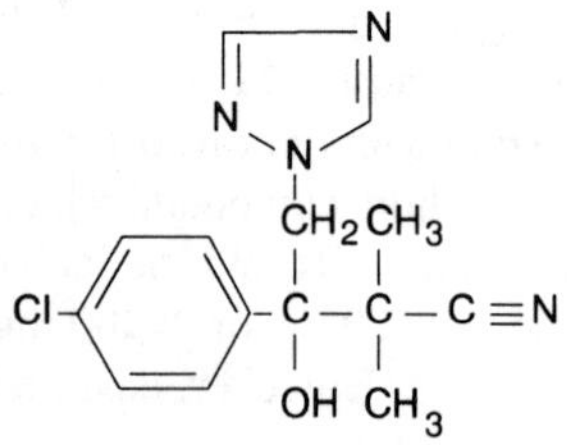

Bay U3625

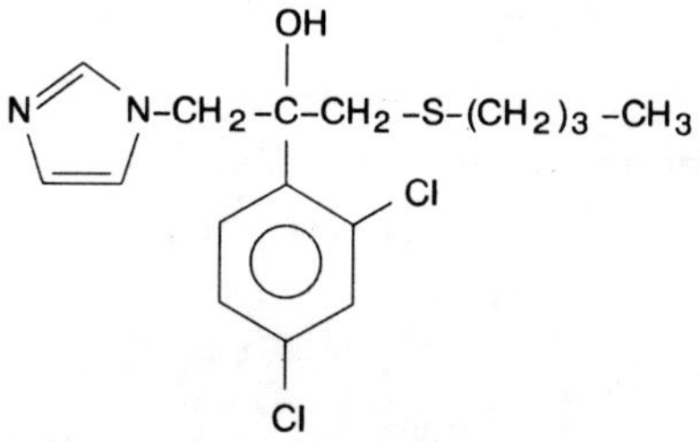

SM-8668 (Sch 39,304)

SM-4470

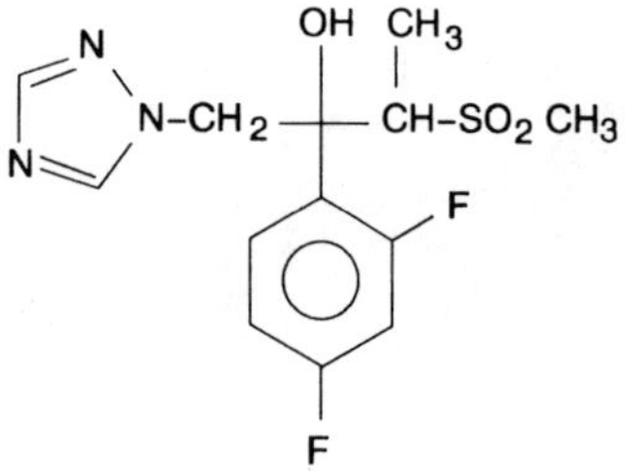

Dupont

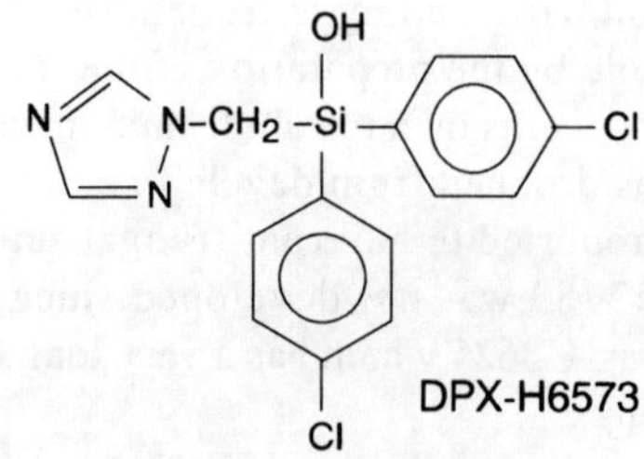

DPX-H6573

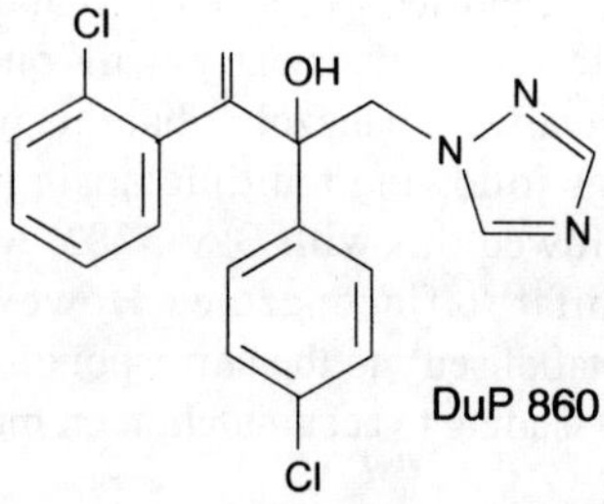

DuP 860

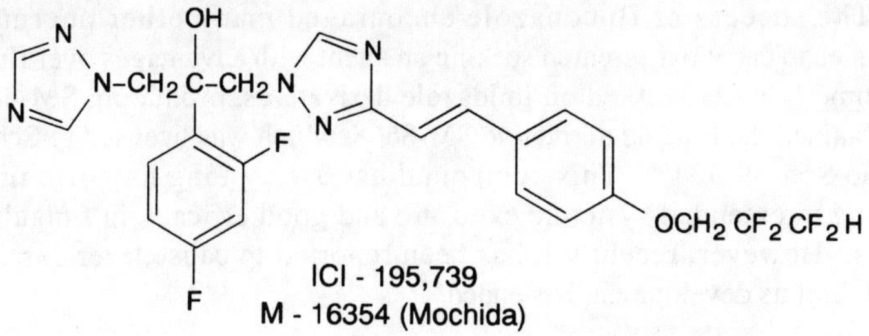

ICI - 195,739
M - 16354 (Mochida)

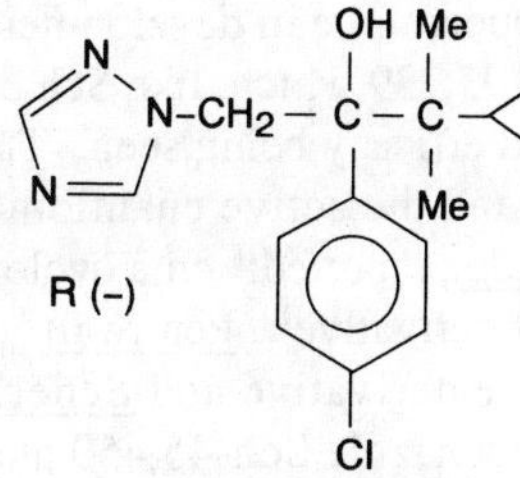

R (−)

SDZ - 89-485

Shionogi

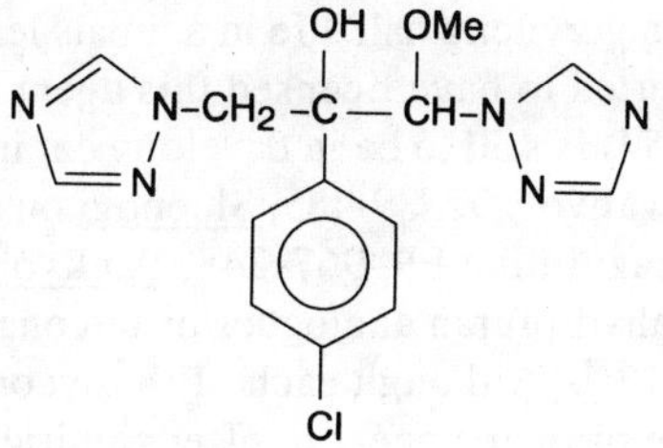

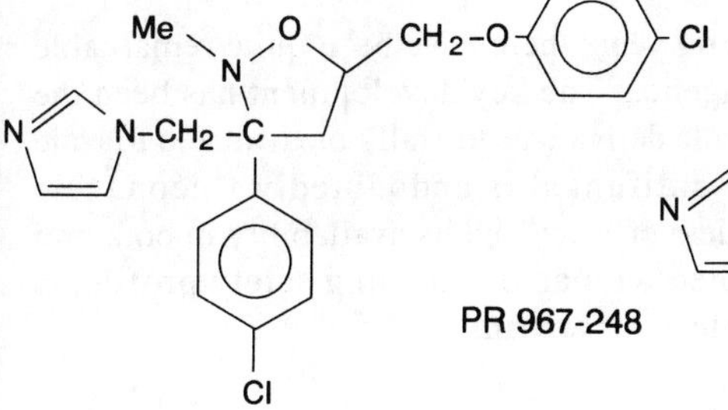

Sankyo

PR 967-248

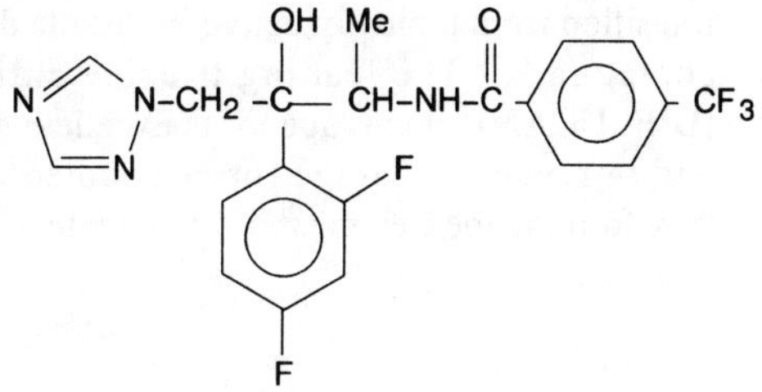

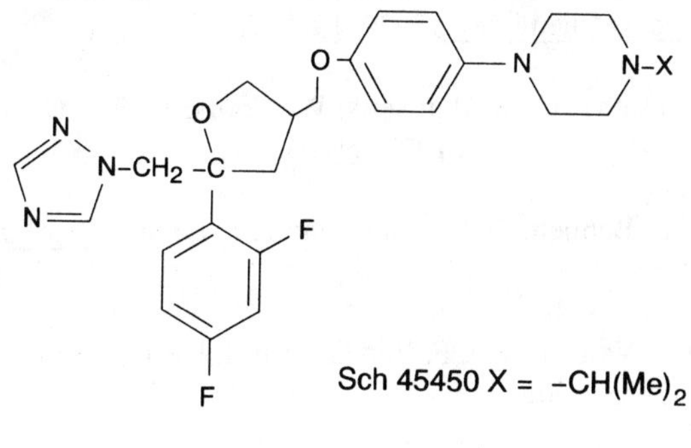

Sch 45450 X = −CH(Me)$_2$

Sch 45012 X =

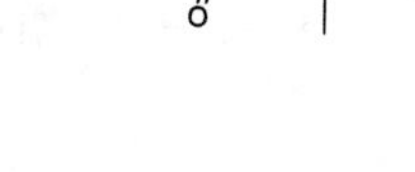

The success of fluconazole encouraged many other pharmaceutical companies to carry out research seeking an agent with advantages over fluconazole. Sumitomo initially worked on imidazole derivatives, producing SM-4470[13], but then obtained the triazole derivative SM-8668, which was licensed to Schering and became Sch-39,304[14]. This compound has a very long half-life in animals, resulting in extended systemic exposure and good efficacy in fungal infection models. However, recently it has been reported to cause liver carcinomas in animals and its development has ended.

Dupont prepared the silicon-containing triazole derivative DPX-H6573[15] and the methylene derivative DuP-860[16] but neither appear to be in development. ICI have reported[17] on the fluconazole derivative ICI-195,739 which, like Sch 39,304, has a very long half-life in animals leading to good efficacy being seen. They are reported to have licensed this agent to Mochida and the active enantiomer (M-16354) is said to be in development in Japan. Sandoz reported[18] on a cyclopropyl derivative SDZ-89-485, Shionogi on a α-methoxy derivative[19], Pennwalt[20] on the isoxazolidine PR-967-248, Sankyo[21] on an amide derivative and Schering on tetrahydrofuran analogues of terconazole and itraconazole Sch-45,450 and Sch-45012[22]. Although each of these compounds have some interesting properties, none of them appear to offer significant advantages over the currently marketed agents.

In summary, over the past several years there have been quite remarkable advances made in azole antifungal agents. The key development has been the transition from topically-active imidazole derivatives to orally bioavailable triazole compounds. The leading triazole antifungal is undoubtedly fluconazole (DIFLUCAN) which, due to its excellent efficacy and its availability in both oral and intravenous dosage forms coupled with an outstanding safety profile, is transforming the treatment of fungal infections in man.

References

1.   S. Jevons, G.E. Gymer, K.W. Brammer, D.A. Cox and M.R.G. Leeming, Antimicrob. Agents Chemother., 1979, 15, 597.

2.   H. Van den Bossche, G. Willemsens, W. Cools, W.F.J. Lauwers and L. LeJeune, Chem. Biol. Interact., 1978, 21, 59.

3.   R.C. Young, J.E. Bennett, G. Geelhoed and A.S. Levine, Ann. Intern. Med., 1974, 80, 605.

4.   D. Thienpoint, J. Van Cutsem, F. Van Gerven, J. Heeres and P.A.J. Janssen, Experientia, 1979, 35, 606.

5.   A series of papers in Rev. Infect. Dis., 1980, 2, N° 4.

6.   J.H. Lewis, H.J. Zimmerman, G.D. Benson and K.G. Ishak,

Gastroenterology, 1984, 86, 503.

7.   K. Richardson, K. Cooper, M.S. Marriott, M.H. Tarbit, P.F. Troke and P.J. Whittle, Ann. N.Y. Acad. Sci., 1988, 544, 4.

8.   J.T. Henderson, 'Recent Trends in the Discovery Development and Evaluation of Antifungal Agents', J.R. Prous, Barcelona, 1987, p.77.

9.   F. Peeters, H. Van den Pas, J. Proost, D. Janssens, E. Snauwaert, J. Van Cutsem and G. Cauwenbergh, Curr. Ther. Res., 1986, 39, 496.

10.   F.C. Odds, J. Antimicrob. Chemother., 1989, 24, 533.

11.   W. Ritter and M. Plempel, B. Soc Antimicrob. Chemother., 1984, p243.

12.   D. Pappagianis, B.L. Zimmer, G. Theodoropoulos, M. Plempel and R.F. Hector, Antimicrob. Ag. Chemother., 1990, 34, 1132.

13.   K. Ichise, T. Tanio, I. Saki and T. Okuda, Antimicrob. Ag. Chemother., 1986, 30, 366.

14.   T. Tanio, K. Ichise, T.Nakajima and T. Okuda, Antimicrob. Ag. Chemother., 1990, 34, 980.

15.   Report in Chem. and Eng. News, 1985, p.43.

16.   J. Cuomo, DuPont patent, EP 251,086.

17.   F.T. Boyle, D.J. Gilman, M.G. Gravestock and J.M. Wardleworth, Ann. N.Y. Acad. Sci., 1988, 544, 86.

18.   M.A. Grassberger Xth Cong. Int. Soc. Hum. Anim. Mycol. (Barcelona), 1988, Abs. P-151.

19.   M. Ogata, Shionogi patent, EP 234,499.

20.   G.B. Mullen, P.A. Swift, D.M. Marinyak, S.D. Allen, J.T. Mitchell, C.R. Kinsolving and V. St. Georgiev, Helv. Chim. Acta, 1988, 71, 718.

21.   T. Konosu, Y. Tajima, N. Takeda, J. Miyaoka, M. Kasahara, H. Yasuda and S. Oida, Chem. Pharm. Bull., 1991, 39, 2581.

22.   A.K. Saksena, V.M. Girijavallabhan, D.F. Kane, R.E. Pike, J.A. Desai, A.B. Cooper, E. Jao, A.K. Ganguly, D. Loebenberg, R.S. Hare and R. Parmegiani, 9th Int. Symp. Fut. Trends in Chemother., (Geneva), 1990, Abs. 128.

13

New Antifungal Agents: Synthesis and Biological Activity

V. Girijavallabhan, A. K. Ganguly, A. K. Saksena,
A. B. Cooper, R. Lovey, D. Loebenberg, D. Rane, J. Desai,
R. Pike, and E. Jao

SCHERING-PLOUGH RESEARCH INSTITUTE, 60 ORANGE STREET,
BLOOMFIELD, 07003, USA

Systemic fungal infections can be debilitating and in some cases fatal. While these diseases are becoming increasingly prevalent for a number of reasons, e.g., the result of immune deficiencies as in AIDS and cancer patients, general aging process, side effects of medications, etc., the available drugs have shown only limited and specific utility. There is, therefore, a pressing, continuous need for more effective antifungal agents. Several approaches to this goal are ongoing at Schering-Plough Research Institute, and for this discussion we have chosen to cover two distinct aspects:

1) Fungal Cell Wall Inhibitors, and 2) New Azole Agents.

Fungal Cell Wall Inhibitors: The cell wall is an essential feature for the viability of yeast and fungi - therefore an effective inhibitor can impart cidal as well as broad spectrum activity. Since the mammalian cells lack the cell wall system, selective inhibitors should be expected to have relatively low risks of side effects in the host.

Among the several unique components of the fungal cell wall, chitin[1] and glucan represent attractive targets. Natural products nikkomycins[2, 3, 4] **1** and polyoxins[5] **2**, known chitin synthesis inhibitors[6] act as competitive inhibitors, presumably due to their structural similarity to the enzyme substrate UDP N-acetyl glucosamine **4.** As part of a systematic investigation we have synthesized a large number of nikkomycin compounds and here we report some of our observations.

Pure polyoxin D, obtained through ion exchange column work-up of the commercially available polyoxin Zn salt was used as the primary source for the synthesis of the key intermediate uracil polyoxin C (UPOC). Reaction of polyoxin D **5** with phenylisothiocyanate afforded the thiourea adduct **6**, which upon acid treatment cleanly generated polyoxin C **7** as a salt. The carboxyl group in the uracil ring in **7**, when needed, could be readily removed by reacting with sodium bisulphite[5] to obtain **8**. This method of preparing **7** and **8** was very convenient for our purpose when compared to some of the total syntheses reported in the literature.[7, 8, 9] In order to make dipeptide derivatives of UPOC, new amino acids were synthesized and were then coupled to 7 and 8.

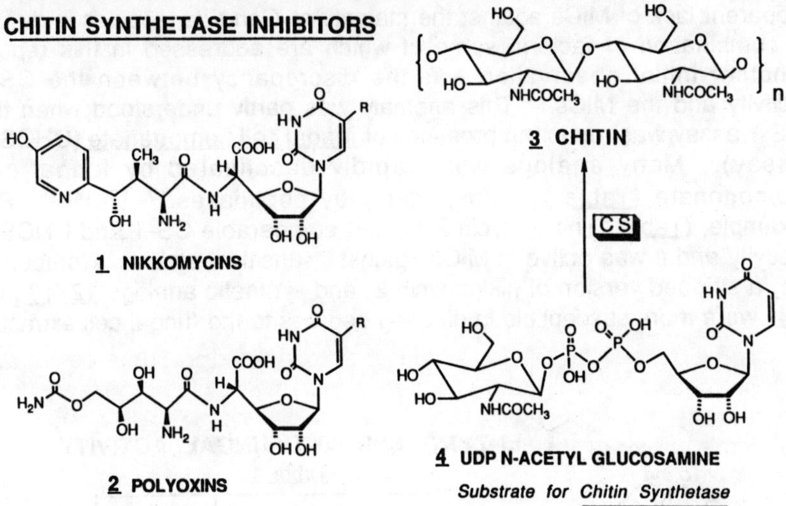

CHITIN SYNTHETASE INHIBITORS

1 NIKKOMYCINS

2 POLYOXINS

3 CHITIN

4 UDP N-ACETYL GLUCOSAMINE
Substrate for Chitin Synthetase

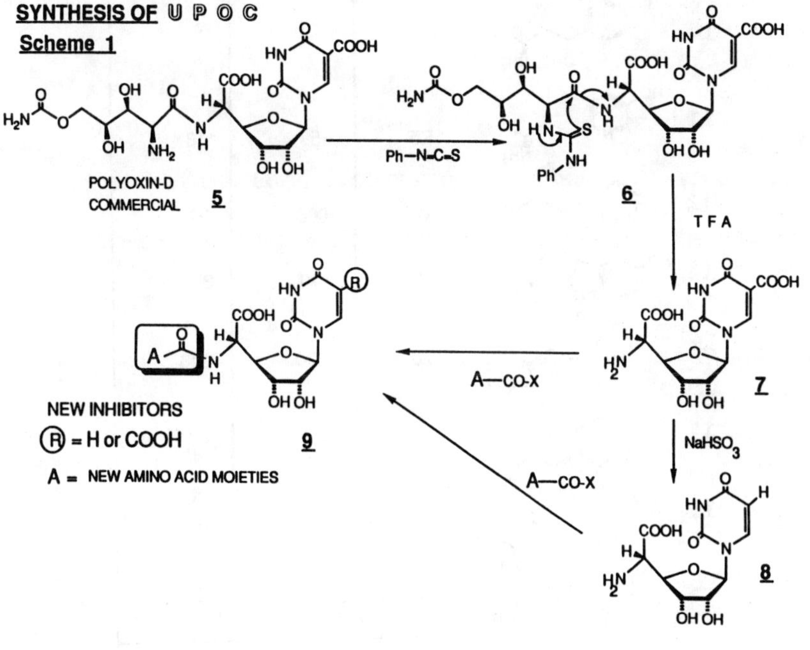

SYNTHESIS OF U P O C
Scheme 1

POLYOXIN-D COMMERCIAL **5**

6

TFA

7

NaHSO₃

8

A—CO-X

A—CO-X

NEW INHIBITORS
(R) = H or COOH **9**
A = NEW AMINO ACID MOIETIES

The biological properties of the new inhibitors were measured in a variety of assays. First, the chitin synthetase enzyme[10] (CS-1) activity was determined using semi-purified CS-1 enzyme, followed by MICs (yeast nitrogen broth medium) against a number of selected *Candida* strains. It is important to note that polyoxins and nikkomycins were active against approximately 20% of the *Candida* strains - termed nikkomycin "sensitive" strains and the rest were considered "resistant".[11] This apparent lack of MICs against the majority of *Candida* species is due to a combination of factors, some of which are addressed in this report. Another initial observation was the discrepancy between the CS-1 activity and the MICs.   This anomaly was partly understood when the CS-1 assay was run in the presence of fungal cell homogenate (CHCS-1 assay).   Many analogs were rapidly deactivated by fungal cell homogenate (Table 1),   presumably by peptidases.[12, 13, 14, 15]   For example, (Table 1) nikkomycin Z 10 had comparable CS-1 and CHCS-1 activity and it was active in MICs against "sensitive" strains. Compound 11, a stripped version of nikkomycin Z, and synthetic analogs 12, 13 and 14, were more susceptible in differing degrees to the fungal cell extract.

ENZYME AND ANTIFUNGAL ACTIVITY
Table 1

A =	CS-1	CHCS-1	MIC *
10	1.2	2.0	6
11	15.5	59	118
12	150	1000	39
13	2.5	10	37
14	120	500	2048
15	-	-	42
16	-	-	91
17	-	-	40

CS-1= CHITIN SYNTHETASE1,I C-50, µg/ml
CHCS-1 = CS-1 + FUNGAL CELL HOMOGENATE ,I C-50, µg/ml
* AGAINST NIKKOMYCIN SENSITIVE CANDIDA STRAINS, µg/ml

Also shown in <u>Table 1</u> are compounds <u>15</u>, <u>16</u> and <u>17</u> having new heterocyclic amino acid moieties. These compounds also had meaningful MICs. Introduction of a methyl group β to the carbonyl moiety as in nikkomycin was found to provide substantial stability to peptidases in a number of synthetic analogs - <u>Table 2</u>. Both <u>18</u> and <u>19</u> are relatively more resistant to cell extract degradation than <u>11</u> and this fact is reflected in the MICs as well. This protecting effect is even more significant when comparing <u>12</u> and <u>20</u>, as well as <u>21</u> and <u>22</u>. We have also investigated a number of functionalaties other than the methyl group in the β position, but most were virtually inactive. It was also observed that the presence of the L-amino group was found to be a requirement for good enzyme activity.

EFFECT OF METHYL SUBSTITUTION

<u>Table 2</u>

A =	CS-1	CHCS-1	MIC *
11	15.5	59	118
18	27	40	9.3
19	61	100	51
12	100	540	53
20	50	50	7.3
21	20	360	128
22	4.6	6.5	35
23	3	7	2048
24	8.6	8.4	1024

CS-1= CHITIN SYNTHETASE1,I C-50, μg/ml
CHCS-1 = CS-1 + FUNGAL CELL HOMOGENATE ,I C-50, μg/ml
* <u>AGAINST NIKKOMYCIN SENSITIVE CANDIDA STRAINS</u> , μg/ml

Compounds <u>23</u> and <u>24</u> portray another problem: while they have good enzyme activity and are reasonably stable in cell extract assay, these analogs were totally inactive in the MIC assay. This lack of activity against the whole cell is thought to be due to poor transport into the fungal cells. Nikkomycins are known to be taken up through an active peptide transport mechanism,[11, 13, 15] and this mode of transport of the inhibitors is blocked in "resistant" strains.

EFFECT OF LIPOPHILIC GROUPS. Table 3

A	MIC (YNB) *	
	SENSITIVE	RESISTANT
25	51	8.6
26	103	17
27	56	10
28	28	28
29	26	18
30	26	17
31	30	19.7

* AGAINST CANDIDA STRAINS, μ g/ml

Therefore we explored other means of transport of these inhibitors into the fungal cells. Since lipophilic compounds tend to be taken up through a passive diffusion process, we made derivatives having lipophilic amide side chains. MICs shown in Table 3 are consistent with the phenomenon that passive diffusion can be relatively uniform in both sensitive and resistant strains. Also observed is that the lipophilic amides are poor substrates for fungal peptidases, thus obviating the need for the β-methyl substitutions in the peptide chain. Even the quinoline derivative **26** had activity against whole cells.

To summarize this section, we have shown that for an inhibitor to be effective, it should have, in addition to the enzyme activity, stability towards fungal proteases. Bypassing the active peptide transport system is another means to achieve broader spectrum activity. More work, however, will be needed to further enhance the potency of these inhibitors.

New Azole Agents:[16] Though fungistatic, new azole agents such as fluconazole and itraconazole continue to provide significant advantages over older drugs such as miconazole and ketoconazole in terms of spectrum, pharmacokinetics, duration, etc. Sch 42427, a recent clinical candidate had a superior spectrum and was efficacious, especially against *Aspergillus* infections in immune deficient patients. However, the results of longer term toxicity studies have precluded its further development in the clinic.

AZOLE ANTIFUNGALS

32 (±) MICONAZOLE

33 FLUCONAZOLE
Most Recent

34 (±) KETOCONAZOLE

35 SCH-42427

36a X= Cl = ITRACONAZOLE
36b X= F = SAPERCONAZOLE
MIXTURE OF FOUR ISOMERS

These agents share a common characteristic feature in their structure as highlighted by the shaded region in 32 - 36. Certain lipophilic moieties present in itraconazole, saperconazole and Sch 42427 appear to have rendered substantial enhancement to the spectrum of activity, particularly against *Aspergillus*. Here we wish to report some of our results from modifying the structure of the trisubstituted five-membered ring system present in 34, 36a and 36b. This theme is highlighted in Scheme 2. For the convenience of synthesis, all the initial compounds were prepared and tested as racemates. However, it will be clear from this work that absolute stereochemistry will be a critical parameter for good biological activity.

STRATEGY FOR MORE ACTIVE AGENTS. Scheme 2

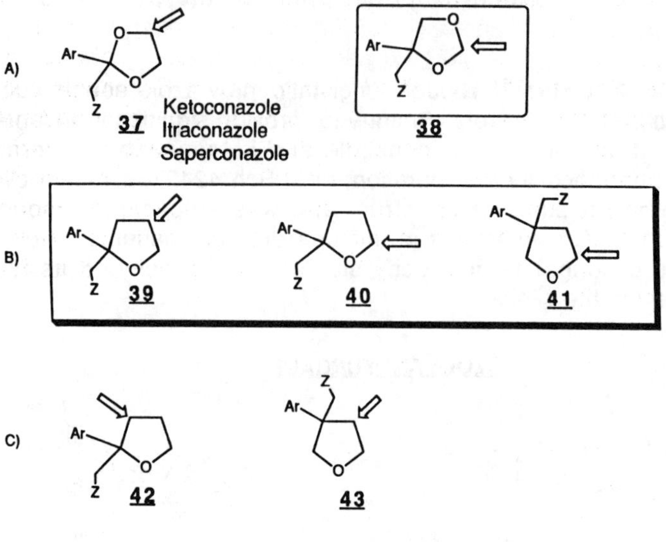

A)

Ketoconazole
37 Itraconazole
Saperconazole

38

B)

39   **40**   **41**

C)

42   **43**

D)   **4&6 MEMBERED RING SYSTEMS**

SYNTHESIS OF RACEMIC TETRAHYDROFURAN MOIETIES. Scheme 3

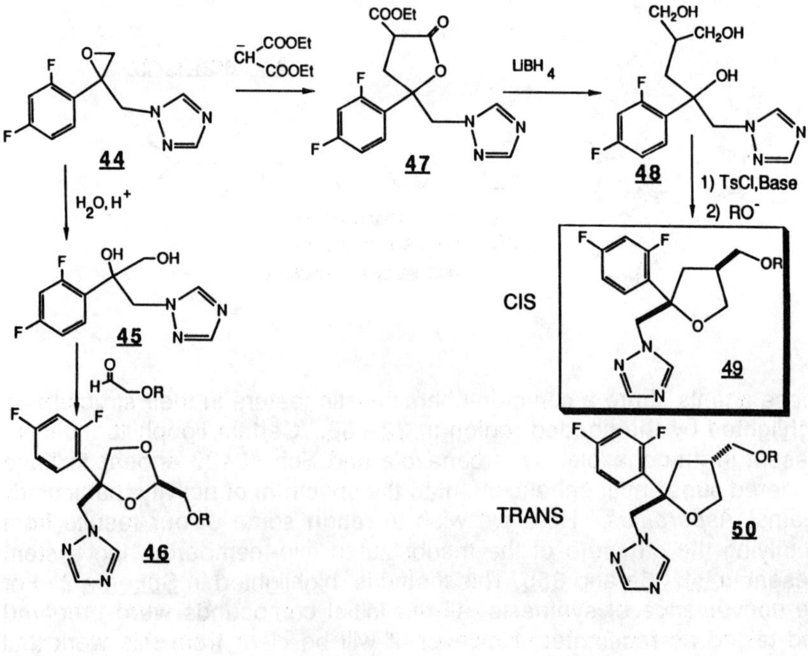

44   **47**   **48**   1) TsCl,Base
                                  2) RO⁻

45

46

CIS   **49**

TRANS   **50**

Compounds of type 46, differing in the R groups were prepared starting from the known[17] epoxide 44 - Scheme 3. The same epoxide when reacted with malonate anion afforded the lactone 47, which upon reduction with lithium borohydride gave the triol 48. Tosylation of the primary alcohol groups in 48, followed by cyclization yielded 49 (cis) and 50 (trans) [approximately 2:3 ratio]. The various R groups were introduced by displacing the remaining tosyl group in 49 (R=tosyl). Cis compounds of type 49 were significantly more active than the corresponding trans analogs 50.

Scheme 4

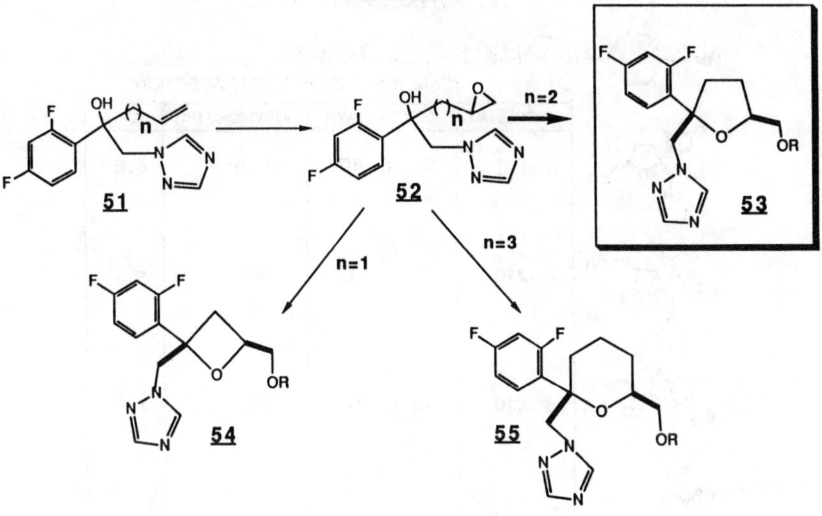

Scheme 5

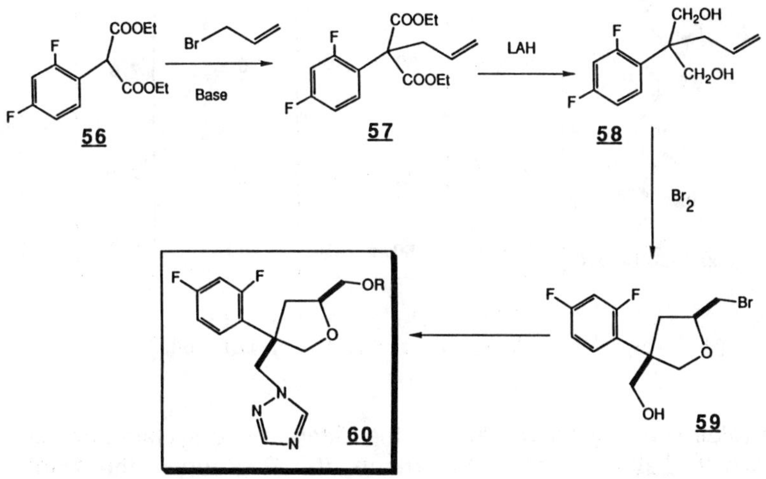

The regioisomer 53 was prepared from the olefin 51 by epoxidation to 52, followed by cyclization - Scheme 4. The R groups were introduced by the established procedure. The same scheme of events was followed to make the corresponding 4 (n=1) and 6 (n=3) membered ring analogs by varying n in 51. It is worthwhile to mention that both types 54 and 55 had significantly less biological activity when compared to the five membered analog 53. The regioisomer type 60 was prepared from the malonate 56 as shown in Scheme 5. Among all the ring analogs, the cis isomers were consistently more potent than the corresponding trans analogs.

RELATIVE ACTIVITIES

Table 4
(CANDIDA ALBICANS - C43 INFECTION)

	INOCULUM	DOSE x DAYS	SURVIVORS (%)	CFU
61	5×10^6	50 x 4D	100	5.6
62	5×10^6	50 x 4D	90	5.3
63	5×10^6	50 x 4D	90	5
64	1×10^6	100 x 10D	90	6
65	5×10^6	50 x 4D	20	7.9
66	5×10^6	50 x 4D	20	8.2
KETOCONAZOLE	5×10^6	50 x 4D	100	5.5

```
CFU = Colony Forming Unit
Dose in mg/Kg, p.o., rat model
```

The relative biological activities of the different ring system analogs are shown in Table 4. For comparison, the R group is the same for compounds 61 through 66. It can be seen from the data that tetrahydrofurans 61, 62 and 63, are clearly better than the dioxolane 64 in a *Candida* infection model.

EFFECT OF SIDECHAINS- Table 5

R =	INOCULUM *	Rx DAYS	SURVIVORS	C F U
61	5×10^6	4	100	5.6
67	1×10^6	4	100	4.2
68	"	4	100	4.8
69	"	10	100	7.6
70	"	10	90	6.8
71	5×10^6	4	100	4.7
72	"	4	80	6.3
KETOCONAZOLE---	"	4	100	5.5

* *CANDIDA ALBICANS - C43;*   Dose: 50mg/Kg, p.o.

The effect of various R groups in a particular ring system is shown in Table 5. Compounds 61, 67, 68, 71 and 72, having well known side chains show very good activity in a *Candida* infection model.

Like ketoconazole and itraconazole, the above analogs were synthesized as racemates or diastereomers. Enantiomerically pure isomers could be prepared by modifying Scheme 3, as depicted in Scheme 6.

ENANTIOSELECTIVE SYNTHESIS - Scheme 6

The striking effect of absolute stereochemistry on the biological activity was revealed when the corresponding optical isomers were tested side by side in a separate, severe infection experiment - Table 6. The (-) enantiomer 73 virtually had all the activity and the (+) isomer 74 provided no protection. This pattern appeared to hold true for the rest of the ring systems as well. It was also noted that compound 73 did not have any appreciable *Aspergillus* activity.

ACTIVITY OF OPTICAL ISOMERS - Table 6

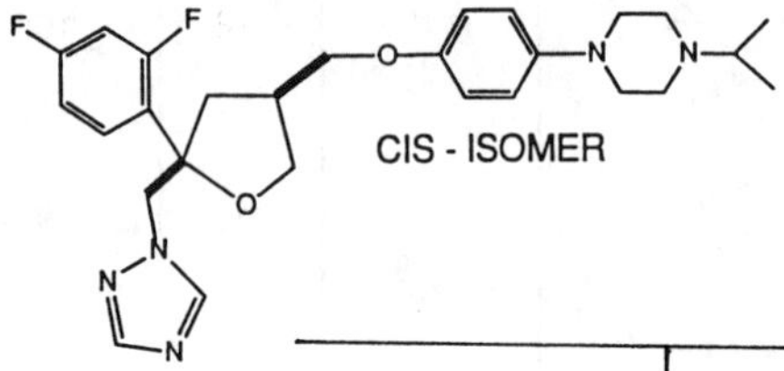

CIS - ISOMER

	SURVIVORS (%)	CFU*
62 (±)	80	6
73 (-)	100	5.5
74 (+)	0	9
KETOCONAZOLE	20	8.1

* *CANDIDA ALBICANS - C43*

Inoculum 5×10^6;   Dose: 10mg/Kg, 4 days

NOT ACTIVE AGAINST *ASPERGILLUS*

However, enhancement of the spectrum to include *Aspergillus* activity was possible by altering the R groups. Compounds 75, 76 and 77 were synthesized and their activities were compared with itraconazole and saperconazole in the same experiment - Table 7. It was clearly evident that compound 76 offered superior efficacy to itraconazole and saperconazole against *Candida albicans*.

Further spectrum advantages of 76 were revealed in an *Aspergillus* infection model. As can be seen from the efficacy graph, we have achieved the synthesis of compounds having broad spectrum antifungal activity. Compound 76 showed superior efficacy when compared to itraconazole and saperconazole.

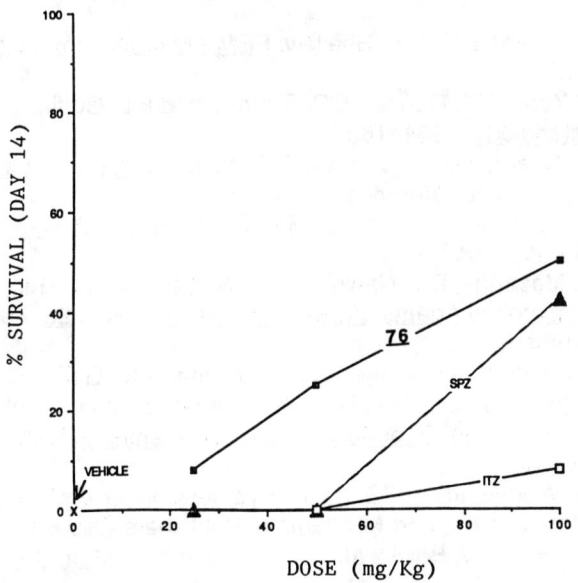

Table 7

R= [structure]

	SURVIVORS (%)	CFU*
75 [structure]	45	7.1
76 [structure]	80	5.4
ITRACONAZOLE	67	6.2
SAPERCONAZOLE	0	9
77 [structure]	80	7.5

Dose: 50mg/Kg, 4 days and
100mg/Kg, 10 days for **77**

**EFFICACY OF 76 IN COMPROMISED CF-1 MICE INFECTED
BY INHALATION WITH *Aspergillus Flavus (ND134)***

Acknowledgements: We thank Dr. R. Hare, Dr. K. Shaw, Dr. R. Parmegiani and Dr. T. Cacciapuoti for their support and effort in this Antifungal work.

REFERENCES

1.   D.J. Adams and G.W. Gooday, Abh. Akad. Wiss. DDR Abt. Math. Naturwiss Tech. IN., 1983, 39.
2.   R. Furter and D.M. Rast, FEMS Microbiology Letters, 1985, 28, 205.
3.   H. Decker, F. Walz, C. Bormann, H. Zahner, H-P Fiedler, H. Heitsch and W.A. Konig, J. Antibiotics, 1990, 43 and cited references.
4.   W.A. Konig, H. Hahn, R. Rathman, W. Hass, A. Keckeisen, H. Hagenmeir, C. Barmann, W. Dehler, R. Kurth and H. Zahner, Liebigs Annalen. Der. Chemie, 1986, 407 and cited references.
5.   K. Isono and S. Suzuki, Heterocycles, 1979, 13, 333.
6.   M. Hori, K. Kakiki and T. Misaio, Agr. Biol. Chem., 1974, 38, 699.
7.   N.P. Damodaran, G.H. Jones and J.G. Moffatt, J. Amer. Chem. Soc. 1971, 93, 3812.
8.   J. Fiander, M-T.G. Lopez, F.G. De Los Heras and P.P. Mendez-Castrillon, Synthesis, 1987, 987.
9.   A.G.M. Barrett and S.A. Lebold, J. Org. Chem. 1991, 56, 4875.
10.   For different chitin synthetases (CS-1, CS-2 and CS-3) see:
      a) S.J. Silverman, A. Sburlati, M.L. Slater and E. Cabib, Proc. Natl. Acad. Sci. USA, 1988, 85, 4735.
      b) M.H. Valdeveso, P.C. Mol, J.A. Shaw, E. Cabib and A. Duran, J. Cell Biol. 1991, 114, 101.
      c) C.E. Bulawa and B.C. Osmond, Proc. Natl. Acad. Sci. USA, 1990, 87, 7424.
      d) E. Cabib, Antimicrobial Agents and Chemotherapy, 1991, 35, 170.
11.   J.W. Payne and D.A. Shallow, FEMS Microbiology Letters, 1985, 28, 55.
12.   J-C Yadan, M. Gonneau, P. Sarthou and F.L. Goffic, J. Bacteriology, 1984, 160, 884.
13.   B.K. Khare, J.M. Becker and F.R. Naider, J. Med. Chem. 1988, 31, 650 and cited references.
14.   G. Emmer, N.S. Ryder and M.A. Grass-Berger, J. Med. Chem. 1985, 28, 278.
15.   P.J. Macarthy, D.J. Newman, L.J. Nisbet and W.D. Kingsbury, Antimicrobial Agents and Chemotherapy, 1985, 28, 494 and cited references.
16.   For a recent review see: A.K. Saksena, V.M. Girijavallabhan, A.B. Cooper and D. Loebenberg, Recent Advances in Antifungal Agents, Annual Reports in Medicinal Chemistry, 1989, 24, 111 and cited references.
17.   P.A. Worthington in 'Sterol Biosynthesis Inhibitors', Part 1, 24, Editors D. Berg and M. Plempel, Publishers Ellis Horwood, 1988.
18.   J. Heeres, L.J. Balckx and J. Van Custem, J. Med. Chem., 1984, 27, 894.

14

Adventures in Polyene Macrolide Chemistry: The Derivatisation of Amphotericin B

D. T. MacPherson,* D. F. Corbett, B. C. Costello,
M. J. Driver, A. R. Greenlees, W. S. MacLachlan,
C. T. Shanks, and A. W. Taylor
SMITHKLINE BEECHAM PHARMACEUTICALS, GREAT BURGH, YEW TREE
BOTTOM ROAD, EPSOM, SURREY KT18 5XQ, UK

1 INTRODUCTION

The isolation of the polyene macrolide amphotericin B (AmB, 1) from *Streptomyces nodosus* was reported in 1956.[1] AmB is now established as a potent, broad spectrum antifungal agent and is particularly useful for the treatment of serious systemic mycoses in immunocompromised patients.[2] However, its use is often limited by its high toxicity, especially nephrotoxicity,[3] and poor aqueous solubility. A fundamental cause of the drug's insufficient antifungal selectivity is its ability to perturb both fungal and mammalian cellular membranes by interaction with endogenous sterols, principally ergosterol in fungi and cholesterol in mammals. This interaction increases the permeability of the cell membrane, possibly through the formation of hydrophilic pores or channels, thus allowing leakage of cellular constituents across the membrane.[4]

AMPHOTERICIN B (AmB, 1)

Two main approaches have been adopted to improve the therapeutic ratio of AmB. Novel formulations of the drug, e.g. liposomes, appear to show promise in reducing toxicity,[5] and one such formulation, AmBisome, has recently been launched by Vestar.[6] Alternatively, several attempts have been made over the last twenty years to derivatise AmB with the primary aim of reducing its toxicity. Chemical manipulation has concentrated almost exclusively on the zwitterionic C-16 carboxylate and the C-3' mycosamine amino functionalities.[7] The C-16 carboxylate has been converted to esters,[8] amides[9] and an acyl hydrazide,[10] while a variety of modifications have been carried out on the C-3' amine, including the preparation of N-alkyl[11] and N-acyl[8,12] derivatives. *In vitro* antifungal activity is retained or slightly reduced, but in the case of C-3' amides activity is substantially reduced unless the amide is substituted with a basic nitrogen atom. Many of these analogues also demonstrate improved antifungal selectivity relative to AmB in studies *in vitro*, and in addition certain derivatives can be converted to water soluble salts.[13] Two compounds in particular, amphotericin B methyl ester (2)[14] and N-(D)-ornithylamphotericin B methyl ester (3),[15] were also markedly less nephrotoxic, both in animal models and in the case of (2), in clinical trials. However, the development of both (2) and (3) was terminated due to unexpected neurotoxicity.[14,16]

R

(2)   H

(3)   (D)-ornithyl

Although these studies did not lead to an antifungal agent of clinical importance, they did indicate that it is possible to retain activity *in vivo,* reduce nephrotoxicity and impart water solubility through chemical derivatisation, particularly at C-16. However, the range of AmB modifications carried out at C-16 and elsewhere has been inhibited by the highly functionalised nature of AmB, together with its insolubility in many common organic solvents. We now wish to report the synthesis of a wide range of new amphotericin B analogues, possessing novel functionality at C-16, C-13 and C-14 respectively.

2 C-16 DERIVATIVES

Preparation and Derivatisation of 16-Hydroxymethyl Analogue.

In 1986, Nicolaou[17] during studies on the total synthesis of AmB described the conversion of (1) in three steps to the mixture (4ab) (Scheme 1). The C-16 methyl ester was then reduced with sodium borohydride to the hydroxymethyl derivative (5ab). Modification of this procedure as follows led to the synthesis of our initial target, 16-decarboxy-16-hydroxymethylamphotericin B.

Before commencing our studies, we realised that a suitable protecting group for the C-3' amine would be required. Accordingly we functionalised the amine with a fluorenylmethoxycarbonyl (Fmoc) group to provide *N*-Fmoc AmB (6) (Scheme 2). After alkylation with methyl iodide we obtained the C-16 methyl ester (7) which was protected at C-13 by acid catalysed anomeric exchange[18] with methanol to give (8) (30% overall from AmB). The C-16 methyl ester of (8) was then reduced with sodium borohydride, without the need for acetonide protection, to give the 16-hydroxymethyl derivative (9) in 58% yield. Removal of the protecting groups from (9) by anomeric exchange at C-13 and treatment with piperidine to remove the Fmoc group gave the desired product (10) [65% from (9)].

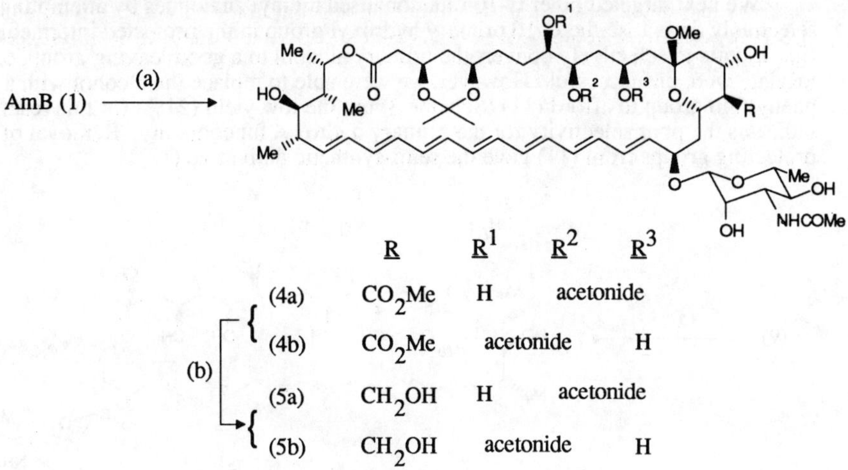
AmB (1)
(a)
OR1
OMe
OH
Me
HO
Me
O
OR2
OR3
R
Me
Me
OH
OH
NHCOMe
R
R1
R2
R3
(4a) CO2Me H acetonide
(4b) CO2Me acetonide H
(b)
(5a) CH2OH H acetonide
(5b) CH2OH acetonide H

Scheme 1 (a) i. Ac_2O; ii. CH_2N_2; iii. 2,2-dimethoxypropane, camphorsulfonic acid (CSA), MeOH (b) $NaBH_4$, MeOH.

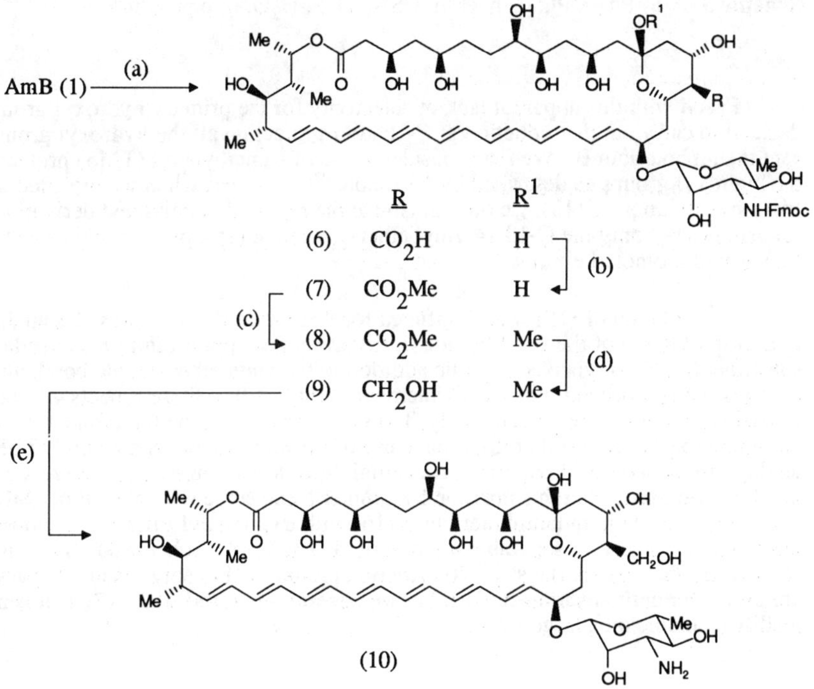
AmB (1)
(a)
OH
OR1
OH
Me
HO
O
Me
OH
OH
OH
OH
R
Me
Me
OH
OH
NHFmoc
R
R1
(6) CO2H H
(b)
(7) CO2Me H
(c)
(8) CO2Me Me
(d)
(9) CH2OH Me
(e)
CH2OH
NH2
(10)

Scheme 2 (a) Fmoc-*O*-succinimide, pyridine (b) *i*-Pr_2NEt, MeI (c) pyridinium-*p*-toluenesulfonate (PPTS), THF, methanol (d) $NaBH_4$, MeOH (e) i. PPTS, THF, H_2O; ii. piperidine.

We next targeted other C-16 functionalised methyl analogues by attempting to selectively derivatise the C-16 primary hydroxyl group in the protected intermediate (9). Efforts to selectively convert the primary alcohol to a good leaving group, e.g. tosylate were unsuccessful. However, we were able to replace the alcohol with a phenylthio group to afford (11) (Scheme 3) but the low yield (21%) for this reaction indicates the poor selectivity for the primary hydroxyl functionality. Removal of the protecting groups from (11) gave the semi-synthetic derivative (12).

	R	R^1
(11)	Fmoc	Me
(12)	H	H

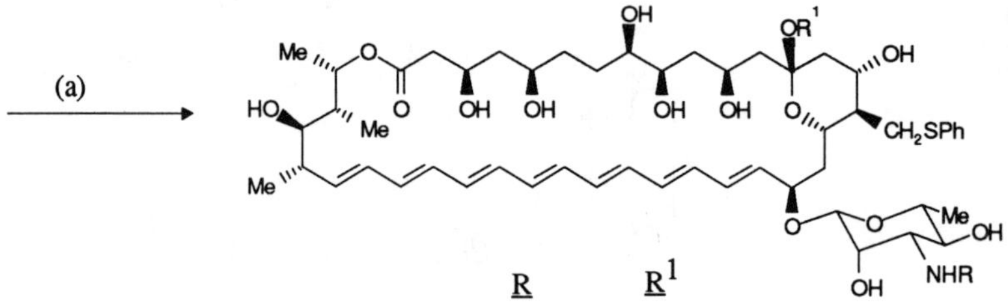

Scheme 3   (a) Ph$_3$P, PhSSPh (b) i. CSA, THF, H$_2$O; ii. piperidine.

Faced with this apparent lack of selectivity for the primary hydroxyl group we decided to carry out the reduction at C-16 after protecting all the hydroxyl groups present in amphotericin B. We first considered use of trimethylsilyl (TMS) protection of the hydroxyl groups as described by Nicolaou.[19] However, when we repeated the trimethylsilylation of (13), we did not isolate the reported persilylated derivative (14) (Scheme 4) but only the C-13,14 vinyl ether derivative (15), presumably as a result of Lewis acid promoted elimination from C-13.[20]

The isolation of (15) was significant for three reasons: it (*i*) provides an alternative for protection of the C-13 hydroxyl, (*ii*) presents an opportunity for introducing substituents at C-14 via electrophilic addition to the vinyl ether double bond, and (*iii*) changes the conformation of the six membered ring. All of these aspects will be discussed later, but for the present study, TMS ethers proved to be too labile to column chromatography and we therefore made use of the more stable triethylsilyl (TES) protecting group. To avoid complex product mixtures arising from differing reactivity at C-13, we first protected this position by anomeric exchange with methanol. Silylation of *N*-Fmoc 13-*O*-methylamphotericin B (16) with triethylsilyl triflate in dichloromethane gave a readily separable mixture of (17) and (18) (Scheme 5). Switching the solvent to *n*-hexane for the silylation reaction produces (17) only. Both (17) and (18) are useful for derivatisation studies, but we generally proceed with (17) as it is more readily available on a large scale.

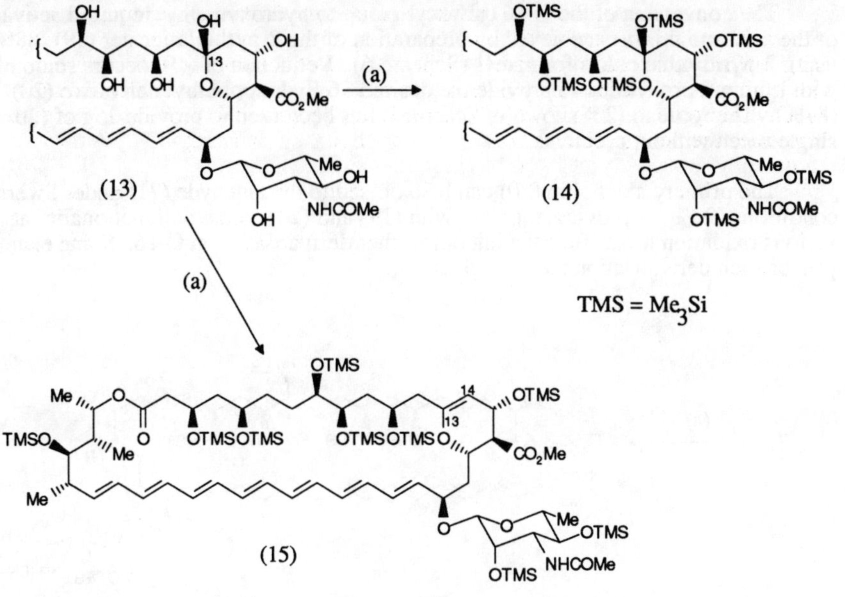

Scheme 4 (a) Me₃SiOSO₂CF₃ (TMSOTf), 2,6-lutidine, CH₂Cl₂ (see text).

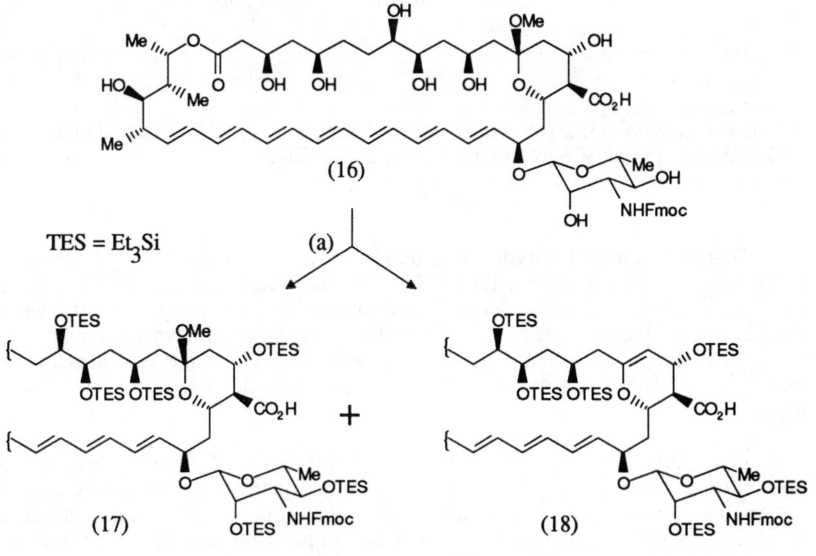

CH₂Cl₂	31%	24%
n-hexane	69%	–

Scheme 5 (a) i. Et₃SiOSO₂CF₃ (TESOTf), 2,6-lutidine; ii. silica gel.

The conversion of the C-16 carboxyl group to hydroxymethyl required activation of the acid and this was achieved by preparation of the 2-pyridylthioester (19) (88%) using 2-pyridylthio chloroformate[21] (Scheme 6). Reduction of (19) occurs smoothly with lithium borohydride to provide the desired 16-hydroxymethyl derivative (20) (87%). The route to (20) shown in Scheme 6 has been used to provide 25g of (20) in a single batch without problems.

The primary alcohol in (20) can be oxidised to the aldehyde (21) under Swern conditions (90%) to provide, together with (19) and (20), versatile functionality at various oxidation levels for carrying out further derivatisation at C-16. Some examples of such derivatisation are now given.

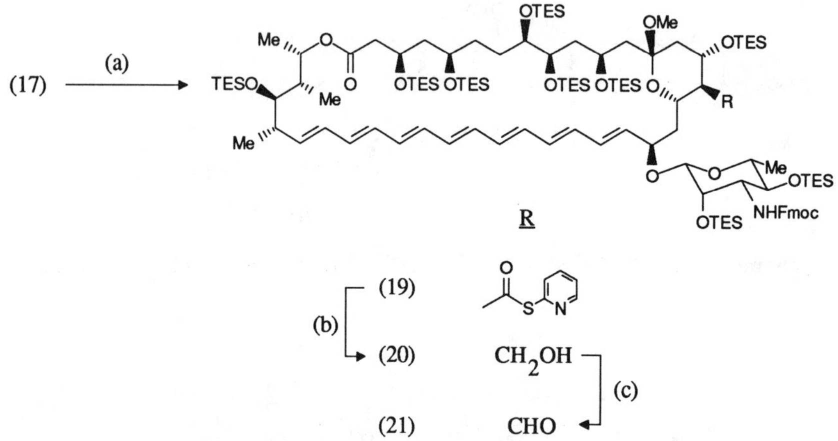

Scheme 6   (a) NEt$_3$, 2-pyridylthio chloroformate (b) LiBH$_4$, diethyl ether (c) DMSO, (CF$_3$CO)$_2$O, tetramethyl urea, then NEt$_3$.

Displacement of the hydroxyl group in (20) with nucleophiles was carried out in one pot *via* the intermediate triflate (22) to provide the azido (23) (79%), iodo (24) (87%) and cyano (25) (77%) derivatives (Scheme 7). Occasionally a small amount of another compound, assigned the structure (28), was isolated from these reactions, presumably as a result of intramolecular ketal exchange. However, (28) appeared to be unstable and readily ring opened at C-13 on attempted deprotection to the AmB analogue.

The iodide (24) was further reacted with lithium triethylborohydride to afford the 16-methyl derivative (26) (65%). The azide functionality was introduced as a precursor to 16-aminomethyl derivatives, but attempted reduction of the azide in (23) under a variety of conditions resulted in extensive decomposition. Reduction with triethyl phosphite[22] produced the phosphoramidate (27) (82%) which could not be induced to react further to produce the amine. All the derivatives (23)-(27) were readily deprotected in three steps (i. HF-pyridine, pyridine, THF, methanol; ii. PPTS, THF, H$_2$O; iii. piperidine) to provide the novel AmB analogues (29)-(33).

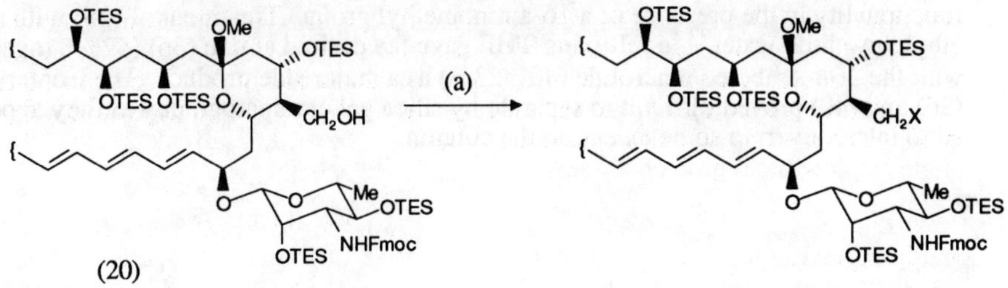

Scheme 7   (a) $(CF_3SO_2)_2O$, pyridine then $R_4N^+X^-$ for X = N_3, I, CN (b) $LiBEt_3H$
(c) $P(OEt)_3$.

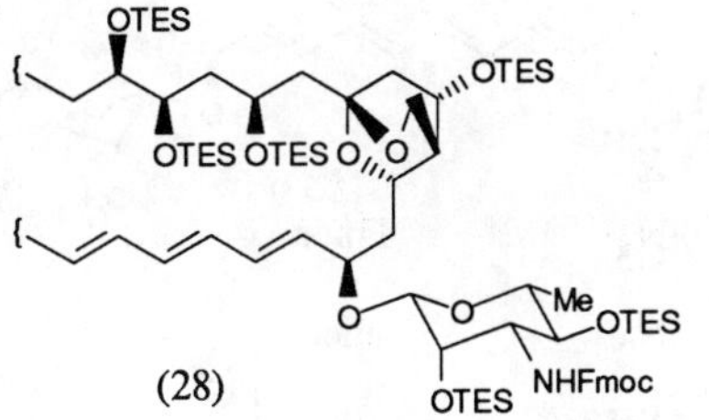

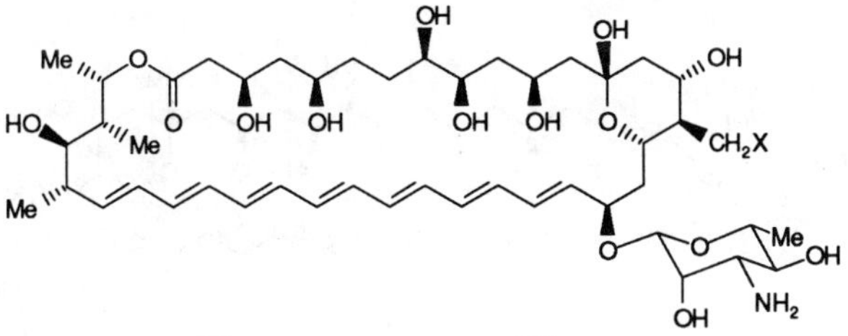

	X		X
(29)	N_3	(32)	H
(30)	I	(33)	$NHPO(OEt)_2$
(31)	CN		

The azide was reduced efficiently after desilylation and exchange of the C-3'
amino protecting group as follows. First (23) was desilylated to (34) in 73% yield
(Scheme 8). However, another product (~10%) was isolated from this reaction, whose
spectral data (IR, UV, MS, 2D-NMR) indicated that it was the isomeric ring contracted
macrolide (39) arising from intramolecular transesterification of the C-35 hydroxyl
group. The chemical shifts of H-35 [3.47ppm in (34); 4.99ppm in (39)] and H-37
[5.46ppm in (34); 4.06ppm in (39)] were particularly diagnostic for this transformation. Exchange of the *N*-Fmoc group in (34) for the allyloxycarbonyl (alloc) protecting group, giving (35) (94%), was necessary because of the instability of the Fmoc

functionality in the presence of a 16-aminomethyl group. Treatment of (35) with tri-ethylphosphine-water[23] in refluxing THF gave the desired amine (36) (57%), together with the 36-membered macrolide (40) (23%) as a major side product. The isomers (36) and (40) proved difficult to separate by silica gel chromatography as they appeared to interconvert to some extent on the column.

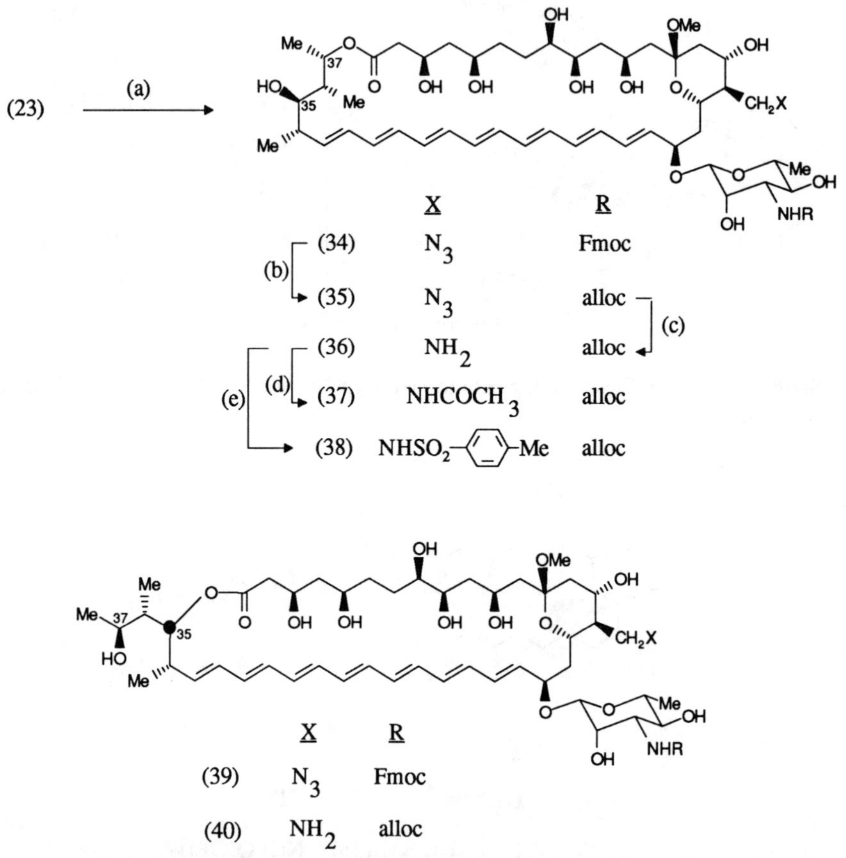

Scheme 8   (a) HF-pyridine, pyridine, THF, MeOH (b) i. piperidine; ii. alloc-*O*-succinimide, pyridine (c) PEt$_3$, THF, H$_2$O, reflux (d) Ac$_2$O (e) *p*-toluenesulfonyl chloride, K$_2$CO$_3$.

The amine in (36) was further derivatised to the acetamide (37) and the sulfon-amide (38), both of which appeared to be less prone to undergo the ring contracting transesterification reaction. Removal of the protecting groups from (37) and (38) (i. PPTS, THF, H$_2$O; ii. Pd[0], *n*-Bu$_3$SnH, AcOH)[24] gave the target AmB analogues (41) [61% from (36)] and (42) [42% from (36)].

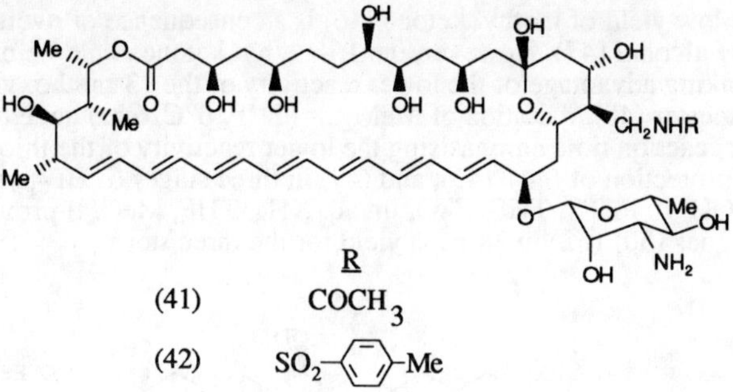

(41) COCH$_3$

(42) SO$_2$—⟨ ⟩—Me

C-16 Ketone Derivatives

The pyridylthioester functionality in (19) (Scheme 6) had been chosen to activate the C-16 carboxyl group as we hoped to use this group as a precursor to C-16 ketone derivatives via the addition of Grignard reagents.[25] However, the reduction substrate (19) was unsuitable for this purpose as the *N*-Fmoc group was unstable to these reagents and its pyridylthioester group was not sufficiently reactive. We discovered that use of the *N*-trifluoroacetyl protecting group[26] and introduction of a C-13,14 double bond to produce (43) (Scheme 9) provided a suitable substrate for ketone synthesis. This substrate illustrates two useful features of the C-13,14 double bond which were alluded to earlier, viz the C-13 hemiketal is protected as a vinyl ether and the conformational change induced by the double bond in the 6-membered ring makes the C-16 pyridylthioester in (43) more sterically accessible and thus more reactive towards Grignard reagents. Addition of phenyl, 2-pyrrolyl,[27] and methyl Grignard reagents to (43) at 0°C produced ketones (44) (73%), (45) (82%), and (46) (20%), respectively. In some reactions partial removal of the *N*-trifluoroacetyl group was observed and in such cases the crude product was reacylated with trifluoroacetic anhydride.

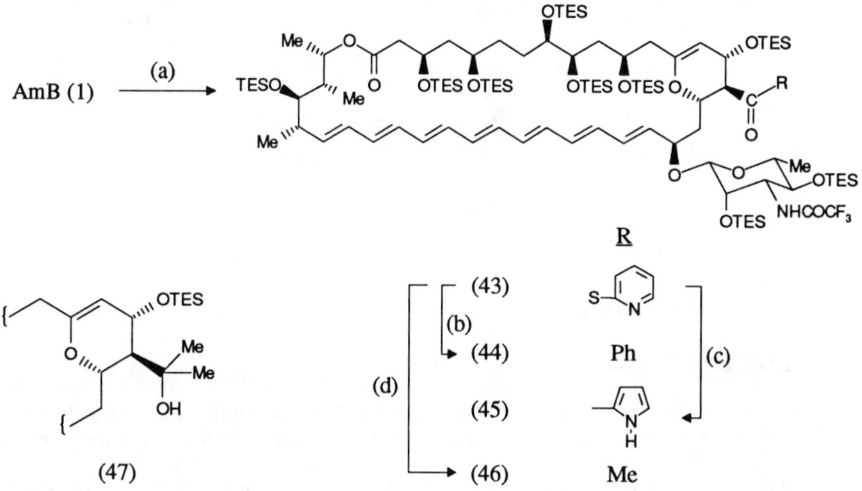

Scheme 9 (a) i. CF$_3$CO$_2$Et, pyridine; ii. PPTS, THF, MeOH; iii. TESOTf, 2,6-lutidine, CH$_2$Cl$_2$; iv. NEt$_3$, 2-pyridylthio chloroformate (b) i. PhMgBr, THF, 0°C; ii. (CF$_3$CO)$_2$O, NEt$_3$ (c) 2-pyrrolylMgBr (d) MeMgBr, THF, 0°C, 15 min.

The low yield of methyl ketone (46) is a consequence of overaddition to produce the tertiary alcohol (47). However, the 16-methyl ketone could be prepared in good yield by taking advantage of the lower reactivity of the 13-methoxy substituted pyridylthioester (48). Addition of MeMgBr (THF, 0°C, 6 hr) gave (49) in 68% yield, the longer reaction time emphasising the lower reactivity of the thioester in this substrate. Deprotection of (44), (45), and (49) in three stages (i. HF-pyridine, pyridine, THF, MeOH; ii. PPTS, THF, H_2O; iii. aq. NH_3, THF, MeOH) provided the C-16 ketone analogues (50)-(52) in 38-51% yield for the three steps.

	R
(48)	S—pyridyl
(49)	Me

	R
(50)	Ph
(51)	pyrrolyl
(52)	Me

3 C-13 DERIVATISATION

C-13 Alkoxy Derivatives

The acid catalysed anomeric exchange of hydroxyl with methoxy at C-13 [(7) → (8), Scheme 2] provided the basis for the preparation of novel C-13 alkoxy derivatives of AmB. In order to facilitate purification of intermediates in this and other studies, we protected the C-16 carboxyl group of AmB as an allyl ester, which was introduced by alkylation of *N*-Fmoc AmB (6) with allyl bromide to give (53) (56%) (Scheme 10).

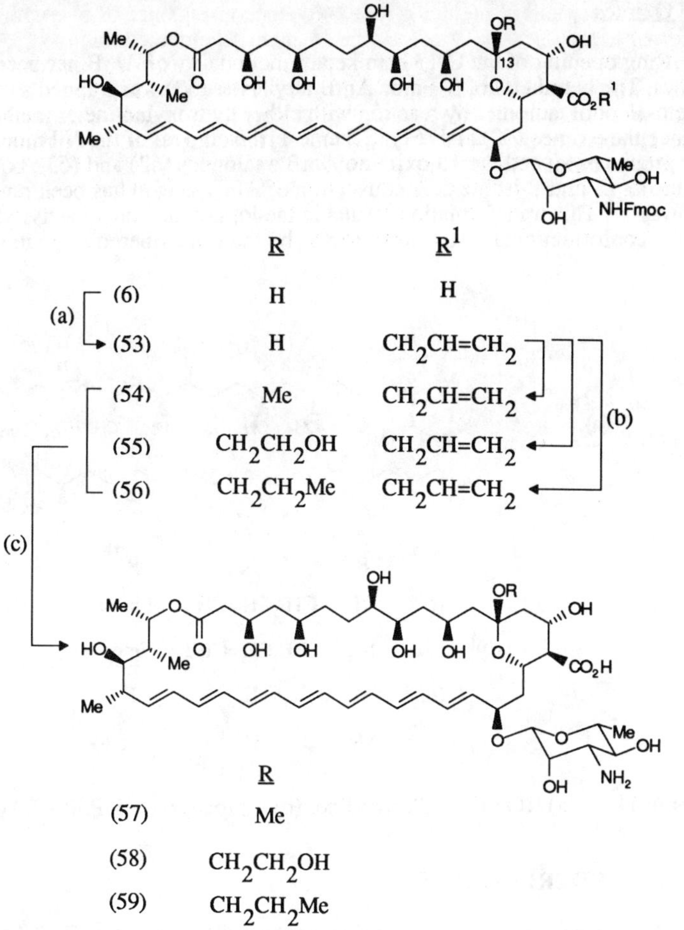

		R	R¹
(a)	(6)	H	H
	(53)	H	$CH_2CH=CH_2$
	(54)	Me	$CH_2CH=CH_2$
(b)	(55)	CH_2CH_2OH	$CH_2CH=CH_2$
	(56)	CH_2CH_2Me	$CH_2CH=CH_2$

(c)

	R
(57)	Me
(58)	CH_2CH_2OH
(59)	CH_2CH_2Me

Scheme 10   (a) *i*-Pr$_2$NEt, allyl bromide (b) ROH, THF, CSA (cat.) (c) i. piperidine; ii. Pd[P(Ph)$_3$]$_4$, pyrrolidine.

Treatment of (53) with camphorsulfonic acid (CSA, 0.4 equiv.) in a 1:1 mixture of THF and the relevant alcohol introduced the desired alkoxy group at C-13. Methanol, ethylene glycol and *n*-propanol were introduced in this manner to give (54), (55) and (56), respectively, in moderate yield (~50%) after column chromatography. Removal of the *N*-Fmoc group with piperidine followed by Pd(0) catalysed deallylation[28] gave the first AmB analogues (57)-(59) to be modified only at the C-13 position. An unexpected property of certain of these derivatives is their inherent water solubility.

C-13 Oximes

Ring opening of the C-13 hemiketal functionality of AmB has been achieved as follows. The hemiketal of N-Fmoc AmB allyl ester (53) was trapped as a derivative of its keto-alcohol tautomer by reaction with either hydroxylamine or methoxylamine to produce the oximes (60) and (61) (Scheme 11). Removal of the N-Fmoc and 16-allyl ester groups produced the 13-oximino AmB analogues (62) and (63), both as a 1:1 mixture of E- and Z-isomers. A conversion of AmB to (63) has been reported by Borowski.[29] This transformation results in the loss of antifungal activity, suggesting that the conformational restraints imparted by the 6-membered ring are vital for activity.

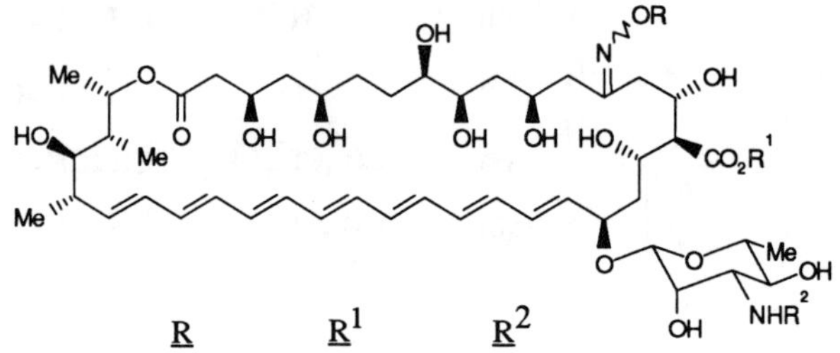

		R	R^1	R^2	
	(60)	H	CH$_2$CH=CH$_2$	Fmoc	
(b)	(61)	Me	CH$_2$CH=CH$_2$	Fmoc	
	(62)	H	H	H	(b)
	(63)	Me	H	H	

Scheme 11   (a) RONH$_2$.HCl, pyridine. (b) i. piperidine; ii. Pd[P(Ph)$_3$]$_4$, pyrrolidine.

4 C-14 DERIVATIVES

The discovery of the Lewis acid promoted dehydration at C-13,14 (described earlier in Scheme 4) was important in allowing us access to a number of C-14 derivatives of AmB. The parent 13,14-anhydro AmB (66) was prepared by treatment of N-Fmoc AmB allyl ester (53) with trimethylsilyl triflate and 2,6-lutidine to give (64), which without purification was desilylated to provide (65) [50% from (53)] (Scheme 12). Removal of the Fmoc and allyl groups gave 13,14-anhydroamphotericin B (66) (78%).

Electrophilic addition to the C-13,14 double bond offered the potential for introducing substituents at C-14 of AmB for the first time. By appropriate choice of substrate we have introduced a hydroxyl group in both the R and S configurations at C-14 via oxidation of the vinyl ether double bond with m-chloroperbenzoic acid (m-CPBA). The hydroxyl group was introduced with the R configuration when (65) was treated with m-CPBA in aqueous THF to provide (67) (75%) (Scheme 13). Presumably the electrophile is directed to the lower face by the 15-OH group. The stereochemistry around the hemiketal ring was assigned from ^{1}H NMR coupling constants (H-14, 3.89 ppm; J$_{14-15}$, 2.9 Hz). Removal of protecting groups from (67) gave 14(R)-hydroxyamphotericin B (68) (58%).

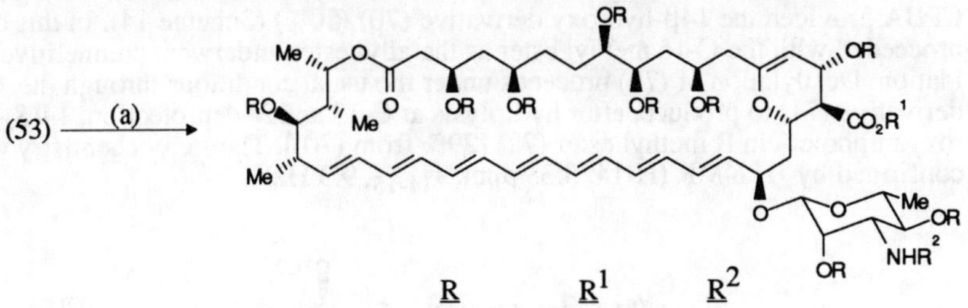

Scheme 12   (a) TMSOTf, 2,6-lutidine, CH_2Cl_2 (b) HF-pyridine, pyridine, THF (c) i. piperidine; ii. $Pd[P(Ph)_3]_4$, pyrrolidine.

Scheme 13   (a) *m*-CPBA, THF, H_2O (b) i. piperidine; ii. $Pd[P(Ph)_3]_4$, pyrrolidine.

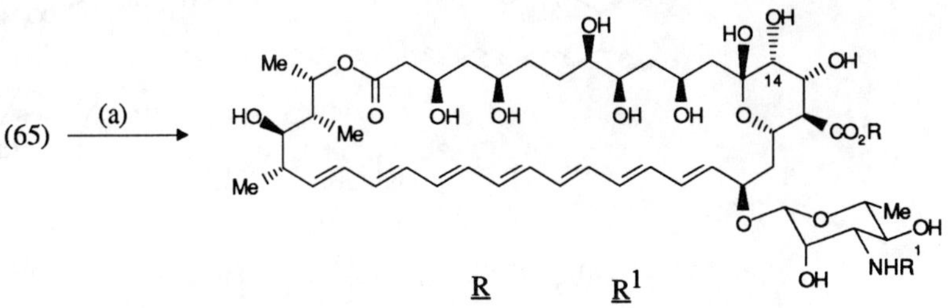

Oxidation of the vinyl ether in pertriethylsilylated derivative (69) with *m*-CPBA provided the 14β-hydroxy derivative (70) (50%) (Scheme 14). In this case we proceeded with the C-16 methyl ester as the allyl ester underwent competitive epoxidation. Desilylation of (70) proceeds under the usual conditions through the 13-fluoro derivative (71) to produce, after hydrolysis at C-13 and *N*-deprotection, 14(*S*)-hydroxyamphotericin B methyl ester (72) [29% from (70)]. The stereochemistry was again confirmed by ^{1}H NMR (H-14, 3.38 ppm; $J_{14\text{-}15}$, 9.3 Hz).

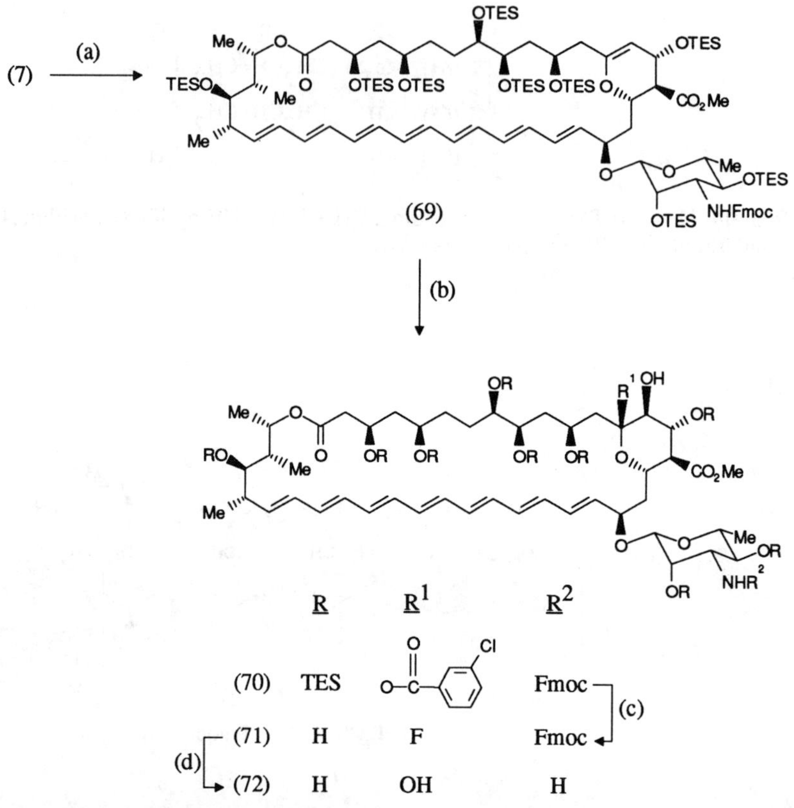

	R	R^1	R^2	
(70)	TES	$O\text{-}C(=O)\text{-}C_6H_4\text{-}Cl$	Fmoc	(c)
(71)	H	F	Fmoc	
(72)	H	OH	H	

Scheme 14   (a) i. TMSOTf, 2,6-lutidine, CH_2Cl_2; ii. HF-pyridine, pyridine, THF; iii. TESOTf, 2,6-lutidine, CH_2Cl_2 (b) *m*-CPBA, THF, *n*-hexane (c) HF-pyridine, pyridine, THF (d) i. H_2O, THF; ii. piperidine.

5 SUMMARY AND BIOLOGICAL PROPERTIES

In summary, we have described versatile methodology which allows novel, selective modification of amphotericin B, not only at the C-16 carboxylic acid but also for the first time at C-13 and C-14. To achieve this we have developed the use of a range of protecting groups for the various functionalities in the molecule. The new C-16 derivatives can be converted to water-soluble salts of the 3'-amine while certain of the C-13 modified compounds have a degree of intrinsic water solubility, thus overcoming one of the disadvantages of the parent molecule. Furthermore, many of the

Table 1 *In Vitro* Antifungal Activity and Rat Serum Chemistry of C-16 Modified Amphotericin B Derivatives

Compound Organism	MIC (μg/ml) in SAB after 2 days, inoculum 10^5 cells/ml (10^4 for *A. fumigatus*)							
	16-CH$_2$OH (10)	16-CH$_2$SPh (12)	16-Me (32)	16-CH$_2$N$_3$ (29)	16-CH$_2$NHAc (41)	16-CH$_2$CN (31)	16-COPh (50)	AmB (1) (mean)
Candida albicans 73/079	1	16	8	8	2	1	1	2
Candida parapsilosis 937A	2	16	8	2	4	2	2	8
Cryptococcus neoformans 451	1	2	4	0.25	0.5	0.5	1	2
Aspergillus fumigatus	2	8	8	2	4	1	1	4

	BUN levels (mmol/l) after 4 days in rats *						
	(10)	(12)	(32)	(29)	(41)	(31)	(50)
Derivative	4.4 ($\pm$0.2)	–	8.0 ($\pm$0.3)	6.4 ($\pm$0.7)	12.1 ($\pm$8.9)	19.4 ($\pm$8.3)	4.5 ($\pm$0.3)
AmB	23.2 ($\pm$3.8)	–	16.6 ($\pm$4.3)	26.7 ($\pm$5.9)	21.0 ($\pm$2.1)	23.9 ($\pm$2.9)	25.8 ($\pm$12.9)
Control	2.6 ($\pm$1.0)	–	8.4 ($\pm$1.2)	4.3 ($\pm$0.5)	3.5 ($\pm$0.2)	4.6 ($\pm$0.6)	3.9 ($\pm$0.8)

*Dosage 6x10 mg/kg i.p. for derivatives and 2x10 mg/kg i.p. for AmB

Table 2 *In Vitro* Antifungal Activity and Rat Serum Chemistry of C-13/14 Modified Amphotericin B Derivatives

Compound Organism	MIC (μg/ml) in SAB after 2 days, inoculum 10^5 cells/ml (10^4 for *A. fumigatus*)						
	13-OMe (57)	13-OCH$_2$CH$_2$OH (58)	13=NOH (62)	13,14= (66)	14(R)-OH (68)	14(S)-OH, 16-CO$_2$Me (72)	AmB (1) (mean)
Candida albicans 73/079	2	4	>32	8	2	2	2
Candida parapsilosis 937A	8	8	>32	32	4	2	4
Cryptococcus neoformans 451	0.5	2	>32	4	1	1	1
Aspergillus fumigatus	2	2	>32	8	8	2	2

	BUN levels (mmol/l) after 4 days in rats*					
	(57)	(58)		(66)	(68)	(72)
Derivative	11.0 ($\pm$2.7)	9.9 ($\pm$1.3)		4.1 ($\pm$0.3)	7.7 ($\pm$3.4)	3.1 ($\pm$0.9)
AmB	20.0 ($\pm$6.7)	23.9 ($\pm$2.9)	–	20.3 ($\pm$9.4)	24.9 ($\pm$9.3)	25.4 ($\pm$11.1)
Control	3.2 ($\pm$0.7)	4.6 ($\pm$0.6)	–	3.2 ($\pm$0.8)	4.2 ($\pm$1.5)	3.9 ($\pm$0.4)

*Dosage 6x10 mg/kg i.p. for derivatives and 2x10 mg/kg i.p. for AmB

analogues we have synthesised have the potential of a higher therapeutic ratio than AmB. This is illustrated in Tables 1 and 2, which show representative minimum inhibitory concentrations (MIC) for certain derivatives against a range of fungi, together with rat blood urea nitrogen (BUN) levels, increases in which are indicative of nephrotoxicity.

The MIC data demonstrate that, in general, a range of substitution at C-13, C-14 and C-16 in AmB is compatible with retention of the *in vitro* antifungal activity of the parent. Although there are some differences in potency, only compound (62) lacking an intact hemiketal ring is without activity. The rat serum chemistry data in Tables 1 and 2 show that the novel derivatives tested are potentially much less nephrotoxic than AmB. Even at the higher dose used for the derivatives (6 x 10 mg/kg compared to 2 x 10 mg/kg for AmB), only compound (31) shows increased BUN levels approaching those of AmB, and several compounds show negligible increases over the controls indicating very low nephrotoxicity.

Several of the semi-synthetic novel polyenes with increased antifungal selectivity are being evaluated further in a variety of efficacy and toxicity studies. The 16-hydroxymethyl compound (10) has shown a particularly promising profile to date and is being investigated as a potential clinical candidate.

ACKNOWLEDGEMENTS

We wish to thank P. A. Hunter and her team of biologists, especially E. A. Carter, P. R. Murdock and L.A. Williams, for the biological data. We are also grateful to S. A. Readshaw for his assistance with the interpretation of NMR spectra.

REFERENCES

1. J. Vandeputte, J.L.Watchtel and E.T. Stiller, <u>Antibiotics Annual</u>, 1956, 587.
2. H.A. Gillis, R.H. Drew and W.H. Pickard, <u>Reviews of Infectious Diseases</u>, 1990, <u>12</u>, 308.
3. R. Sabra and R.A. Branch, <u>Drug Safety</u>, 1990, <u>5</u>, 94.
4. J. Brajtburg, W.G. Powderly, G.S. Kobayashi and G. Medoff, <u>Antimicrob. Ag. Chemother.</u>, 1990, <u>34</u>, 183.
5. J. Brajtburg, W.G. Powderly, G.S. Kobayashi and G. Medoff, <u>Antimicrob. Ag. Chemother.</u>, 1990, <u>34</u>, 381.
6. D.C.E. Speller and D.W. Warnock (Editors), <u>J. Antimicrob. Chemother.</u>, 1991, <u>28</u>, Supplement B, 1-118.
7. For a review see C.P. Schaffner in " Recent Trends in the Discovery, Development and Evaluation of Antifungal Agents " ed. R.A. Fromtling, J.R. Prous Science Publishers, Barcelona, 1987, p. 595.
8. W. Meclinski and C.P. Schaffner, <u>J. Antibiotics</u>, 1972, <u>25</u>, 256.
9. A. Jarzebski, L. Falkowski and E. Borowski, <u>J. Antibiotics</u>, 1982, <u>35</u>, 220.
10. J. Grzybowska and E. Borowski, <u>J. Antibiotics</u>, 1990, <u>43</u>, 907.
11. A. Czerwinski, W.A. Konig, T. Zieniawa, P. Sowinski, V. Sinnwell, S. Milewski and E. Borowski, <u>J. Antibiotics</u>, 1991, <u>44</u>, 979.
12. A. Czerwinski, J. Grzybowska and E. Borowski, <u>J. Antibiotics</u>, 1986, <u>39</u>, 1025 ; J.J.K. Wright, J.A. Albarella, L.R. Krepski and D. Loebenberg, <u>J. Antibiotics</u>,

1982, 35, 911 ; C.P. Schaffner and E. Borowski, Antibiot. Chemother., 1961, 11, 724.

13.   see for example USP 4,041,232/ 1977.

14.   P.D. Hoeprich, N.M. Flynn, M.M. Kawachi, K.K. Lee, R.M. Lawrence, L.K. Heath and C.P. Schaffner, Annals of the New York Academy of Sciences, 1988, 544, 517.

15.   R.M. Parmegiani, D. Loebenberg, B. Antonacci, T. Yarosh-Tomaine, R. Scupp, J.J. Wright, P.J.S. Chiu and G.H. Miller, Antimicrob. Ag. Chemother., 1987, 31, 1756.

16.   T. Massa, D.P. Sinha, J.D. Frantz, M.E. Filipek, R.C. Weglein, S.A. Steinberg, J.T. McGrath, B.F. Murphy, R.J. Szot, H.E. Black and E. Schwartz, Fundamental and Applied Toxicology, 1985, 5, 737.

17.   K.C. Nicolaou, T.K. Chakraborty, R.A. Daines and N.S. Simpkins, J. Chem. Soc.,Chem. Commun., 1986, 413 ; K.C. Nicolaou, T.K. Chakraborty, Y. Ogawa, R.A. Daines, N.S. Simpkins and G.T. Furst, J. Amer. Chem. Soc., 1988, 110, 4660.

18.   R.M. Kennedy, A. Abiko and S. Masamune, Tetrahedron Lett., 1988, 447.

19.   K.C. Nicolaou, T.K. Chakraborty, R.A. Daines and Y. Ogawa, J. Chem. Soc., Chem. Commun., 1987, 686.

20.   M.J. Driver, W.S. MacLachlan, D.T. MacPherson and S.A. Readshaw, J. Chem. Soc., Chem. Commun., 1990, 636.

21.   E.J. Corey and D.A. Clark, Tetrahedron Lett., 1979, 2875.

22.   A. Koziara, K. Osowska-Pacewicka, S. Zawadzki and A. Zwierzak, Synthesis, 1985, 202.

23.   M. Bartra, F. Urpi and J. Vilarrasa, Tetrahedron Lett., 1987, 5941.

24.   O. Dangles, F. Guibé, G. Balavoine, S. Lavielle and A. Marquet, J. Org. Chem., 1987, 52, 4984.

25.   M. Araki, S. Sakata, H. Takei and T. Mukaiyama, Bull. Chem. Soc. Jpn., 1974, 47, 1777.

26.   T.J. Curphey, J. Org. Chem., 1979, 44, 2805.

27.   K.C. Nicolaou, D.A. Claremon and D.P. Papahatjis, Tetrahedron Lett., 1981, 4647.

28.   R. Deziel, Tetrahedron Lett., 1987, 4371.

29.   A. Czerwinski, W.A. König, P. Sowinski, L. Falkowski, J. Mazerski and E. Borowski, J. Antibiotics, 1990, 43, 1098.

15

2,3-Oxidosqualene–Lanosterol Cyclase: An Attractive Target for Antifungal Drug Design

S. Jolidon, A. Polak-Wyss, P. G. Hartman, and P. Guerry

PHARMA DIVISION, PRECLINICAL RESEARCH, F. HOFFMANN-LA ROCHE LTD, CH-4002 BASEL, SWITZERLAND

1 INTRODUCTION

The number and severeness of mycotic infections is still increasing due to the growing population of immunosuppressed and immunocompromised patients, but none of the available drugs fulfils the need for a genuinely broad spectrum fungicidal agent. Almost all known antifungals act on enzymes involved in the biosynthesis of ergosterol,[1] the most important sterol constituent of the fungal plasma cell membrane (Figure 1). The disturbance of its formation is often lethal for the fungus and the enzymes involved in the biosynthesis of this sterol are therefore major targets for antifungal chemotherapy.

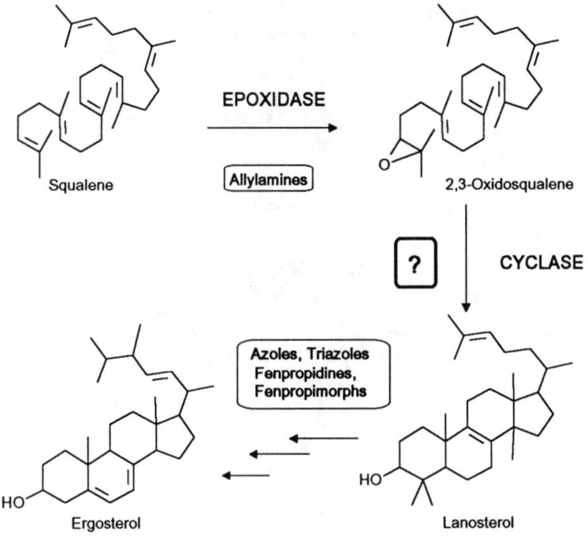

Figure 1 Sterol biosynthesis in fungi

Azoles or triazoles as well as fenpropidines or fenpropimorphs act on enzymatic steps involved in the biochemical processing of the sterol skeleton; allylamines, which block squalene-epoxidase, are the only marketed antifungals acting on a pre-sterol enzymatic step.

2,3-oxidosqualene-lanosterol cyclases [EC 5.4.99.7] are key enzymes which, through cyclization of 2,3-oxidosqualene give rise to lanosterol, the first cyclic compound to appear in the biosynthesis of sterols. These ubiquitous enzymes have been proposed as suitable targets for the design of several types of drugs, including antifungals [2,3]. Unfortunately, these enzymes are poorly characterized, partly due to their microsomal and membrane bound nature which renders isolation and purification rather tedious. Partial purification of hog liver enzyme [4] and purification of rat liver enzyme [5] has been carried out. Concerning yeasts, purification of the enzyme of *Saccharomyces cerevisiae* [6] should be mentioned, where even affinity chromatography steps have been used.[7] Furthermore, a 2,3-oxidosqualene-lanosterol cyclase coding gene from the medically important fungus *Candida albicans* has been cloned and characterized.[8]

Knowledge of the tertiary structure of a 2,3-oxidosqualene-lanosterol cyclase would be highly desirable for a better understanding of its mechanism and optimization of inhibitors, but so far this information is not available. Thus, design and optimization of cyclase inhibitors have been based

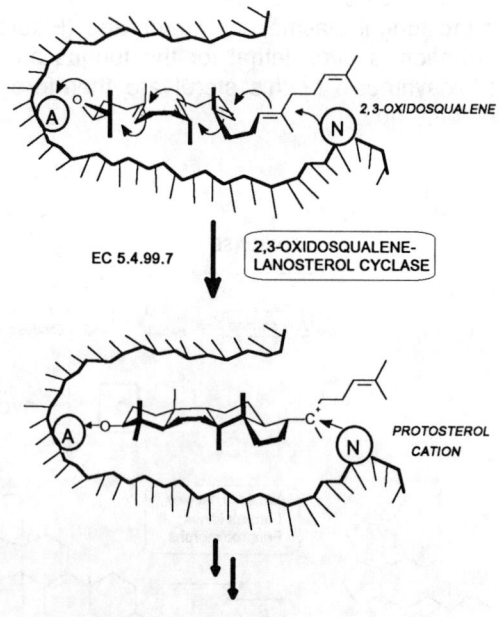

Lanosterol

Figure 2 Postulated mechanism of action of the 2,3-oxidosqualene-lanosterol cyclase [EC 5.4.99.7]

on lead structures arising from a more mechanistical or even empirical approach.[9,10] The postulated mechanism of action of this enzyme has been described several times [9,11,12] and is shown in a simplified manner in Figure 2.

The enzyme is thought to induce cyclization of the prefolded 2,3-oxidosqualene (chair/boat/chair-conformation) by protonation and ring opening of the epoxide. The subsequent cyclization process is probably non-concerted [12] and may even proceed through intermediates with a 5-membered C-ring.[13] At the end of the cyclization, the protosterol cation at C-20 is probably stabilized by an enzymatic nucleophile. A series of methyl group shifts then occur, preceding release of the final product lanosterol. Nothing is known about the nature of the active site acid or nucleophile, but it has been suggested that an -SH group might play a key role as the enzymes are inactivated by cysteine modifiying agents.[4,6]

2 DESIGN OF INHIBITORS AND STRUCTURE / ACTIVITY RELATIONSHIP

The mechanistic approach towards inhibitors of 2,3-oxidosqualene-lanosterol cyclase has led to different classes of compounds :

The concept of mimics of presumptive carbocationic high energy intermediates formed during cyclization of 2,3-oxidosqualene led to the synthesis of a series of lipophilic amines, quaternary ammonium salts and N-oxides [3,14], which proved to be potent inhibitors of the target enzyme. In our hands, some of the most potent compounds of this type were simple tertiary amines bearing two small alkyl or alkenyl groups and a longer lipophilic chain (i.e. 2,3-dihydro-2-azasqualene or N,N-dimethyldodecylamine; Figure 3). It can easily be envisaged that such compounds strongly interact with the acidic group **A** of the active site through their basic nitrogen atom.

A somewhat different approach aims at an interference with the nucleophilic group **N** of the active site, which will trap the C-20 protosterol cation. At least two categories of compound belong to this class of inhibitors, enol-ethers [15] and vinyl-derivatives of 2,3-oxidosqualene [16] (Figure 3). Enol-ethers can be thought of as giving stabilized ketal-like intermediates by interaction with the enzyme nucleophile, whereas the vinyl-derivatives might act as an electrophilic trap, leading to covalently bound inhibitors.

We thought it would be interesting to combine both approaches by designing and synthesizing inhibitors which interfere with both the active site acid and nucleophile. Such compounds, acting at the same time as a base and an electrophile, were obtained in the form of the aminoketones shown in Figure 3.[17,18] According to the aforementioned postulated mechanism of action of these inhibitors, their inhibitory activity should be dependent on the distance between the basic nitrogen atom of the tertiary amine and the electrophilic carbonyl-C-atom of the benzophenone moiety. This distance

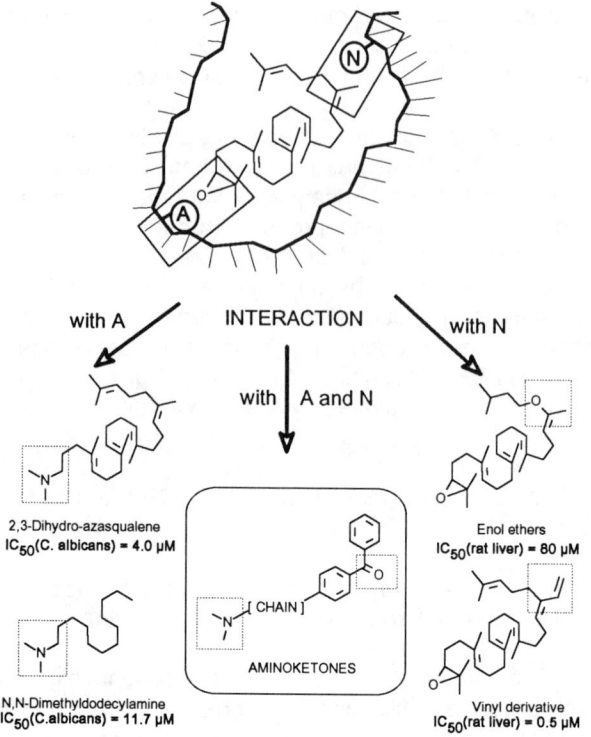

Figure 3 Different classes of cyclase inhibitors

can be estimated to be approximately 10.7 Å as shown by molecular modelling studies on the intermediate protosterol cation, where it represents the distance between the oxygen atom and the cation formed at C-20. A similar distance is found in the prefolded chair/boat/chair-conformation of 2,3-oxidosqualene between the oxygen atom of the epoxide and the future position of the C-20 in the sterol skeleton.

This hypothesis is supported by the data shown in Figure 4. Whereas Ro-02-9467 (distance N to C=O : 6.7 Å) has a rather poor IC_{50}-value, Ro-43-6300 (distance N to C=O : 10.7 Å) is a powerful enzyme inhibitor. Ro-47-2459 (distance N to C=O : 14.7 Å) on the other hand behaves almost like a simple lipophilic tertiary amine (see IC_{50}-values for 2,3-dihydro-2-aza-squalene or N,N-dimethyl-dodecylamine in Figure 3), as the ketone is positioned too far away for an interaction with the active site nucleophile. Figure 2 also shows that the prefolded 2,3-oxidosqualene as well as the protosterol cation are rather flat and stretched molecules; this could explain the very low activity of Ro-02-9467, where the bent nature of the molecule no longer fits into the enzyme cavity. This explanation is strengthened by the observation that analogues of Ro-43-6300 where the benzylamine- and the benzoyl-group, instead of being linked to the 4,4'-po-

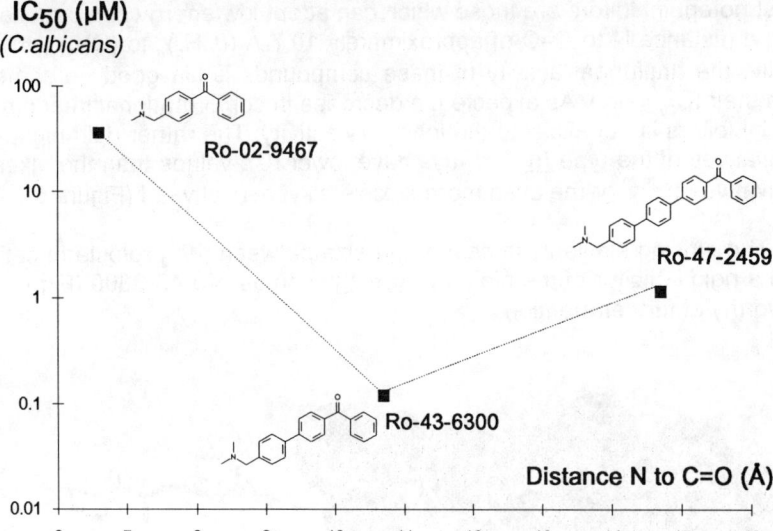

Figure 4 IC_{50}-values of rigid inhibitors plotted as a function of the distance between N and C=O

sition of the biphenyl skeleton, are linked in a 3,4'- or 2,4' - arrangement, have comparably low IC_{50}-values; these molecules also have a bent rather than a stretched molecular shape. A similar structure / activity relationship is found for the more flexible aminoketones **I** [19,20] (Figure 5), where the

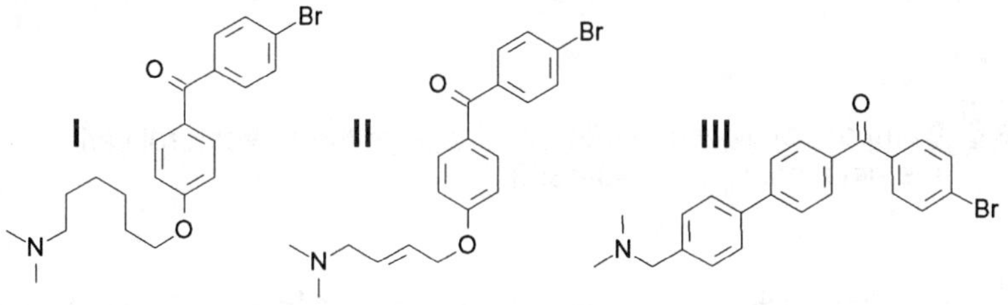

I	**II**	**III**
Ro-43-8212	Ro-44-4281	Ro-44-2103

$$IC_{50} (\mu M)$$
(Candida albicans)

1.09	**0.32**	**0.11**

Figure 5 Different types of bifunctional cyclase inhibitors and their IC_{50}-values

most potent inhibitors are those which can adopt low energy conformations with a distance N to C=O of approximately 10.7 Å ($(CH_2)_5$ to $(CH_2)_7$). Generally, the antifungal activity of these compounds is in good agreement with their IC_{50}-value. As expected, a decrease in conformational freedom of the inhibitors is beneficial to the inhibitory activity. The rather rigid biphenyl derivatives of the type III [21] always have lower IC_{50}-values than the alkenyl derivatives II [19,20] or the even more flexible alkyl derivatives I (Figure 5).

The striking similarity in shape and size between the protosterol cation and a rigid inhibitor of the biphenyl type III such as Ro-43-6300 (Figure 6) is worthy of further mention.

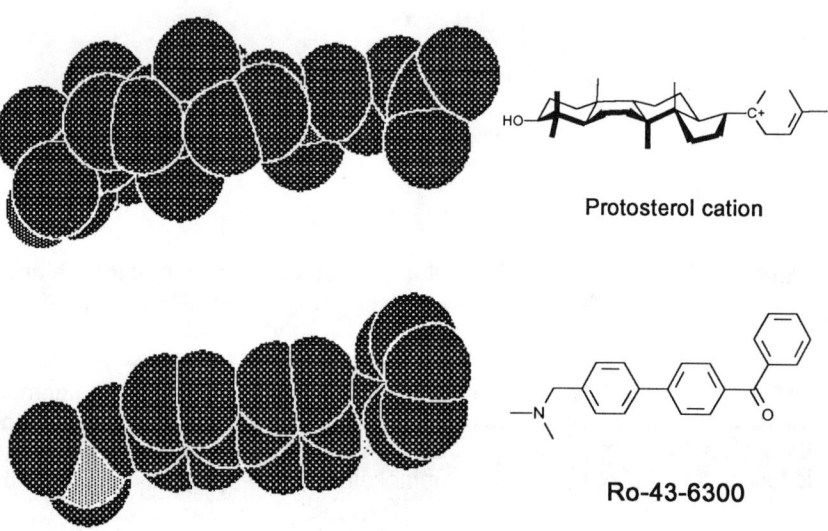

Protosterol cation

Ro-43-6300

Figure 6 Comparison of the molecular shape of the protosterol cation and the rigid inhibitor Ro-43-6300

Whereas the influence of the N to C=O distance on IC_{50}-values measured is easily explained, the influence of other structural modifications is more difficult to interpret. For example, we would expect strongly electron withdrawing substituents on the benzoyl-moiety to increase the electrophilic nature of the carbonyl group; according to the postulated mechanism of action of these aminoketones, this should lead to a stronger binding of the inhibitor and thus to lower IC_{50}-values. The influence of p-substituents is shown in Figure 7, where σ_p°-constants [22] are plotted against IC_{50}-values. Strong electron withdrawing substituents such as NO_2 or CN show surprisingly high IC_{50}-values against the 2,3-oxidosqualene-lanosterol cyclase of *Candida albicans*. The best inhibitors bear halogen- or simply hydrogen-substituents; this activity pattern is also found among the more flexible inhibitors of the type **I** or **II**.

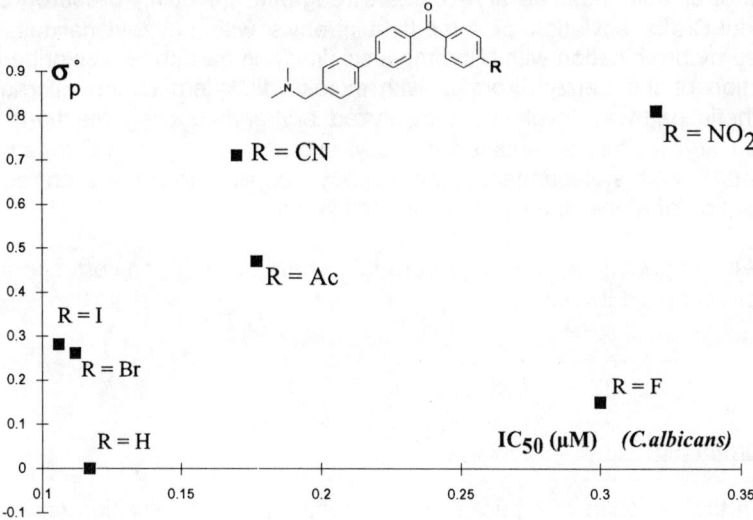

<u>Figure 7</u> Hammet-$\sigma_p°$-constants plotted against IC_{50}-values

Furthermore, the IC_{50}-value seems to be dependent on the size of the amino-group substituents; replacement of the N,N-dimethyl-groups of the type **I** to **III** - inhibitors by larger substituents leads to a gradual increase in the IC_{50}-value, the N,N-dibutyl derivative being almost inactive. Again, the antifungal activity of these compounds roughly parallels their enzyme inhibitory activity. N-Allyl-N-methyl derivatives generally show a 5 to 10 times better enzyme inhibition than their N,N-dimethyl counterparts, which might suggest some correlation between activity and basicity of the amine.

3 CHEMISTRY

4-[[6-(Dialkylamino)-hexyl]-oxy]-benzophenones of the type **I** or 4-[[4-(dialkylamino)-2-butenyl]-oxy-]-benzophenones of the type **II** can easily be prepared by a phase transfer catalyzed alkylation of 4-hydroxybenzophenones with an excess of 1,6-dibromohexane or 1,4-dibromo-2-butene. Treatment of the intermediate bromo-ether with an excess of N,N-dialkylamine leads to the desired compounds **I** or **II** (X = O). An alternative method involves Mitsunobu-type[23] coupling of 4-hydroxybenzophenones with ω-(N,N-dialkylamino)-1-alkanols or -1-alkenols. A possibility of obtaining the carbon analogs of **I** and **II** (X = CH_2) is given, for example, by the copper catalyzed coupling of aryl-Grignard reagents with α,ω-dihalo-alkanes or -2-butenes.[24] The intermediate α-aryl-ω-haloalkane or -alkene is then treated with excess dialkylamine in ethanol or acetone. The aryl-Grignard reagents are easily prepared by a Friedel-Crafts acylation of 4-methyl-biphenyls with benzoyl halides, followed by bromination with N-bromosuccinimide in carbon tetrachloride and reaction of the benzyl bromide with excess dialkylamine. An alternative synthetic pathway involves Pd-catalyzed aryl-aryl-coupling reactions between aryl halides or -triflates and aryl stannanes [25] or aryl-Grignard reagents.[26] With aryl stannanes, the biphenyl coupling does not even require protection of a possible intermediate aryl ketone.

All compounds synthesized were fully characterized and corresponding spectroscopic data obtained.

4 BIOCHEMISTRY AND BIOLOGY

Determination of IC_{50}-values

In the absence of a pure enzyme preparation, the inhibition constants of the test compounds against the 2,3-oxidosqualene-lanosterol cyclase of *Candida albicans* are measured in a cell free assay. 1.0 g of fungal cells are treated with 1 mg Zymolase 100T (Seikagaku Kogyo, Japan) and 12.5 µl of ß-mercaptoethanol in 5 ml digestion buffer for 30 min. at 30°. The protoplasts so formed are lysed by addition of 2 ml 100 mM phosphate buffer. This method may be scaled up as required. Subsequent centrifugation at 15000 g yields a cell free extract which retains full cyclase activity as shown by a 42% incorporation of racemic [1-^{14}C]-2,3-oxidosqualene synthesized from squalene in analogy to a published method.[27] It should be kept in mind that only the (3S)-enantiomer of racemic 2,3-oxidosqualene is cyclized by the enzyme.[12] Interestingly, use of the non ionic detergent Decyl Poe (n-decylpentaoxyethylene; Bachem, Switzerland) prevents further conversion of lanosterol to other fungal sterols, thus allowing accurate measurements of IC_{50}-values if the assay is done in presence of varying amounts of the test compound. After treatment with aqueous potassium hydroxide, the non saponifiable lipids are extracted and applied to silica gel TLC-plates, which are run twice in dichloromethane. The radio-labelled spots are quantified with an automatic TLC-scanner (Rita-3200, Raytest).

A similar assay can be performed with whole cells of *Candida albicans* or *Histoplasma capsulatum* if ^{14}C-radiolabelled mevalonate is used instead of [1-^{14}C]-2,3-oxidosqualene. This experiment allows monitoring of other radioactive non-saponifiable lipids formed in presence of the test compounds. The assay showed an almost exclusive accumulation of radiolabelled 2,3-oxidosqualene, thus confirming the specificity of our inhibitors against 2,3-oxidosqualene-lanosterol cyclase, even in living fungal cells.

Antifungal activity

The antifungal activity of the test compounds was obtained by determination of the minimum inhibitory concentrations (MIC-values) on Rowley agar against strains of *Candida albicans* and *Aspergillus fumigatus* (after 2 days) and against strains of *Histoplasma capsulatum* and *Trichophyton mentagrophytes* (after 7 days).

5 CONCLUSIONS

Our approach toward structurally new inhibitors of 2,3-oxidosqualene-lanosterol cyclase shows that it is possible to rationally design and synthesize potent inhibitors based on mechanistic considerations even in the absence of a detailed knowledge of the 3-dimensional structure of the enzyme or the nature of its active site. These aminoketones thus represent the first cyclase inhibitors which show an impressive *in vitro* activity against a wide range of medically important pathogenic and opportunistic fungi; the inhibition constants of some approach the nanomolar range and their *in vitro* antifungal activity can be favourably compared with several marketed antifungals (Table 1).

Studies of lipids accumulated in whole fungal cells indicate that these compounds exert their antifungal activity through inhibition of 2,3-oxidosqualene-lanosterol cyclase. Structure / activity studies within this series of compounds in general show a rough correlation between anti-

	IC_{50} (µM)	Average MIC (µg/ml)				
	(C. albicans)	C.albic. (16 strains)	C.spec. (10 strains)	Crypt. (8 strains)	Derm. (6 strains)	Dimorph. (6 strains)
ITRACONAZOLE	----	1.25	0.07	0.15	5.00	1.25
Ro-43-6913	0.66	0.60	0.11	0.35	0.30	0.30
Ro-46-6523	0.03	1.68	0.81	1.25	0.23	0.15
Ro-44-4082	0.11	1.25	0.37	3.40	1.24	0.44

Table 1 Antifungal activity of selected cyclase inhibitors in comparison to the marketed antifungal Itraconazole

fungal activity and enzyme inhibition, even if some aspects are not completely understood. However, it is quite clear that many aspects of the inhibition of the target enzyme remain to be clarified. Furthermore, it still remains unclear whether the antifungal activity of these aminoketones is due to the accumulation of 2,3-oxidosqualene or to a depletion in ergosterol in the fungal cells.

It would appear, that our approach could lead to interesting, structurally new and simple antifungals with a novel mode of action.

REFERENCES

1. D. Berg and M. Plempel, "Sterol Biosynthesis Inhibitors; Pharmaceutical and Agrochemical Aspects", Verlag Chemie, Weinheim, 1988.
2. L. Cattel, M. Ceruti, F. Viola, L. Delprino, G. Balliano, A. Balliano, A. Duriatti and P. Bouvier-Navé, Lipids, 1986, 21, 38.
3. M. Ceruti, G. Balliano, F. Viola, L. Cattel, N. Gerst and F. Schuber, Eur. J. Med. Chem. , 1987, 22, 199.
4. A. Duriatti and F. Schuber, Biochem.Biophys.Res.Commun., 1988, 151, 1378.
5. M. Kusano, I. Abe, U. Sankawa and Y. Ebizuka, Chem. Pharm.Bull., 1991, 39, 239.
6. T. Hoshino, H.J. Williams, Y. Chung and A.I. Scott, Tetrahed. 1991, 47, 5925.
7. E.J. Corey and S.P.T. Matsuda, J.Amer.Chem.Soc., 1991, 113, 8172.
8. R. Kelly, S.M. Miller, M.H. Lai and D.R. Kirsch, Gene, 1990, 87, 177.
9. E.I. Mercer, Lipids, 1991, 26, 584.
10. A.C. Oehlschlager and E. Czyzewska in "Emerging Targets in Antibacterial and Antifungal Chemotherapy", Chapman and Hall, New York, 1992, p.449.
11. C. Walsh, "Enzymatic Reaction Mechanism", W.H. Freeman and Co., San Francisco, 1979, p.625.
12. E.E. van Tamelen, J.Amer.Chem.Soc., 1982, 104, 6480.
13. A. Krief, J.-R. Schander, E. Guittet, C. Herve du Pentoat and J.-Y. Lallemand, J.Amer.Chem.Soc. 1987, 109, 7910.
14. A. Rahier, P. Bouvier, L. Cattel, A. Narula and P. Benveniste, Biochem. Soc. Trans., 1983, 11, 537.
15. M. Ceruti, F. Viola, F. Dosio, L. Cattel, P. Bouvier-Navé and P. Ugliengo, J. Chem. Soc. Perkin Trans. I, 1988, 461.
16. X.-Y. Xiao and G.D. Prestwich, J.Amer.Chem.Soc., 1991, 113, 9673.
17. S. Jolidon, A. Polak, P. Guerry and P.G. Hartman, Pestic.Sci., 1991, 31, 588.
18. S. Jolidon, A. Polak, P.G. Hartman and P. Guerry in "Proceedings of the Third International Symposium on Molecular Aspects of Chemotherapy, Gdansk, Poland, June 19-21, 1991", Springer Verlag, Berlin, 1992, p.143.

19. F. Hoffmann-La Roche and Co., Eur.Pat.Appl. No 401798, 12 Dec. 1990.
20. F. Hoffmann-La Roche and Co., Eur.Pat.Appl. No 410359, 30 Jan. 1991.
21. F. Hoffmann-La Roche and Co., Eur.Pat.Appl. No 464465, 8 Jan. 1992.
22. C.G. Swain and E.C. Lupton, J.Amer.Chem.Soc., 1968, 90, 4328.
23. M.S. Manhas, W.H. Hoffman, B. Lal and A.K. Bose, J.Chem.Soc. Perkin Trans. I, 1975, 461.
24. H. Andringa, J. Hanekamp and L. Brandsma, Synth. Commun., 1990, 20, 2349.
25. A.M. Echavarren and J.K. Stille, J.Amer.Chem.Soc., 1987, 109, 5478.
26. D.A. Widdowson and Y.-Z. Zhang, Tetrahed., 1986, 42, 2111.
27. R.G. Nadeau and R.P. Hanzlik, Methods Enzym., 1969, 15, 346.

16
Synthetic Studies on the Papulacandins

A. G. M. Barrett,* M. Peña, and J. A. Willardsen

DEPARTMENT OF CHEMISTRY, COLORADO STATE UNIVERSITY, FORT
COLLINS, CO 80523, USA

1 INTRODUCTION

There exist legions of bioactive natural products that contain spiroketal ring systems.[1] Examples include the avermectins, a class of potent insecticidal and anthelmintic agents,[2] and the calyculins, a group of highly selective phosphatase inhibitors.[3] Representative examples of these natural products are avermectin A_{2b} (1) and calyculin A (2). The papulacandins are a group of antifungal agents isolated from the fermentation broths of *Papularia sphaerosperma*[4] and *Dictyochaeta simplex*.[5] Structurally, these substances are derivatives of the 1,7-dioxaspiro[5.4]decane skeleton, a common spiroketal residue. These carbohydrate antifungal agents are exemplified by the papulacandins A (3) and D (4) and L-687,781(5). Chaetiacandin is a structurally related natural product from *Monochaetia dimorphospora* that lacks the spiroketal ring system.[6]

Several of the papulacandins show potent *in vitro* activities against various clinical isolates of *Candida albicans*. For example, L-687,781(5) has an MIC of 1µg/mL against *Candida albicans* MY1055. Other susceptible fungi include *C. tropicalis*, and *Microsporum canis*, etc. The mode of action of these substances has been shown to involve the inhibition of 1,3-β-D-glucan synthases. These enzymes are crucially important for the biosynthesis of various 1,3-β-D-glucans such as pachyman (6). These biopolymers are vital constituents of fungal cell walls and inhibition of their biosynthesis represents a viable strategy for the design of novel antifungal agents.[7] Such enzymes are important for the life cycle of *Pneumocystis carinii*, a major opportunistic infection of AIDS victims.

Recently, it has been shown that papulacandins are effective against *P. carinii* induced pneumonia.[5] In the rat acute PCP model, L-687,781(5) brought about 84% cyst reduction at a dosage of 5.0mg/kg. Ciba Geigy chemists have studied structure activity relationships on semi-synthetic papulacandins by changing arene ring substitution and by hydrogenating or hydrolysing the fatty acyl side chains. These results unequivocally demonstrate the importance of the long C-3' ester for activity.[4]

It is clear from these biological activity profiles that the papulacandins are important lead compounds for the design of new antifungal agents. As such, they are excellent targets for the development of simple synthetic methodology. None of the papulacandins have been prepared by total synthesis. Indeed there are still stereochemical ambiguities in all the structures. Neither the C-7 nor the C-14 stereocenters of the long fatty acyl side chain (see structure **3**) have been defined. In addition, the absolute stereochemistry of the galactose *O*-6 acyl fragment of L-687,781(**5**) and related papulacandins is also unknown. Several groups have reported the construction of the spirocyclic core of papulacandin D.[8,9,10,11,12] Three distinct strategies have been employed to elaborate such simple spiroketals from D-glucose precusors. Firstly, both Schmidt and Danishefsky have employed the homologation of appropriately protected D-glucose derivatives, oxidation, and spirocyclization to reveal the papulacandin D spirocycle.[8] Secondly, Beau and Friesen have used the palladium(0) catalyzed coupling of protected 1-tributylstannyl-glucal with various aryl bromides as the key process in constructing the skeleton of papulacandin D.[9] However, both of these approaches are multistep. There is a third strategy which is considerably more concise that has been independently reported by Bihovsky,[10] Czernecki,[11] and our own laboratories.[12] The process involves the condensation of protected derivatives of D-gluconic acid δ-lactone with aryllithium reagents and acid mediated spirocyclization. In contrast to this extensive work on the synthesis of the spirocyclic core of papulacandin D, only very limited studies on the synthesis of the fatty acyl side chains of the papulacandins have appeared.[12]

2 RESULTS AND DISCUSSION

Synthesis of the Papulacandin Acyl Side Chains

The synthesis of the diene acid residue **10** of papulacandin A (**3**) is summarized in Scheme 1. Thus, the *cis*-addition of di-isobutylaluminum hydride to 1-heptyne and iodine-metal exchange[13] gave the corresponding E-vinyl iodide **8** (53%).¶ This substance was smoothly condensed with methyl (E)-3-(tributylstannyl)acrylate,[14] under Stille conditions,[15] to provide the corresponding (E,E)-dienoate **9** (60%). Finally saponification gave the corresponding crystalline (E,E)-dienoic

¶ All new compounds were fully characterized by spectral data and microanalyses or HRMS.

acid **10** (67%) (mp 46 - 47 °C). This substance (mp 49 - 50 °C) has previously been prepared by alternative methods.[16]

In addition to the elaboration of acid **10**, we have also undertaken the synthesis of the more complex, long fatty ester **24**. There is an added complication in this venture: neither the absolute stereochemistry of the C-7 nor the C-14 centers in this unit is known. However, it is reasonable to speculate that L-iso-leucine (**11**) may be the biosynthetic precursor[†][17] for the terminal $EtCH(Me)CH_2$ unit. Thus, arbitrarily, the synthesis of

Scheme 1

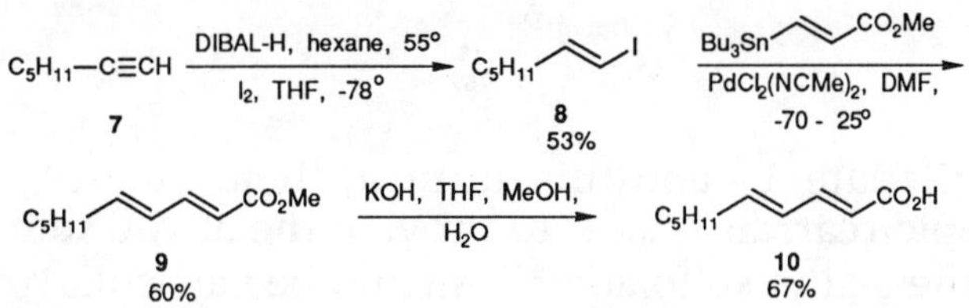

ester **24** was started from L-iso-leucine (**11**). Our synthetic endeavours are summarized in Schemes 2 - 4. Diazotization of **11** using sodium nitrite and hydrochloric acid[18] gave the known α-chloro-acid.[18] This substance was formed as the (2S 3S)-diastereoisomer since the substitution proceeded via the α-lactone and the well known double inversion process. Chloro-acid **12** was converted into the corresponding bromide **13** via lithium aluminum hydride reduction, toluene-4-sulfonylation, and displacement using lithium bromide.

Bromide **13** was converted into the unsaturated aldehyde **15** using transition metal mediated carbonylation and vinylsilane chemistry. Thus, the bromide **13** was homologated to produce (4S)-methylhexanal via Grignard reagent formation and carbonylation using (pentacarbonyl)iron (0).[19] Subsequent addition of (E)-2-(trimethylsilyl)vinyllithium[20] gave the corresponding allylic alcohol **14**. Reaction of alcohol **14** with benzenesulfenyl chloride gave, on silver nitrate work up, the corresponding E-enal **15** (76%) and this substance was formed exclusively as the E-geometric isomer. The proposed mechanism for this transformation is summarized in Scheme 3. Thus alcohol **14** is presumably initially converted into the

---

† L-isoleucine is a common biosynthetic precursor of *sec*-butyl containing natural products; for example see reference 17.

Scheme 2

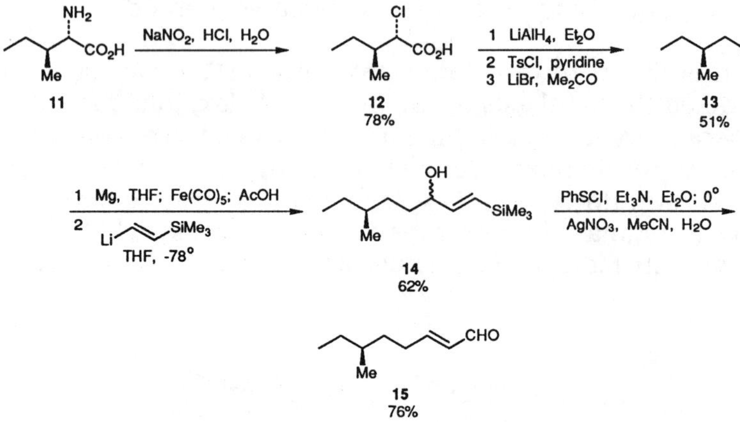

allylic sulfenate **16** and this, in turn, should undergo a [2,3]-sigmatropic rearrangement to provide the allylic sulfoxide **17**. Finally, the α-silyl sulfoxide **17**, should be particularly prone to undergo a silyl-Pummerer rearrangement to provide the acetal **18**. Work up, with silver nitrate, should result in hydrolysis and thus provide the enal **15**. Our strategy for the conversion of the alcohol **14** into the enal **15** is an adaption of some very elegant methodology originally reported by Parsons.[21]

Scheme 3

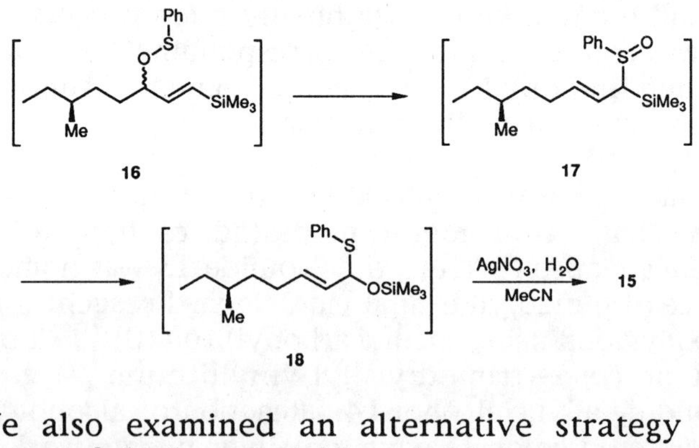

We also examined an alternative strategy for the elaboration of the key enal **15** and this is summarized at the start of Scheme 4. Nitrile **19** was readily prepared in good overall yield (65%) from the chloro-acid **12** via lithium aluminum hydride reduction, toluene-4-sulfonylation and potassium cyanide displacement in the presence of 18-crown-6. Subsequent di-isobutylaluminum hydride reduction of nitrile

19 and direct condensation with ethyl (triphenyl-phosphorylidene)acetate gave the α, β-unsaturated ester **20** (85%). As expected, this Wittig reaction with the stabilized ester ylide proceeded with excellent geometric selectivity (E : Z > 95:5). Di-isobutyl-aluminum hydride reduction of ester **20** and subsequent pyridinium chlorochromate oxidation in the presence of silica[22] gave the enal **15**. A second Wittig reaction of **15** and ethyl (triphenylphosphorylidene)propanoate gave the dienoate ester **21** (88%). Again, as expected, the geometric selectivity in this process was excellent.

Scheme 4

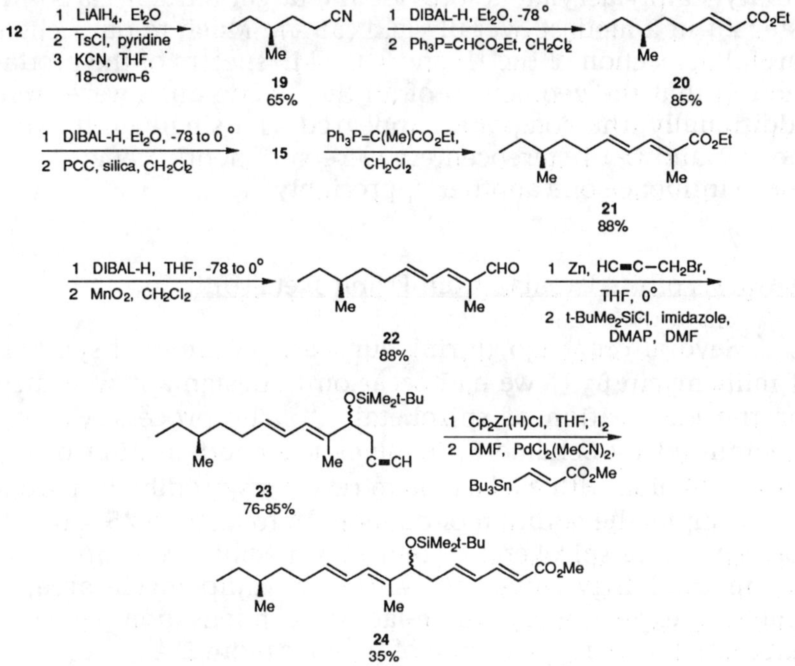

Di-isobutylaluminum hydride reduction of the ester **21** and subsequent oxidation of the resultant allylic alcohol over manganese dioxide gave the delicate (E, E)-dienal **22**. This substance was smoothly propargylated using 3-bromopropyne and zinc[23] to provide, on protection,[24] the silyl ether **23** (76 - 85%). Clearly, in the process, **23** was produced as a mixture of diastereoisomers and, in the subsequent chemistry, the (7-R, S)-mixture was taken on without separation. It is germane, at th-is point, to briefly comment upon the synthesis of the separate (R, S)- and (S, S)- diastereoisomers corresponding to silyl ether **23**. Firstly, the enantioselective synthesis of either diastereoisomer of the ether **23**, using chiral propargylating reagents,[25] proceeded with only modest diastereoselectivities. However, the two diastereoisomers of **23** were obtained by esterification of the corresponding alcohol with (S)-*O*-methylmandelic acid[26], chromatographic separation, ester hydrolysis and silylation

   The synthesis of the tetraene ester **24** was completed with the mixture of C-7 diastereoisomers (papulacandin numbering). Hydrozirconation[27] of the acetylene **23** and iodine quenching gave the corresponding *trans*-vinyl iodide. Without isolation, this was Stille coupled[15] with methyl 3E-tributylstannylacrylate to provide the target tetraenecarboxylic ester **24** in a modest overall yield (35%). Much to our delight, careful inspection of the ^{1}H and ^{13}C NMR spectra of the product showed that the geometries of all the alkene units were *trans*. Additionally, the compound appeared as a single entity since the C-7 and C-14 stereocenters were sufficiently remote so as not to influence one another appreciably.

Synthesis of the Papulacandin D Spiroketal Unit

   Several years ago, during our work on the total synthesis of milbemycin β_3,[28] we had occasion to design a new strategy for the elaboration of spiroketals.[29] The process, which is exemplified by Scheme 5, involves the condensation of a β-ketone dianion with a δ-lactone to reveal a spirodihydropyrone. For example, the addition of dianion **26** to lactone **25** gave the corresponding spiroketal system **27** on acidic work up. Since this methodology served us well in the milbemycin area, we sought to explore a carbanion-lactone condensation process to elaborate the spirocyclic core of papulacandin D.[12]

Scheme 5

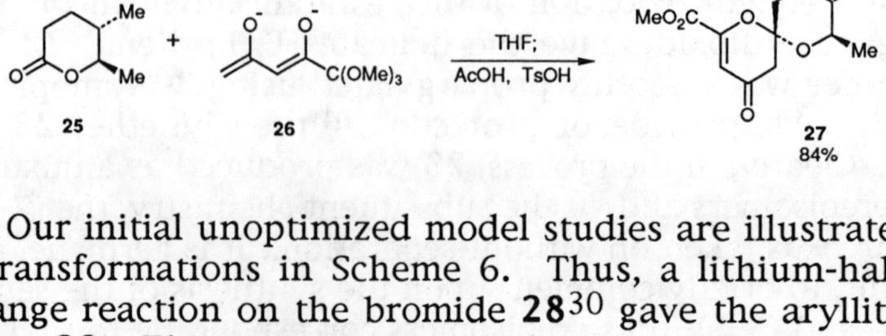

   Our initial unoptimized model studies are illustrated by the transformations in Scheme 6. Thus, a lithium-halogen exchange reaction on the bromide **28**[30] gave the aryllithium reagent **29**. This reactive intermediate was allowed to react

with δ-valerolactone to provide the lactol **30**. Subsequent acidification using hydrogen fluoride in pyridine resulted in desilylation[24] and spirocyclization to reveal the simple spiroketal **31** (67%).

Scheme 6

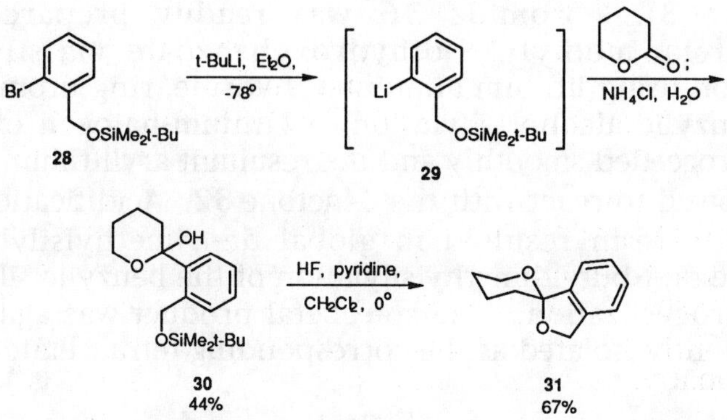

In order to employ this chemistry for papulacandin D (**4**), it is necessary to carefully choose appropriate protection for D-gluconic acid δ-lactone. In contrast to benzyl and benzylidene protected derivatives,[10,11] 2,3,4,6-tetrakis-O-trimethylsilyl-D-gluconic acid δlactone (**32**)[31] is directly available in one step

Scheme 7

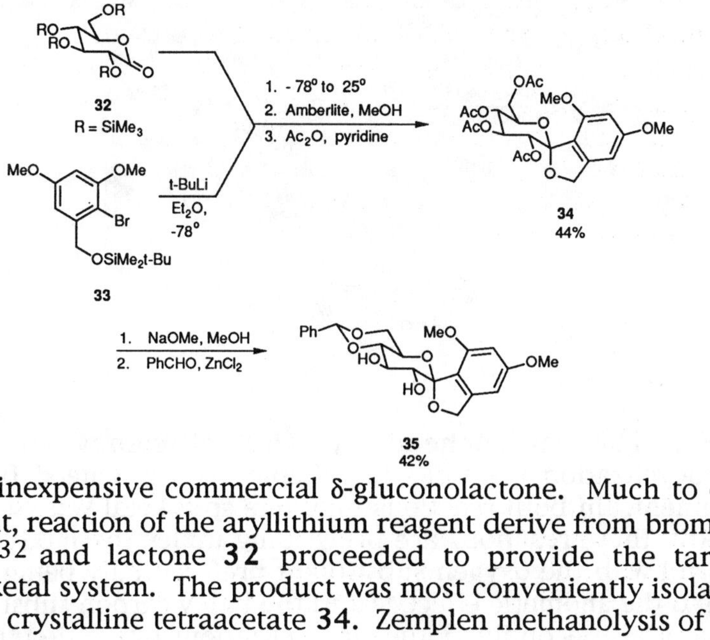

from inexpensive commercial δ-gluconolactone. Much to our delight, reaction of the aryllithium reagent derive from bromide **33**[24,32] and lactone **32** proceeded to provide the target spiroketal system. The product was most conveniently isolated as the crystalline tetraacetate **34**. Zemplen methanolysis of the tetraester **34** and subsequent condensation of the resultant tetraol with benzaldehyde[24] gave the benzylide protected diol **35** (42% unoptimized).

The aryllithium condensation strategy was successfully extended to the corresponding papulacandin D spiroketal **38** (Scheme 8). Bromide **36** was readily prepared from commercial methyl 3,5-dihydroxybenzoate via silylation, reduction using lithium aluminum hydride, ring bromination, and benzylic alcohol silylation.[24] Lithium-halogen exchange again proceeded smoothly and the resultant aryllithium reagent was allowed to react with the δ-lactone **32**. Acidification using Amberlite resin resulted in global de-trimethylsilylation,[24] selective de-t-butyldimethylsilylation of the benzylic silyl ether and spirocyclization. The spiroketal product was again most conveniently isolated as the corresponding tetraacetate **37** (38-

Scheme 8

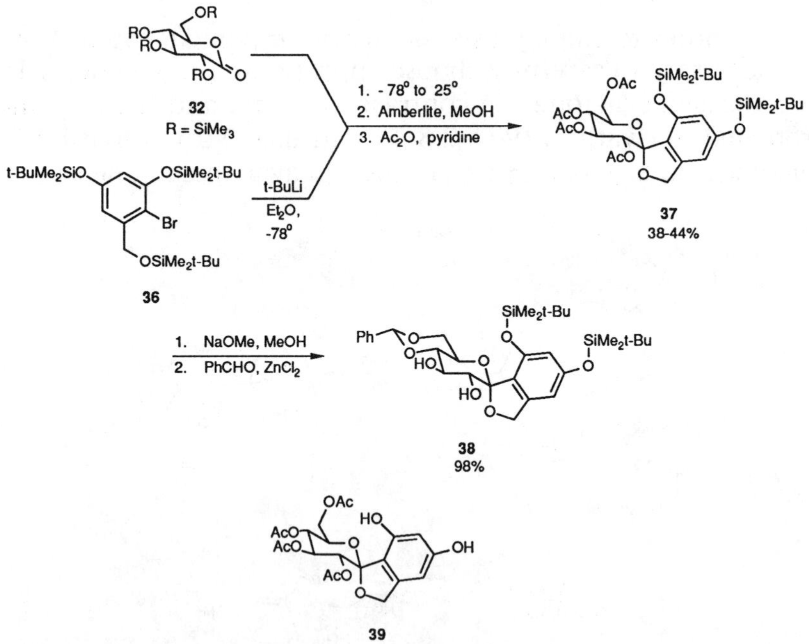

44%). The stereochemistry and efficiency of both spirocyclization reactions in Schemes 7 and 8 need further comment. In both reactions only one spiroketal was isolated. Clearly these reactions are proceeding under thermodynamic control with the oxygen substituent preferentially being axial, due to the anomeric effect,[33] and the bulky carbon substituent equatorial. Secondly, both spirocyclizations proceeded in only modest overall yields. However, this inefficiency is more than offset by the simplicity of spiroannulation using the trimethylsilyl protected lactone **32**. The reactions in both Schemes 7 and 8 scale well to multigram batches.

In exactly the same way as in Scheme 7, spiroketal **37** was converted into the benzylidene protected diol **38**. In this case, the yield of reaction was optimized to 98%. We have briefly examined the ease of desilylation of the ester-silyl ether **37**. Reaction of ether **37** with tetrabutylammonium fluoride[24] in THF rapidly gave the highly crystalline diphenol **39** (~100%).

We next turned our attention to the problem of selective C-3 monoesterification. In a model study, reaction of the diol **35** with cinnamyl chloride (**40**) in the presence of triethylamine and 4-(dimethylamino)pyridine gave an inseparable mixture of *O*-3 and *O*-2 esters (50%, 4:1). Subsequent hydrolysis of the benzylidene protecting group[24] by reaction with acid Dowex resin in methanol and chromatography gave the isomerically pure crystalline monoester **41**. In these reactions, the *O*-3 and *O*-2 esters were readily distinguished by ^{1}H NMR spectroscopy. The major *O*-3 ester **41** showed *inter alia* δ (CD$_3$OD) 4.44 (d, 1H, J = 10 Hz, C$_2$-*H*) and 5.47 (app t, 1H, J = 9.6 Hz, C$_3$-*H*). In contrast, the minor crude *O*-2 isomer showed *inter alia* δ (CD$_3$OD) 4.05 (app t, 1H, J = 9.3 Hz, C$_3$-*H*) and 5.80 (d, 1H, J = 10 Hz, C$_2$-*H*).

Scheme 9

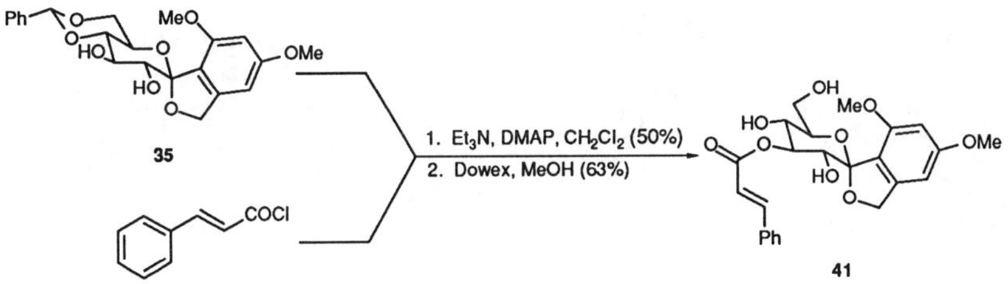

We were suitably encouraged by these results and directly examined the esterification of diol **38** with the fatty acid derived from ester **24**. Saponification of the (7RS, 14S-) ester **24** using potassium hydroxide in aqueous ethanol proceeded smoothly and without elimination. Careful acidification using citric acid gave the corresponding unstable carboxylic acid. Without purification, this was converted into the corresponding mixed anhydride **42** by reaction with 2,4,6-

trichlorobenzoyl chloride and triethylamine in dichloromethane solution. Again, no attempt was made to purify and isolate this delicate substance. Yamaguchi has previously reported the use of mixed anhydrides derived from 2,4,6-trichlorobenzoyl chloride for the synthesis of esters, lactones and thiol esters.[34] The anhydride 42 and 4-(dimethylamino)pyridine were added sequentially to the diol 38 in dichloromethane. Work up gave the O-3-monoester contaminated by the corresponding O-2-monoester (70%, ~3:1). Unfortunately, these substances could not be easily separated at this point. However, reaction of the mixture of adducts with tetrabutylammonium fluoride in THF resulted in clean desilylation to provide the benzylidene derivative (64%). Subsequent treatment of this compound with acid Dowex resin was a reaction of mixed blessings. Unfortunately, this deprotection step proceeded in only low yield. However, chromatography gave a compound (10%) that was identical by TLC and ^{1}H NMR spectroscopy with an authentic sample of papulacandin D (4). In addition, the sample showed a ^{13}C NMR spectrum in excellent agreement with published data.[35] There is, however, a problem in this rosy scenario. The synthetic material 43, which was homogeneous by chromatography, certainly consisted of two side chain C-7

Scheme 10

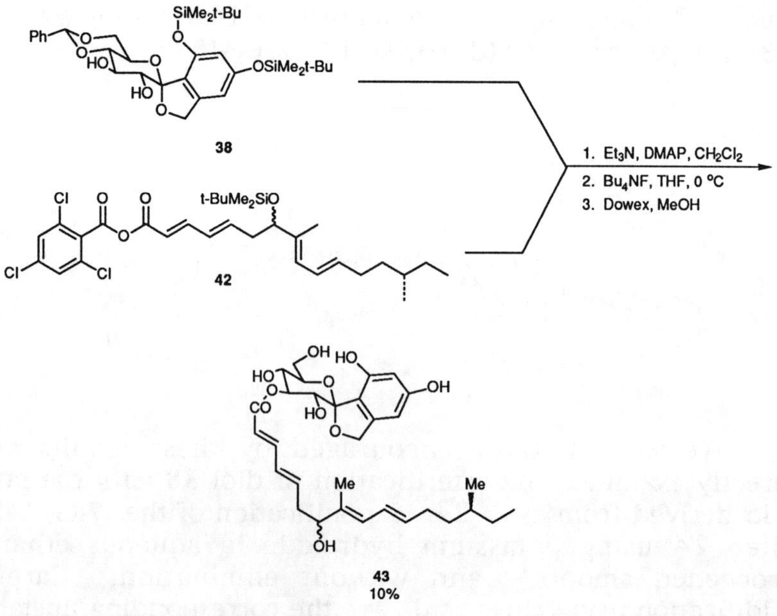

epimers. The identity of this material with natural papulacandin D (4) is both gratifying and troubling. It is not yet clear if we have been successful in preparing the natural product admixed with its C-7 epimer or some other side chain diastereoisomer. We are currently seeking to confirm the identity of the fatty acid C-7 and C-14 stereochemistries by degradation. Once this information is available, we shall adapt Scheme 10 to provide the actual natural product stereoisomer.

Conclusion

It is clear that the strategy outlined in this paper is very appropriate for the synthesis of the papulacandins. The methods should additionally be valuable for the synthesis of various analogues for biological evaluation. The selective *O*-3 esterification reaction in Scheme 10 needs to be optimized by variation in protecting group methodology. Additionally, the fatty acid stereochemistry needs to be first established by degradation before a total synthesis of papulacandin D can be claimed.

Acknowledgment

We thank the National Institutes of Health for support (AI-22252), the Ciba-Geigy company for providing authentic samples of papulacandins A and D, Professor Jeffrey M. Becker at the University of Tennesse for providing compound testing, and Dr. Jason M. Hill for preparing several intermediates on a large scale.

References

1.   F. Perron and K.F. Albizati, <u>Chem. Rev.</u>, 1989, <u>89</u>, 1617.

2.   H.G. Davies and R.H. Green, <u>Chem. Soc. Rev.</u>, 1991, <u>20</u>, 271 and   references therein.

3.   S. Matsunaga, H. Fujiki, D. Sakata and N. Fusetani, <u>Tetrahedron</u>, 1991, <u>47</u>, 2999 and references therein.

4.   P. Traxter, W. Tosch and O. Zak, <u>J. Antibiotics</u>, 1987, <u>40</u>, 1146 and references therein.

5.   F. VanMiddlesworth, M.N. Omstead, D. Schmatz, K. Bartizal, R. Fromtling, G. Bills, K. Nollstadt, S. Honeycutt, M. Zweerink, G. Garrity and K. Wilson, <u>J. Antibiotics</u>, 1991, <u>44</u>, 45.

6.   T. Komori and Y. Itoh, <u>J. Antibiotics</u>, 1985, <u>38</u>, 544.

7.   G. Römmele, P. Traxter, and W. Wehrli, <u>J. Antibiotics</u>, 1983, <u>36</u>, 1539.

8.   S. Danishefsky, G. Phillips and M. Ciufolini, <u>Carbohydrate Res.</u>, 1987, <u>171</u>, 317. R.R. Schmidt and W. Frick, <u>Tetrahedron</u>, 1988, <u>44</u>, 7163.

9.   R.W. Friesen and A.K Daljeet, <u>Tetrahedron Lett.</u>, 1990, <u>31</u>, 6133. E. Dubois and J.-M. Beau, <u>Ibid.</u>, 1990, <u>31</u>, 5165. R.W. Friesen and C.F. Sturino, <u>J. Org. Chem.</u>, 1990, <u>55</u>, 5808. E. Dubois and J.-M. Beau, <u>Carbohydrate Res.</u>, 1992, <u>223</u>, 157.

10.   S.B. Rosenblum and R. Bihovsky, <u>J. Am. Chem. Soc.</u>, 1990, <u>112</u>, 2746.

11.   S. Czernecki and M.- C. Perlat, <u>J. Org. Chem.</u>, 1991, <u>56</u>, 6289.

12.   A.G.M. Barrett, D. Dhanak, S.A. Lebold, M. Peña, and D. Pilipauskas, <u>Pestic. Sci.</u>, 1991, <u>31</u>, 581.

13.   G. Zweifel and C.C. Whitney, <u>J. Am. Chem. Soc.</u>, 1967, <u>89</u>, 2753.

14.   A.J. Leusink, H.A. Budding, and J.W. Marsman, <u>J. Organomet. Chem.</u>, 1967, <u>9</u>, 285.

15.   J.K. Stille, <u>Angew. Chem. Internat. Ed. Engl.</u>, 1986, <u>25</u>, 508.

16.   G. Rickards and L. Weiler, <u>J. Org. Chem.</u>, 1978, <u>43</u>, 3607 and references therein.

17.   D.E. Cane, T.-C. Liang, L. Kaplan, M.K. Nallin, M.D. Schulman, O.D. Hensens, A.W. Douglas, and G. Albers-Schönberg, <u>J. Am. Chem. Soc.</u>, 1983, <u>105</u>, 4110.

18.   V. Schurig, U. Leyrer, and D. Wistuba, <u>J. Org. Chem.</u>, 1986, <u>51</u>, 242.

19.   M. Ryang, I. Rhee, and S. Tsutsumi, <u>Bull. Chem. Soc. Jpn.</u>, 1964, <u>37</u>, 341.

20.   R.F. Cunico and F.J. Clayton, <u>J. Org. Chem.</u>, 1976, <u>41</u>, 1480.

21. I. Cutting, and P.J. Parsons, <u>Tetrahedron Lett.</u>, 1981, <u>22</u>, 2021.

22. G. Piancatelli, A. Scettri, and M. D'Auria, <u>Synthesis</u>, 1982, 245 and references therein.

23. R. Baker and M.A. Brimble, <u>J. Chem. Soc., Chem. Commun.</u>, 1985, 78.

24. T.W. Greene, "<u>Protective Groups in Organic Synthesis,</u>" J. Wiley & Sons, New York, 1981. H. Kunz, and H. Waldmann, "Protecting Groups" in "<u>Comprehensive Organic Synthesis,</u>" Eds. B.M. Trost, I. Fleming, E. Winterfeldt, Pergamon Press, Oxford, 1991, <u>vol. 6</u>, p. 631-701.

25 E.J. Corey, C.-M. Yu and D.-H. Lee, <u>J. Am. Chem. Soc.</u>, 1990, <u>112</u>, 878 and references therein.

26. B.M. Trost, J.L Belletire, S. Godleski, P.G. McDougal, J.M. Balkovec, J.J. Baldwin, M.E. Christy, G.S. Ponticello, S.L. Varga and J.P. Springer, <u>J. Org. Chem.</u>, 1986, <u>51</u>, 2370.

27. J.S. Temple, M. Riediker and J. Schwartz, <u>J. Am. Chem. Soc.</u>, 1982, <u>104</u>, 1310 and references therein.

28. S.V. Attwood, A.G.M. Barrett, R.A.E. Carr, G. Richardson and N.D.A. Walshe, <u>J. Org. Chem.</u>, 1986, <u>51</u>, 4840.

29. S.V. Attwood, A.G.M. Barrett and J.-C. Florent, <u>J. Chem. Soc., Chem. Commun.</u>, 1981, 556. A.G.M. Barrett, R.A.E. Carr, M.A.W. Finch, J.-C. Florent and N.D.A. Walshe, <u>J. Org. Chem.</u>, 1986, <u>51</u>, 4254.

30. D.R. Buckle, A.E. Fenwick, D.J. Outred and C.J.M. Rockwell, <u>J. Chem. Research</u>, 1987, 3144.

31. D. Horton and W. Priebe, <u>Carbohydrate Research</u>, 1981, <u>94</u>, 27.

32. P.D. Noire and R.W. Franck, <u>Synthesis</u>, 1980, 882.

33. A.J. Kirby, "<u>The Anomeric Effect and Related Stereoelectronic Effects at Oxygen</u>", Springer-Verlag: New York, 1983. P. Deslongchamps, "<u>Stereoelectronic Effects in Organic Chemistry</u>," Pergamon: Toronto, 1983; Chapter 2.

34.   Y. Kawanami, Y. Dainobu, J. Inaga, T. Katsuki and M. Yamaguchi, Bull. Chem. Soc. Japan, 1982, 54, 943 and references therein.

35.   P. Traxler, H. Fritz, H. Fuhrer and W.J. Richter, J. Antibiotics, 1980, 33, 967.

Chemistry of Antivirals and Antiparasitics

17

Carbocyclic Nucleoside Analogues as Antiviral Agents

Richard Storer,[1,*] Anthony D. Baxter,[1] Richard E. Boehme,[2] Ian R. Clemens,[1] Graham J. Hart,[2] Martin F. Jones,[1] Clara L. P. Marr,[2] Andrew M. Mason,[1] Chi-Leung Mo,[1] Stewart A. Noble,[1] Ian L. Paternoster,[1] Colin Robertson,[1] D. Michael Ryan,[3] Michael A. Sowa,[3] Niall G. Weir,[1] and Christopher Williamson[1]

[1]DEPARTMENT OF MEDICINAL CHEMISTRY
[2]DEPARTMENT OF VIROLOGY
[3]DEPARTMENT OF CHEMOTHERAPY
GLAXO GROUP RESEARCH LTD, GREENFORD ROAD, GREENFORD, MIDDLESEX UB6 0HE, UK

During the last twenty years a great deal of effort has been devoted to the synthesis and biological evaluation of nucleoside analogues in the search for effective agents for the treatment of diseases resulting from viral infection. Herpes simplex viruses type 1 (responsible for oral herpes infection) and type 2 (responsible for genital herpes infection) have been major targets. Of the other Herpes viruses, there is a particular need for an effective agent for the treatment of infection resulting from Varicella Zoster Virus which is the cause of chicken pox as the primary infection and shingles as the recurrent infection. Cytomegalovirus also represents an expanding market due to increasing incidence as an opportunistic infection in AIDS patients. In the area of respiratory viruses, the major targets have been Influenza A and B although others such as Respiratory Syncytial Virus (RSV) represent significant causes of illness. The discovery of Human Immunodeficiency Viruses as the probable cause of AIDS prompted renewed effort in the area of nucleoside chemistry in the last eight years and for the future, Hepatitis B may provide the next major target. Although much current effort is being devoted to non-nucleoside compounds directed at a variety of molecular targets, to date it is the nucleoside analogues as a class which have provided the compounds of most potential as antiviral agents.

Two nucleoside analogues dominate the antiviral market at present. These are the anti-herpetic agent Acyclovir (ACV) and Retrovir (AZT) for the treatment of AIDS. Among the newer generation of nucleoside analogues which are reaching the market, Ganciclovir (DHPG) is licensed for the treatment of CMV in immunocompromised patients and Ribavirin is used primarily for the treatment of RSV infection in hospitalized infants. Recently ddI and ddC have received limited approval for use in the treatment of HIV infection: the former for use in AZT-intolerant patients and the latter for combination therapy with AZT.

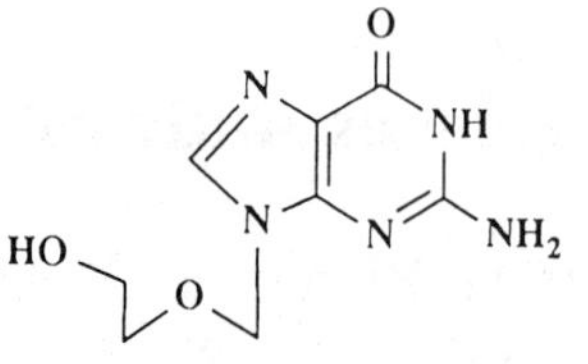

ACV

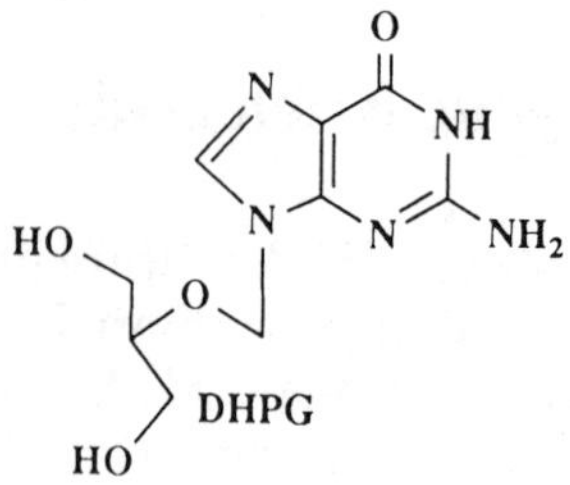

DHPG

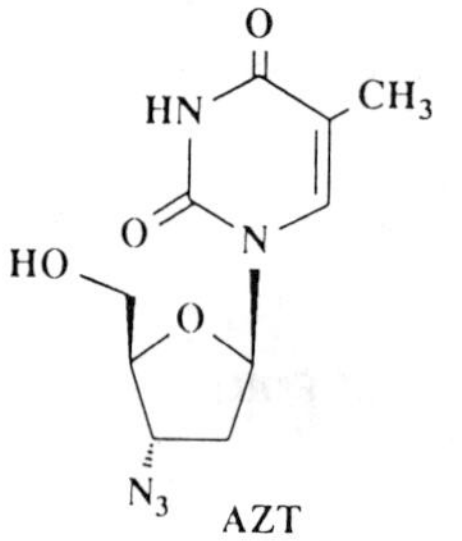

AZT

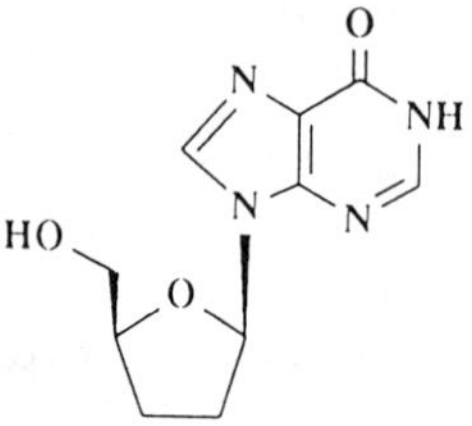

ddI

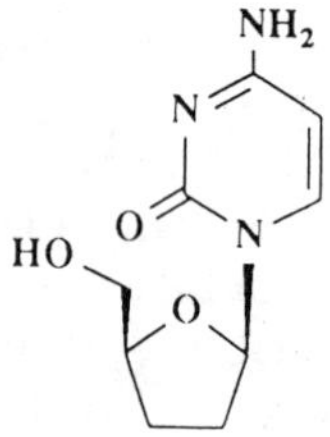

ddC

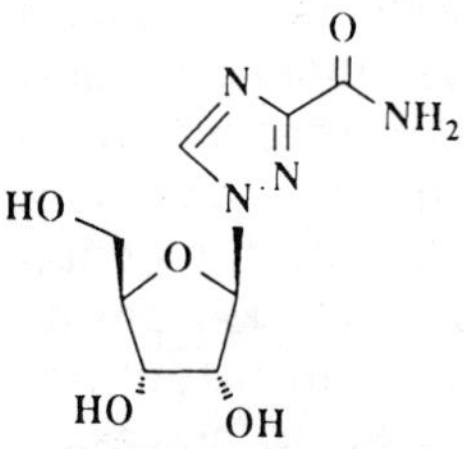

Ribavirin

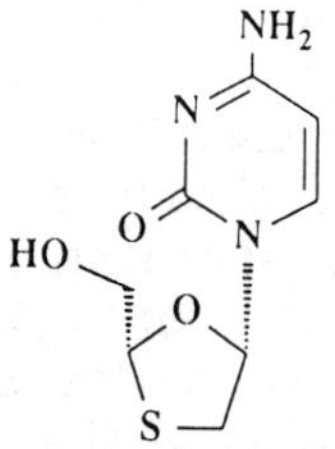

3TC

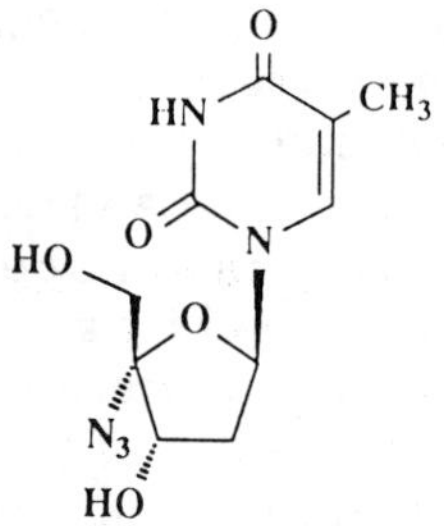

ADRT

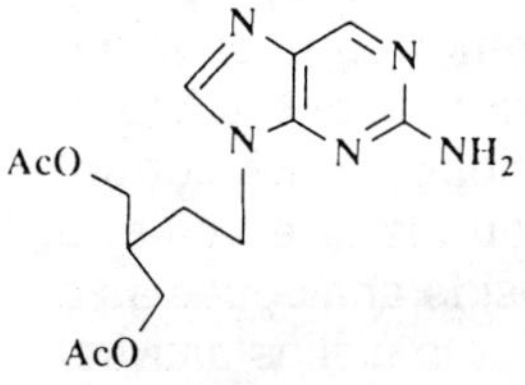

Famciclovir

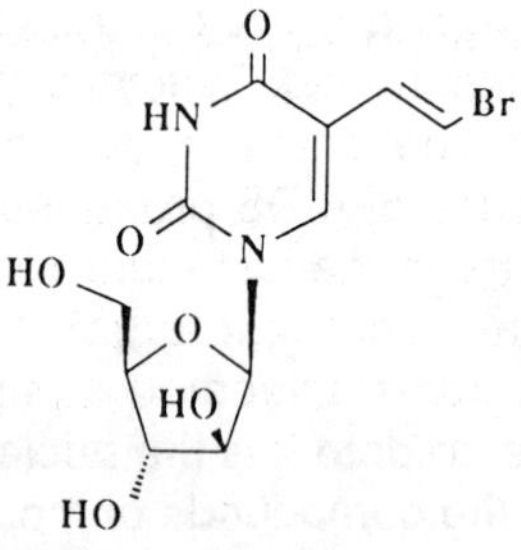

BVaraU

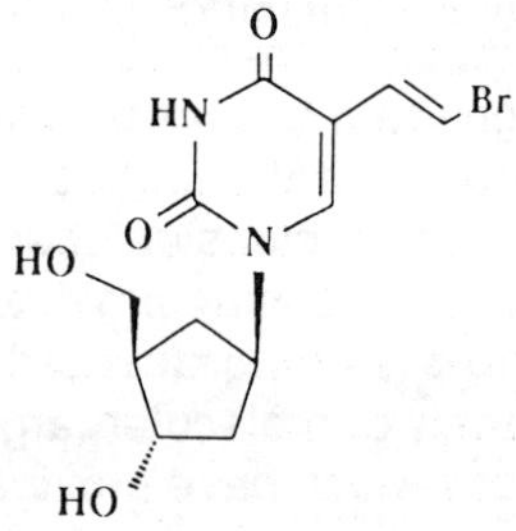

(+)-c-BVdU

The small number of licensed compounds is testament to the difficulty of finding suitably effective agents which display the required degree of selectivity and represents a poor reward for the amount of effort which has been devoted to this area worldwide.

A variety of other nucleoside analogues are currently in various stages of development. In the HIV area, 3TC is being developed by Glaxo/Biochem Pharma and ADRT represents a different structural class being evaluated by Syntex. In the Herpes area, Famciclovir is being developed by Smithkline-Beecham as a possible replacement for ACV and two compounds are being actively developed for the VZV market: BVaraU by Bristol-Myers/Squibb and carbocyclic BVdU by Glaxo. The latter is an example of a carbocyclic nucleoside analogue i.e. where the ribose oxygen has been replaced by a methylene group. Compounds of this type[1,2] have been much studied during the last decade by a number of groups worldwide and have been a particular interest to Glaxo.

Work on carbocyclic nucleoside analogues was first reported in 1966 when Shealy and Clayton described[3] a synthesis of the racemic carbocyclic analogue of adenosine. Glaxo interest[4,5,6] in the area was established as a result of our synthesis of the racemic carbocyclic analogue of BVdU and the demonstration that the compound had similarly potent activity against HSV-1 *in vitro* and *in vivo* and against VZV *in vitro* to the parent sugar BVdU.[7] It was anticipated that the advantage of such carbocyclic analogues of nucleosides would lie in the improved metabolic stability to the phosphorylase enzymes which cleave the glycosidic linkage in normal nucleoside analogues such as BVdU. The carbocyclic analogue of BVdU proved to be a compound of considerable interest to other groups in addition to ourselves and it was Griengl et al[8] who first reported that the activity resided primarily in the "natural" enantiomer i.e. that having the same absolute stereochemistry as the natural nucleosides. In the case of nucleoside analogues prepared from natural nucleosides or sugars, the analogues were generally prepared and tested in the single optical form dictated by the stereochemistry of the starting material. In the carbocyclic series however, compounds had generally been made and tested initially in racemic form and the observed difference in potency of the enantiomers of c-BVdU highlighted the need for obtaining chiral material. Our initial approach to the synthesis of the "natural" (+)-enantiomer is summarized in Scheme 1[9] and employs methodology previously described[10] for the synthesis of related compounds. The route allows chirality to be introduced at the beginning and also benefits from direct introduction of the uracil moiety prior to construction of the bromovinyl substituent.

Studies including these described herein with the chiral carbocycle have provided evidence for the suitability of this compound as a drug candidate. The improvement in stability of the carbocyclic analogue relative to BVdU was evidenced by the data shown in table 1. Under conditions where BVdU itself was rapidly degraded and where even BVaraU showed some breakdown, the carbocyclic analogue was found to be completely stable to phosphorolysis by intact human platelets. The level and profile of activity of the carbocycle *in vitro* (table 2) was found to parallel that of the sugar unlike many carbocyclic pyrimidine nucleoside analogues which have been found to show reduced potency compared to their sugar counterpart. Support for the theory that the profile of activity results from the pattern of phosphorylation was provided

Scheme 1

Synthesis of (+)-Carbocyclic BVdU

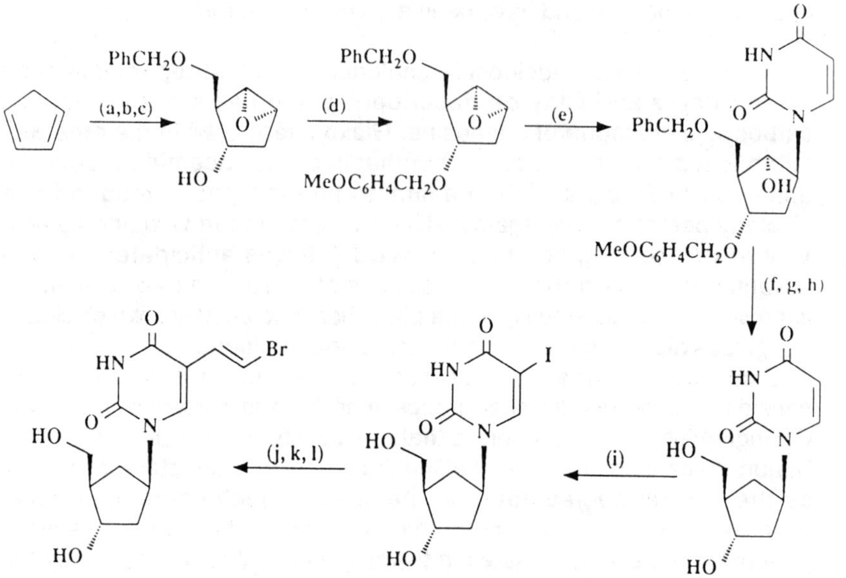

a) benzyl chloromethyl ether b) (-)-diisopinocampheylborane/H₂O₂
c) BuᵗOOH/VO(acac)₂ 50% d) p-MeO-benzyl chloride/NaH 100%
e) uracil/NaH 40% f) PhOCSCl/DMAP g) AIBN - Bu₃SnH 70%
h) H₂Pd-C 86% i) I₂/HNO₃ 100% j) Pd(OAc)₂/Ph₃P/CH₂=CHCO₂CH₃/Et₃N 87%
k) 2M-NaOH 77% l) NBS/KHCO₃ 85%

by the study detailing inhibition of the various DNA polymerases by the triphosphate of (+)-c-BVdU (table 3). The lack of activity against HSV-2 is consistent with lack of phosphorylation in HSV-2 infected cells which in turn results from the absence of a virally-encoded thymidylate kinase. The activity of the compound against HSV-1 *in vivo* in the mouse model where it shows similar potency to BVdU and only a slight reduction in potency compared with ACV anticipates good activity against VZV *in vivo*. Although there is no rodent model for VZV infection, the simian virus provides a good model for the human virus and it has now been shown that (+)-c-BVdU shows good activity against the simian virus both *in vitro* and *in vivo*. The compound represents an excellent drug candidate for VZV infection and its development is being actively pursued.

Table 1

PHOSPHOROLYSIS OF NUCLEOSIDES BY INTACT HUMAN PLATELETS

COMPOUND	10MIN	1HR	2HR	4HR	17HR	20HR	26HR	45HR	93HR
				INCUBATION TIME					
Thymidine	81.3		7.8			0			
Deoxyuridine	39.1		16.9						
BVdU	25.4	22.0		4.5		0			
BV*ara*U				100	99.4	98.8	98.3		
(+)-c-BVdU							100	100	100

Data are the percent of compound remaining as nucleosides after the indicated incubation times.

Table 2

IN VITRO ACTIVITY OF (+)-c-BVdU

COMPOUND	VZV (Ellen)	VZV (Webster)	HSV-1 (KOS)	HSV-2 (186)
	PD_{50} (μg/ml)			
(+)-c-BVdU	0.0033	0.005	0.044	20
BVdU	0.006	0.0029	0.03	3.9
BV*ara*U	0.0003	0.0002	0.07	>30
Acyclovir	6.3	7.8	0.22	1.5

(+)-c-BVdU is about 1000-fold more active than acyclovir against VZV. Similar results were obtained with other lab strains and a number of clinical isolates (data not shown). Against HSV-1, (+)-c-BVdU is about 5-fold more active than acyclovir *in vitro*, but is only weakly active against HSV-2. (+)-c-BVdU is inactive against CMV.

Table 3

INHIBITION OF DNA POLYMERASES BY (+)-c-BVdU TRIPHOSPHATE

ENZYME	K_m for TTP (μM)	I_{50} at [S]=K_m (μM)
HSV-1 DNA Polymerase	0.2	1.4
	0.2	2.5
HSV-2 DNA Polymerase	0.24	0.93
	0.25	0.55
VZV DNA Polymerase	0.25	2.2
	1.5[a]	0.27[a]
HeLa DNA Polymerase α	5.3	>1000[b]
	4.1	>1000[c]
L1210 Mitochondrial DNA Polymerase	2.1	>1000[d]
	1.8	
HeLa DNA Polymerase γ	0.15	37
	0.16	16

Activated calf thymus DNA was used as template except [a]poly dA oligo T_{12-18}
[b]1.04 mM gives 29% inhibition
[c]1 mM give 34% inhibition
[d]1 mM gives 28% inhibition
K_i's were not calculated due to non-linear Dixon Plots

A major objective of our carbocyclic nucleoside programme at Glaxo was to identify compounds having a spectrum of activity which included HSV-2 ideally in addition to HSV-1 and VZV. Our work leading to the discovery and establishment of the profile of activity of carbocyclic 2′ ara-fluoroguanosine (c-AFG) has been described in the literature.[11,12] Racemic c-AFG showed potent activity *in vivo* although the compound proved to be unacceptably toxic in repeat dose studies. The demonstration that the activity resided primarily in the "natural" enantiomer following enzyme-mediated resolution prompted us to devise synthetic approaches to the required optical form. (-)-Aristeromycin, readily available as a fermentation product of *Streptomyces citricolor*, provided a suitable chiral starting material and the required synthesis was readily accomplished[13] employing the adenine to guanine base interconversion initially worked out for the synthesis[14] of (-)-carbocyclic guanosine (c-G) from Aristeromycin. Unfortunately, both activity and toxicity present in the racemic form of c-AFG were found to be due to the same "natural" enantiomer and development of the compound was discontinued.

Until this time, most of the efforts in this area at Glaxo had concentrated on modifications at the 2' and 6' positions in the carbocyclic ring. Work in the carbocyclic series was seen to offer additional possibilities for further substitution compared with the normal sugar analogues and this was used to advantage in the preparation of 4'-substituted compounds concentrating initially on the introduction of hydroxyl and fluoro substituents. Some of the initial work directed to the synthesis of racemic 4'-substituted compounds has been described in the literature.[15] Several analogues having fluoro or hydroxyl substituents at the 4'-position and hydrogen or a 2'-*ara*-fluoro substituent at the 2'-position were found to have activity *in vitro* against both HSV-1 and HSV-2 comparable to that shown by ACV. Compounds having fluorine at the 4'-position, particularly where a 2'-*ara*-fluoro substituent was also present, were found to be more cytotoxic than the corresponding 4'-unsubstituted compounds. In a number of cases however, introduction of a hydroxy substituent at the 4'-position appeared to result in a reduction in cytotoxicity compared with the 4'-unsubstituted compound. A preliminary comparison *in vivo* of the 4'-substituted compounds identified carbocyclic 4'-hydroxy-2'-deoxyguanosine as the best candidate for progression. This compound was one of the few carbocyclic nucleoside analogues first prepared in the "natural" optical form rather than as a racemic mixture having been obtained from Aristeromycin using the procedure outlined in scheme 2.[16]

The synthetic route required us to gain selective access to the unprotected 2'-OH group to allow specific deoxygenation. This was achieved *via* initial silylation of the 5'-hydroxyl group, preparation of the 2',3'-cyclic stannane derivative by treatment with di-n-butyl tin oxide and reaction with benzoyl chloride to give the required 3',5'-protected 2'-hydroxyl compound. Deoxygenation of the 2'-hydroxyl group was followed by fluoride-mediated deprotection at the 5'-position, reaction of the resultant hydroxyl function with Rydon's reagent to form the iodo derivative and elimination to generate the exocyclic olefin at the 4'-position. As expected, we were able to use the 3'-hydroxyl group to direct the epoxidation of the exocyclic olefin to the α-face and thus avoid the problems of stereochemical selectivity encountered in earlier work in the racemic series. This reaction and subsequent cleavage of the epoxide using benzoate anion were performed at the stage shown in the scheme to take advantage of the improved solubility in organic solvents of compounds possessing the intermediate modified base. The use of potassium hydrogen carbonate to mediate the Dimroth rearrangement represented an additional improvement in this route which further exemplified the general applicability and versatility of the adenine to guanine base interconversion. Biological evaluation to define the overall antiviral and pharmacokinetic profile of this compound is continuing. In standard systemic tests in mice, the compound has activity similar to that

Scheme 2

Synthesis of (+)-Carbocyclic 4′-Hydroxy-2′-deoxyguanosine

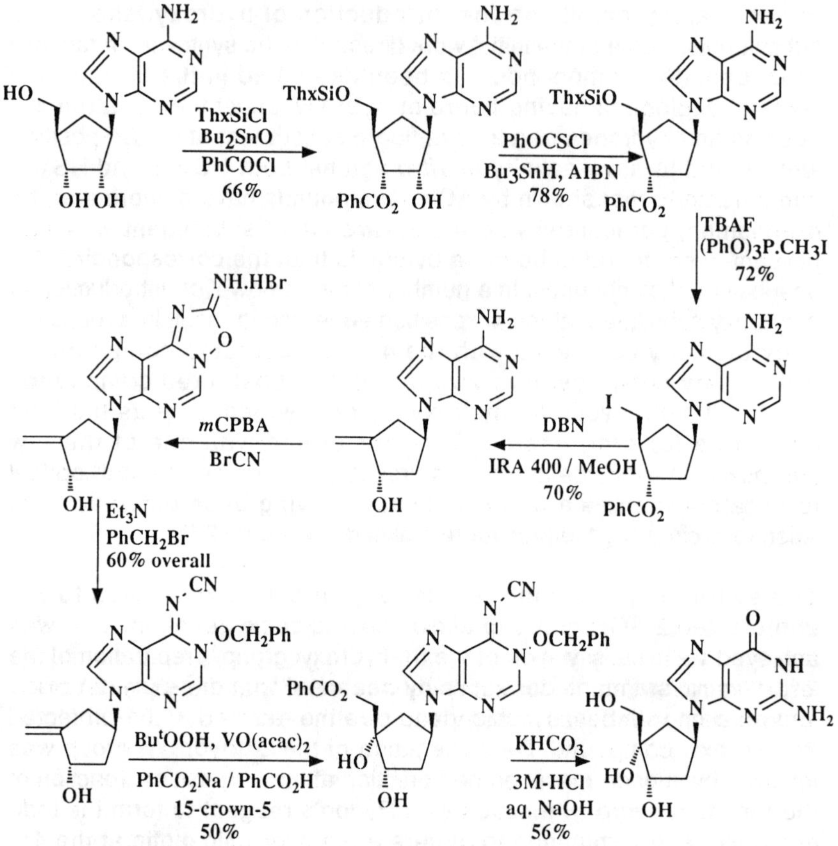

shown by ACV against HSV-1 and approximately ten-fold greater than ACV against HSV-2. The compound appears to have an additional advantage over ACV in terms of duration of action as evidenced by frequency of dosing studies and further work to investigate this is in progress as part of the development studies with this compound.

Whilst the 4′-substituted series had yielded compounds possessing potent activity against HSV-1 and HSV-2, none had shown additional activity against VZV without unacceptable levels of toxicity. Indeed, of all the compounds prepared in the course of our work on carbocyclic nucleoside analogues, only one appeared to have the desired profile having good activity against HSV-1, HSV-2 and VZV without apparent cytotoxicity. This compound, initially prepared in racemic form, had the same 2'-*ara*-fluoro carbocyclic ring present in c-AFG but had 8-aza-guanine as the base.[17] The *in vitro* activity of the compound (8-aza-c-AFG) is shown in table 4. Whilst the compound displayed high levels of activity against the three viruses of most interest, it is perhaps significant that the it showed no activity *in vitro* against CMV. Many other compounds claimed to have broader-spectrum anti-herpetic activity show activity against CMV. Since CMV does not encode a TK enzyme, phosphorylation of such compounds must result exclusively from the action of cellular enzymes with the resultant likelihood of toxicity. The initial synthesis of 8-aza-c-AFG in racemic form was from the same pyrimidine triamine used in the initial synthesis of c-AFG. In order to examine the activity of the two enantiomers, we investigated the use of adenosine deaminase for the separation of small quantities of material.

Adenosine deaminase has been found to be particularly useful for such resolutions of racemic nucleoside analogues. Whilst in many cases it allows completely specific hydrolysis of a 6-amino function in the "natural" enantiomer of a purine nucleoside analogue to the corresponding hydroxyl group as in the deamination of adenosine, the enzyme can also accommodate a wide variety of structural diversity in the substrate. The successful extension of the application of adenosine deaminase mediated resolution to the present case is shown in Scheme 3. Preparation of the racemic 2,6-diaminopurine precursor was effected in the expectation that this would be preferable to the 2-amino-6-chloropurine for recognition by the enzyme. In the event lack of adequate aqueous solubility of the 2,6-diaminopurine compound precluded effective resolution. It was subsequently found that the more soluble 6-chloro compound was recognised by the deaminase and could be resolved directly. As with other carbocyclic nucleoside analogues tested, although not the case for all nucleoside analogues, the *in vitro* activity was found to reside principally in the "natural" (+)-enantiomer which was greater than two orders of magnitude more potent than the "unnatural" (-)-enantiomer. Whilst we were unable to detect "natural" enantiomer impurity in the "unnatural" enantiomer, we were unable to eliminate this possible explanation for the weak activity observed in the latter.

Our attention was then directed to the synthesis of the preferred optical form. Although a variety of possible approaches to the synthesis of this compound may be envisaged, we were attracted by the opportunity to extend the chemistry described earlier to the development of a route to (+)-8-aza-cAFG from Aristeromycin. The overall transformation required was the introduction of a 2'-*ara* fluoro substituent at the expense of the 2'-hydroxyl coupled with synthesis of an 8-aza-guanine from adenine. Various options were available for the order in which these transformations were effected and a number of combinations were investigated and found to be viable. The completed route is shown in

Scheme 3

Resolution of (±)-8-aza-c-AFG

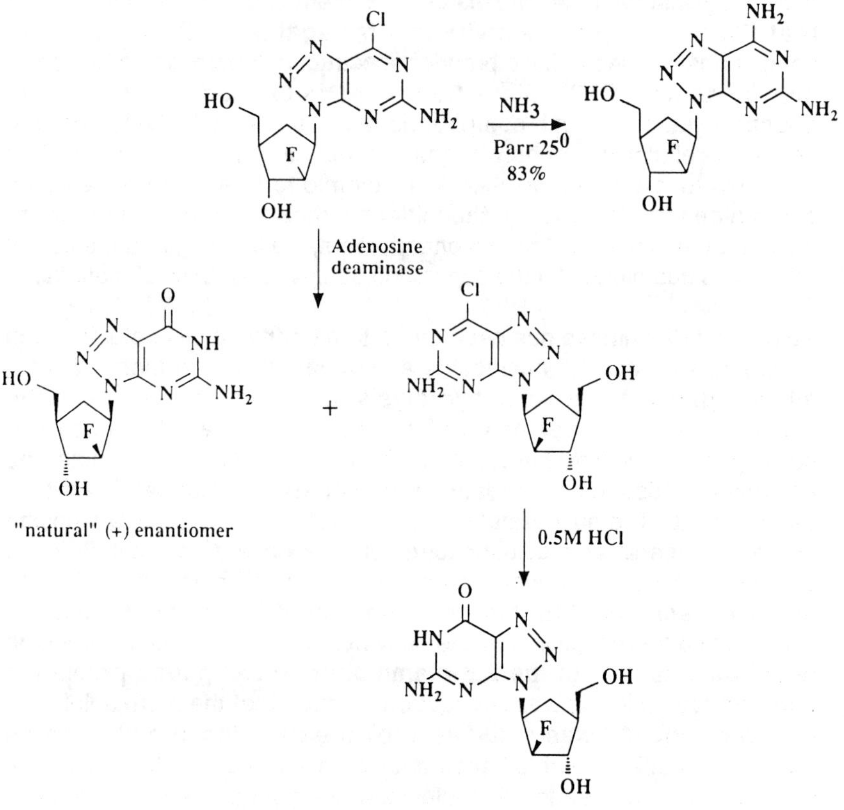

"natural" (+) enantiomer

"unnatural" (-) enantiomer

Table 4

IN VITRO ACTIVITY OF (±)-8-aza-c-AFG

Plaque Redn. ID_{50} (µg/mL)

Compound	HSV1	HSV2	VZV	TOX
(±)-8-aza-c-AFG	0.057-0.25	0.19	0.006-0.054	100
Acyclovir	0.20	0.60	22.0	

The compound has no activity against CMV.

Scheme 4

Synthesis of (+)-8-aza-c-AFG from (−)-Aristeromycin

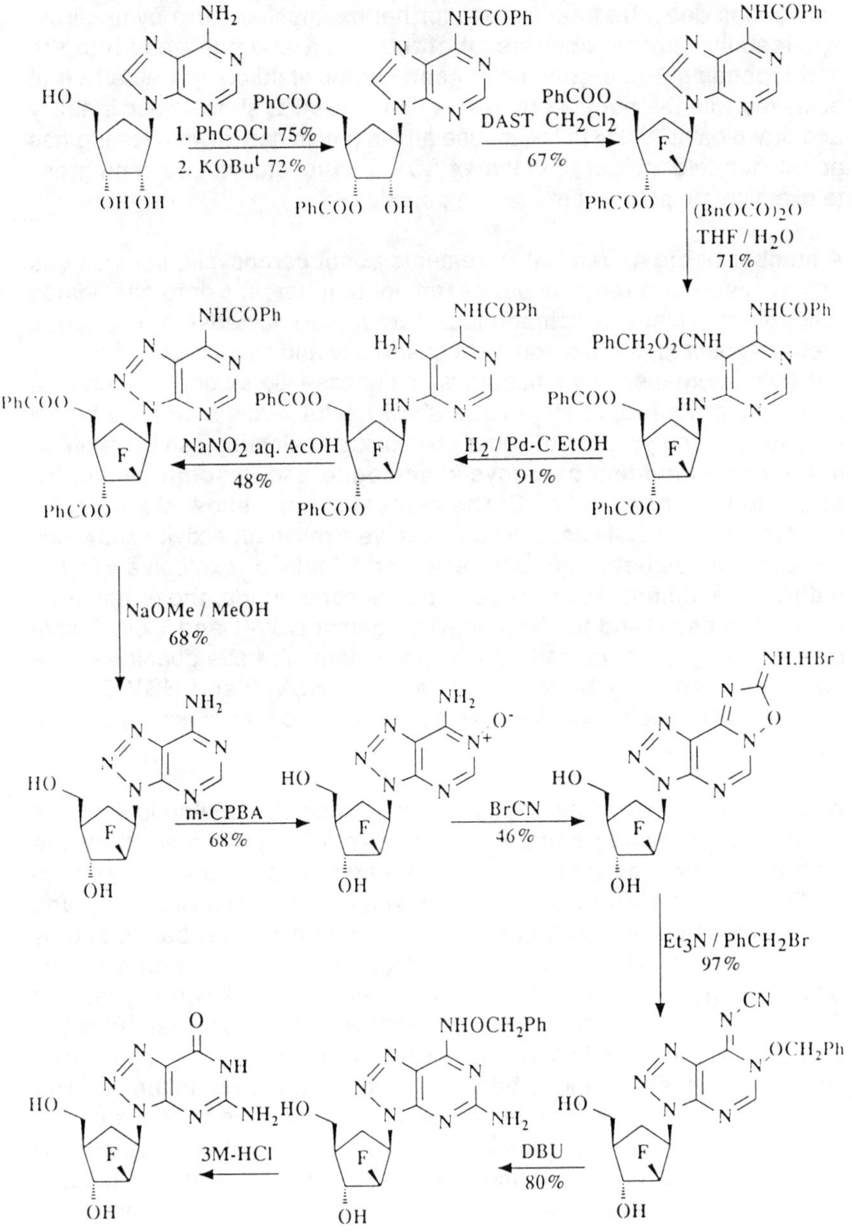

Scheme 4. We first introduced the fluoro substituent having gained access to the unprotected 2′-hydroxyl group as previously reported.[18] Cleavage of the imidazo ring of the purine was effected using dibenzylpyrocarbonate *en route* to diamine precursor of the corresponding 8-aza-adenine. The robustness of the adenine to guanine conversion described earlier was further exemplified here by its direct applicability to the conversion of the 8-aza-adenine base into the corresponding 8-aza-guanine in good overall yield comparable to that achieved with the normal purines. (+)-8-aza-c-AFG shows good stability and oral bioavailability in the mouse and in preliminary in vivo testing has activity ten-fold greater than that of ACV. Further studies are in progress to examine the potential of this compound.

A number of broad general statements about carbocyclic nucleosides can be made as a result of our search for anti-herpetic drug candidates amongst this class of compounds. Carbocyclic nucleoside analogues are, as expected, more stable both chemically and biologically, than their sugar counterparts. Of the hundreds (in our case alone) of compounds of this class prepared, relatively few show useful levels of activity below toxic levels. The potency of sugar nucleoside analogues can be retained in the corresponding carbocyclic analogue and in some cases, for example the case of c-AFG, the carbocycle can show significantly greater activity. In all cases which we have examined, activity shown by a racemate has been found to reside principally or exclusively in the natural enantiomer. In the case of active compounds, those having a pyrimidine base tend to show activity against HSV-1 and VZV. Active purine analogues in the carbocyclic series normally have guanine as the base and generally have activity against HSV-1 and HSV-2. One compound (+)-8-aza-c-AFG shows a broader spectrum of activity without apparent toxicity.

As with other areas of investigation in the nucleoside analogue field, the number of promising compounds discovered is poor reward for the amount of effort which has been devoted to this area worldwide. In addition to the Herpes area, carbocycles have been prepared and evaluated as potential agents for the treatment of disease caused by some of the other viral classes mentioned at the beginning of this presentation. In the respiratory viruses area, carbocyclic analogues[19] of the lead compound ribavirin failed to show activity against influenza viruses. The carbocyclic analogue[20] of clitocine showed anti-influenza A activity *in vitro* but proved to be too toxic for progression. In the anti-HIV area, carbovir which was first described in racemic form by Vince et al.[21] is the only carbocyclic nucleoside analogue reported to have significant activity. The activity of this compound was found to reside in the "natural" optical form which was also available in good yield from Aristeromycin.[22] Recent series of sugar-modified nucleoside analogues which can be considered to be carbocyclic analogues include the 2′- and 3′-

heteroatom substituted analogues, the latter having become known as the iso-nucleosides.[23-27] Although two compounds in the 3'-oxa series having as the base guanine or adenine showed some activity against HIV, neither was sufficiently active or selective to warrant progression. Pyrimidine analogues of the iso-nucleosides were inactive against HIV although compounds having heteroatoms at both the normal ribose oxygen position and at the 3'-position, the dioxolanes and oxathiolanes, show significant anti-HIV activity and will be the subject of a later contribution from Dr Mansour of Biochem Pharma.

ACKNOWLEDGEMENTS

The authors would like to acknowledge contributions to the work described herein by Jane Angier, Janet Cameron, Janet Horne, Amanda Jowett, David Knight and Claire Viner of the Virology Department and Angela Collard of the Chemotherapy Department of Glaxo Group Research, Greenford. The preparation of the manuscript was undertaken by Jackie Halvey and Anne Oakley-Smith and the authors are grateful for their skilled assistance.

REFERENCES

1. V.E. Marquez and M.-I. Lim, Med. Res. Rev.,1986, 6, 1.

2. A.D. Borthwick and K. Biggadike, Tetrahedron, 1992, 48, 571.

3. Y.F. Shealy and J.D. Clayton, J. Amer. Chem. Soc.,1966, 88, 3885.

4. D.I.C. Scopes, R.F. Newton, P. Ravenscroft and R.C. Cookson, Eur. Pat. No. 104066 (1968).

5. R.C. Cookson, P.J. Dudfield, R.F. Newton, P. Ravenscroft, D.I.C. Scopes and J.M. Cameron, Eur. J. Med. Chem., 1985, 20, 375.

6. P. Ravenscroft, R.F. Newton, D.I.C. Scopes and C. Williamson, Tetrahedron Lett., 1986, 27, 747.

7. R.T. Walker, P.J. Barr, E. De Clercq, J. Descamps, A.S. Jones and P. Serafinowski, Nucleic Acids Res., Special Publ. 1978, 4, S103.

8. J. Balzarini, H. Baumgartner, M. Bodenteich, E. De Clercq and H. Griengl, J. Med. Chem., 1989, 32, 1861.

9. P.L. Myers, Eur. Pat. No. 418007 (1991).

10.   K. Biggadike, A.D. Borthwick, A.M. Exall, B.E. Kirk, S.M. Roberts and P.M. Youds, J.Chem. Soc. Chem. Commun.,1987, 1083.

11.   A.D. Borthwick, S. Butt, K. Biggadike, A.M. Exall, S.M. Roberts, P.M. Youds, B.E. Kirk, B.R. Booth, J.M. Cameron, S.W. Cox, C.L.P. Marr and M. Shill, J. Chem. Soc. Chem. Commun., 1988, 656.

12.   A.D. Borthwick, B.E. Kirk, K. Biggadike, A.M. Exall, S. Butt, S.M. Roberts, D.J. Knight, J.A.V. Coates and D.M. Ryan, J. Med. Chem., 1991, 34, 907.

13.   A.D. Borthwick, K. Biggadike, S. Holman and C-L. Mo, Tetrahedron Lett., 1990, 31, 767.

14.   C-L. Mo, R. Storer and P.J. Turnbull, Eur. Pat. No. 409595 (1991).

15.   K. Biggadike and A.D. Borthwick, J. Chem. Soc. Chem. Commun., 1990, 1380.

16.   R. Storer, I.L. Paternoster, A.D. Borthwick and K. Biggadike, Eur. Pat. No. 430518 (1991).

17.   R. Storer, N.G. Weir, A.D. Baxter, C-L. Mo, A.D. Borthwick, K. Biggadike and B.E. Kirk, Eur. Pat. No. 410660 (1991).

18.   K.Biggadike, A.D. Borthwick, A.M. Exall, B.E. Kirk and R. Ward, J. Chem. Soc. Chem. Commun., 1988, 898.

19.   S.A. Noble, N.E. Beddall, A.J.Beveridge, C.L.P. Marr, C-L. Mo, P.L. Myers, C.R. Penn, R. Storer and J.M. Woods, Nucleosides Nucleotides,1991, 10(1-3), 487.

20.   A.D. Baxter, C.R. Penn, R. Storer, N.G. Weir and J.M. Woods, Nucleosides Nucleotides, 1991, 10(1-3), 393.

21.   R. Vince, M. Hua, J. Brownell, S. Daluge, F. Lee, W.M. Shannon, G.C. Lavelle, J. Qualls, O.S. Weislow, R. Kiser, P.G. Canonico, R.H. Schultz, V.L. Narayanan, J.G. Mayo, R.H. Shoemaker and M.R. Boyd, Biochem. Biophys. Res. Commun., 1988, 156, 1046.

22.   A.M. Exall, M.F. Jones, C-L. Mo, P.L. Myers, I.L. Paternoster, H. Singh, R. Storer, G.G. Weingarten, C. Williamson, A.C. Brodie, J. Cook, D.E. Lake, C.A. Meerholz, P.J. Turnbull and R.M. Highcock, J. Chem. Soc. Perkin Trans. 1, 1991, 2467.

23.   M.Bamford, D.C. Humber and R. Storer, Tetrahedron Lett., 1991, 32, 271.

24.   M.F. Jones, S.A. Noble, C. Robertson and R. Storer, Tetrahedron Lett., 1991, 32, 247.

25.   M.F. Jones, S.A. Noble, C. Robertson, R. Storer, R.M. Highcock and R.B. Lamont, J. Chem. Soc. Perkin Trans. 1, 1992, 1427.

26.   D.M. Huryn, B.C. Sluboski, S.Y. Tam, L.J. Todaro and M. Wiegele, Tetrahedron Lett., 1989, 30, 6259.

27.   K.B. Frank, E.V. Connell, M.J. Holman, D.M. Huryn, B.C. Sluboski, S.Y. Tam, L.J. Todaro, M. Weigele, D.D. Richman, H. Mitsuya, S. Broder and I.S. Sim, Ann. N.Y. Acad. Sci., 1990, 616, 408.

18

Discovery and Development of TIBO as a Potential Anti-AIDS Therapeutic

M. J. Kukla,* H. J. Breslin, E. De Clercq, R. Pauwels, K. Andries, and P. A. J. Janssen

JANSSEN RESEARCH FOUNDATION, SPRING HOUSE, PA 19477, USA; REGA INSTITUTE FOR MEDICAL RESEARCH, KATHOLIEKE UNIVERSITEIT LEUVEN, MINDERBROEDERSSTRAAT 10, B-3000 LEUVEN, BELGIUM; AND JANSSEN RESEARCH FOUNDATION, TURNHOUTSEWEG 30, B-2340 BEERSE, BELGIUM

At this point, one doesn't need to describe the need for additional anti-AIDS drug therapy. The recent approval of DDI (Videx) to Bristol Myers Squibb added a second drug to join AZT (Zidovidine) on the pharmacy shelf as treatment modalities for AIDS. Unfortunately, both of these agents act by the same mechanism of action and share some common potential side effect liabilities.

The major causative agent in AIDS is the HIV-1 retrovirus which initially attacks the lymphocytes of the immune system. The replicative cycle of the virus within the lymphocyte has been elucidated to the point that several potential targets for drug intervention have been identified. Some of the areas that have seen reported research effort include: blockade of attachment and entry of the virus into the cell via a specific cell receptor; blockade of the viral reverse transcriptase which transcribes viral genetic RNA into a DNA duplex which can be integrated into the lymphocyte genome; inhibition of the viral TAT regulatory protein which increases viral replication; and blockade of the specific viral protease which processes the large viral protein chain, generated from the transcription process, into the smaller functional and structural proteins necessary for the assembly of a new virus particle.

Our approach to search for a new anti-HIV-1 agent was a broad screening paradigm that would allow us to find compounds that might be useful at any point in the life cycle of the virus. We collaborated with Erik De Clercq's group of the Rega Institute to test compounds in a whole cell assay using MT-4 cells as described previously.[1] The cells were either infected with HIV-1 or mock infected and incubated in the presence of various concentrations of the test compounds. The number of viable cells was then determined 5 days after infection. The determinations for each compound included the concentration required to protect 50% of the cells against the HIV-1 cytopathic effects (IC_{50}) and the concentration of test compound which was cytotoxic (CC_{50}) to uninfected cells. The ratio of these two values was reported as a selectivity index (SI).

R14458
$IC_{50} = 70 \ \mu M$

AZT $IC_{50} = 0.004 \ \mu M$

Our file of potential drug candidates numbered >80,000 compounds; thus an initial choice of testing candidates was made by selecting representative structures and requiring that the compounds had previously shown little or no biological or toxicological effect. From this resultant group of 600 compounds, R14458 was found to weakly but consistently inhibit the HIV-1 virus from killing the MT-4 cells. As shown, it had an IC_{50} of 70 µM as compared to AZT which had an IC_{50} of 0.004 µM in this test. Thus we began structural modification to improve on the lead compound which was 1/17,500 as active as the standard. The acronym TIBO was adopted for the series based on the general chemical nomenclature (indicated by bolding and underling): **4,5,6,7-t**etrahydro-5-methyl**i**midazo[4,5,1-jk][1,4] **b**enzodiazepin-2(1H)-**o**ne.

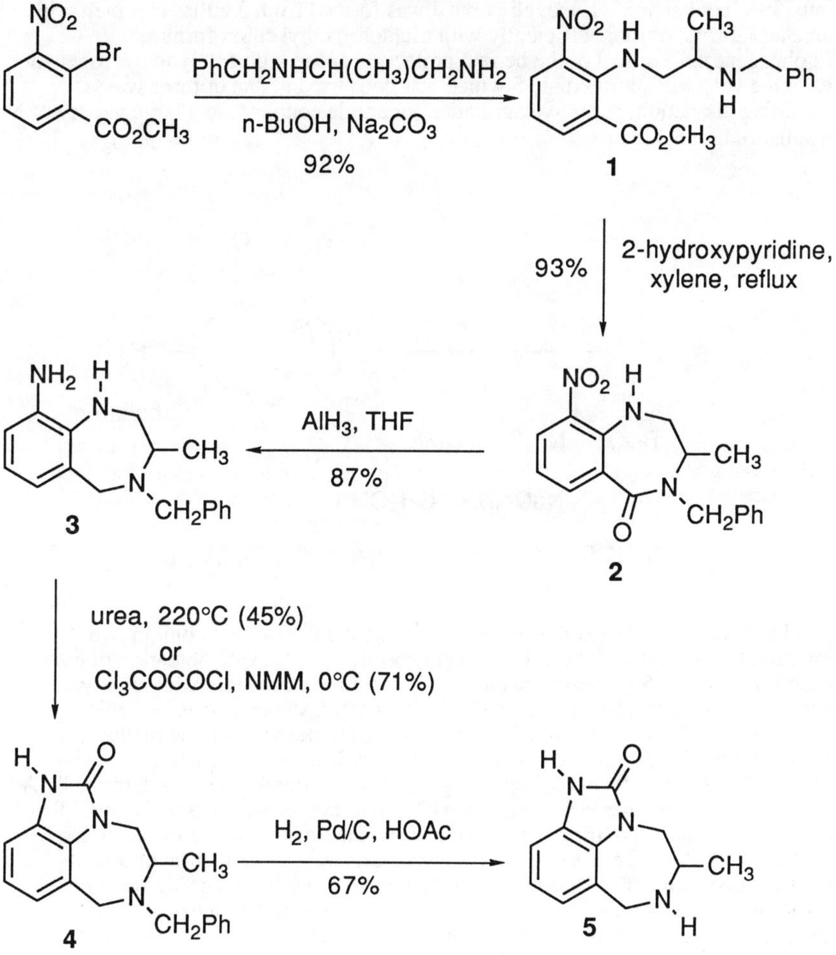

Scheme 1

A number of synthetic approaches to the tricyclic TIBO structures were developed in order to make variations in all parts of the molecule to generate structure activity relationships (SAR) for lead optimization.

As indicated in Scheme 1, methyl-2-bromo-3-nitrobenzoate was heated with benzylated 1,2-diaminopropane in a halogen displacement to give **1**. Cyclization to benzodiazepine **2** was accomplished in 93% yield by heating in refluxing xylene in the presence of 2-hydroxypyridine catalyst. Reduction to triamine **3** was carried out smoothly with aluminum hydride at ambient temperature after initial attempts to carry out this reduction with lithium aluminum hydride were less successful. With the latter reagent, after 2.5 h in a mixture of THF and toluene at reflux, there was complete loss of starting material but only partial (35%) conversion to **3**. The main product resulted from reduction of the nitro to an amine to yield the diamino lactam. Continued reflux led to increasing amounts of **3**, but even after 5 days there was only 79% conversion. The cyclic urea **4** was formed from **3** either in a melt with urea at >200°C, or more efficiently with trichloromethyl chloroformate (diphosgene). Removal of the benzyl protecting group with catalytic hydrogenation gave the versatile intermediate **5** which was converted in one of three ways including alkylation, reductive amination, or amide coupling, to a large variety of N-6 substituted analogs (**6**).[2]

5 ⟶ **6**

R-X, KI, Na$_2$CO$_3$, DMF

R-CHO, NaCNBH$_3$, CH$_3$OH

R-COOH, 1-hydroxybenzotriazole, DCC

The second synthesis, described in Scheme 2, allowed us to obtain two important structural varieties. It is enantiospecific and allowed obtention of either enantiomer of the 5-position substituents in the diazepine ring. Additionally, it was versatile enough to allow replacement of the methyl, shown in this example, with other groups. By using the readily available enantiomers of alanine methyl ester in a standard peptide coupling reaction with 2-amino-3-nitrobenzoic acid, the stereocenter is fixed from the beginning. Catalytic reduction gave o-diamine **8**. A number of methods were explored to cyclize **8** to benzodiazepine **9**. Heating **8** neat at ca. 200°C under vacuum, to remove extruded methanol, turned out to be the best, with an overall yield of 58% from **7**. The results of LAH reduction of **9** were solvent dependent. The lactam carbonyl adjacent to the benzene ring was very resistant to reduction and required reflux temperatures of dioxane or glyme to complete reduction to **10**. As shown previously, the carbonylation to **5** was accomplished with diphosgene. The unprotected secondary amine was converted to a carbamoyl chloride derivative during the course of the reaction and a hydrolytic workup was required to regenerate **5**. This reaction scheme was carried out with a variety of available α-substituted amino esters to give a number of 5-substituted analogs.

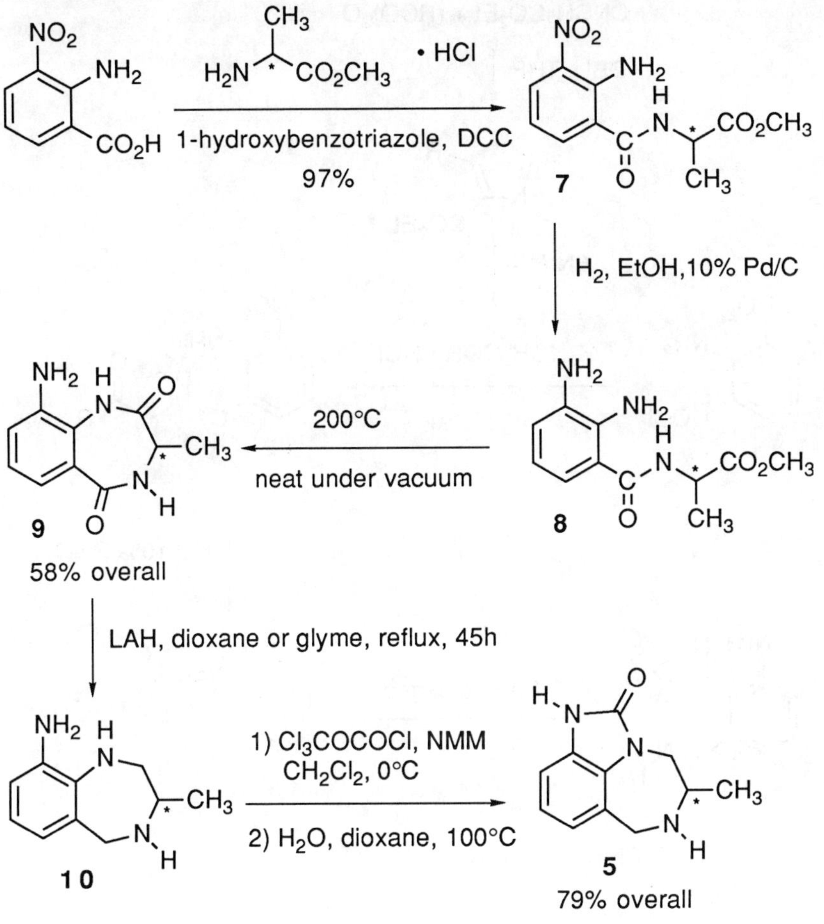

Scheme 2

Alternative methods were required to synthesize analogs with substituents in the other positions on the 7-membered ring. Scheme 3 details the variation used to obtain 4-position analogs. A nitro anthranilic acid derivative **11** was activated as an acid chloride to react with α-amino ketones as shown. The latter were available via a published method from acid hydrolysis of oxazoles.[3] Reduction of the nitro group in **12** by catalytic hydrogenation rendered the ortho amine nucleophilic. Consequently, it took part in ring closure by reductive amination with the ketone to form imine **13** as the major product. The crude material was further reduced with LAH to triamine **14** which was carried forward by previously described methods to analogs of interest with 4-position substituents.

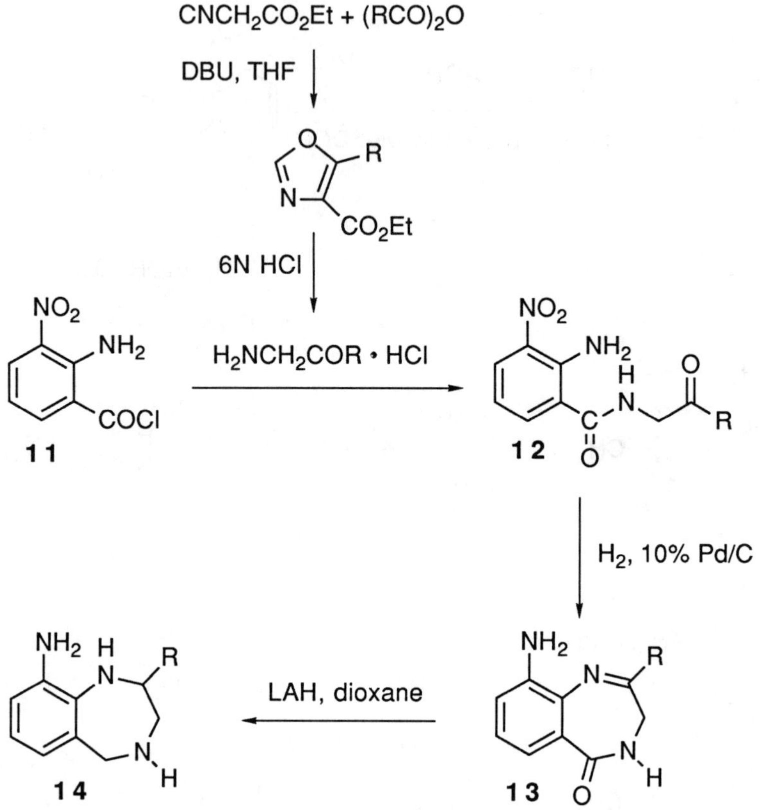
CNCH2CO2Et + (RCO)2O
DBU, THF
R
O
N
CO2Et
6N HCl
NO2
NH2
COCl
11
H2NCH2COR • HCl
NO2
NH2
H
N
O
C
O
R
12
H2, 10% Pd/C
NH2
N
R
N
H
O
13
LAH, dioxane
NH2
H
N
R
N
H
14

Scheme 3

Substitution on the final position of the seven membered ring portion of the TIBO molecule is pictured in Scheme 4. The key intermediate in this sequence is acetophenone derivative **15**. 2-Chloro-3-nitro benzoic acid is converted to an acid chloride prior to reaction with the magnesium salt of diethylmalonate. The resultant α,α-ketodiester was hydrolyzed with a mixture of acetic and sulfuric acids to the diacid which spontaneously bis-decarboxylated to **15**. Reaction with 1,2-diaminoethane gave aromatic halide displacement and condensation of the ketone to form an imine. This was reduced with sodium cyano borohydride to **16**. Alkylation of the secondary amine followed by reduction of the nitro group with LAH gave **17**. The final ring formed in this example, the cyclic thiourea **18**, was generated by reaction of the ortho diamine with carbon disulfide.

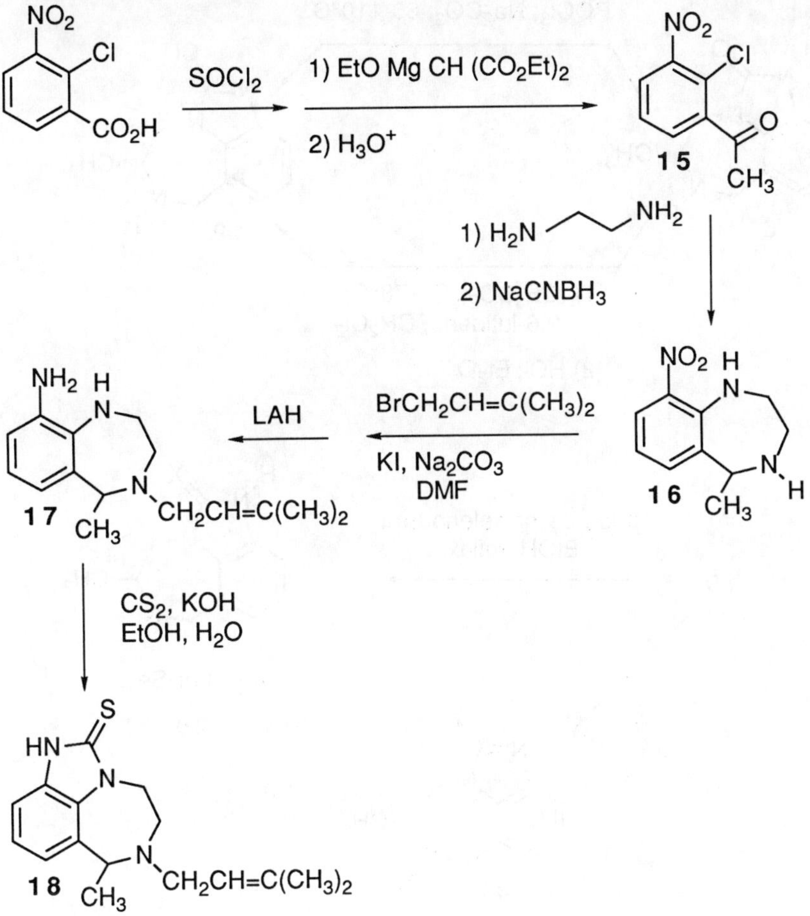

Scheme 4

Variation of the cyclic urea portion of the molecule[4] was accomplished basically through two key intermediates, **19** and **23**. Imino chloride **19** was generated from **6** by two different methods. Treatment of **6** with phosphorus oxychloride at elevated temperatures gave **19** in ca. 60% yields. Alternatively, **6** was treated with triflic anhydride at low temperature to generate the O-triflate which was easily converted to **19** by addition of ethereal HCl. Reaction of **19** with a variety of nucleophiles led to several analogs which are exemplified by structures **20** and **21**.

Another versatile intermediate **23** was synthesized by selective alkylation of triamine **22**. Reaction of **23** with dielectrophilic reagents, as pictured in Scheme 6, yielded even further analogs such as **24** and **25**.

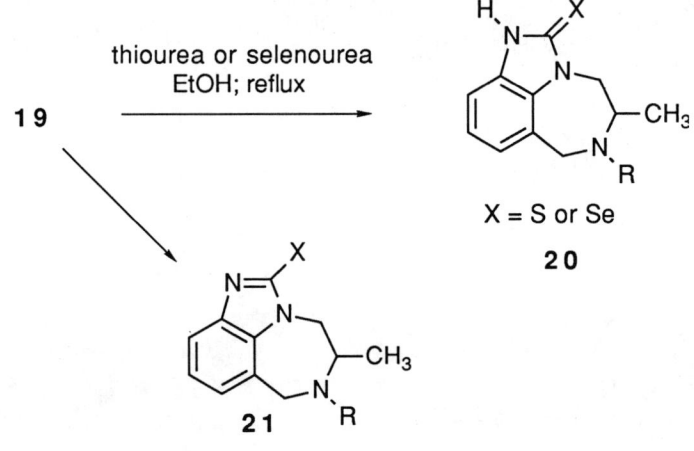

$$POCl_3; Na_2CO_3; 60\text{-}110°C$$

1) $(CF_3SO_2)_2O; -78°C;$
   2,6-lutidene; CH_2Cl_2

2) HCl; Et_2O

thiourea or selenourea
EtOH; reflux

19

X = S or Se

20

21

Conditions	Product (X)
40% aq H_2NCH_3; MeOH	-NHCH$_3$
40% aq $HN(CH_3)_2$; MeOH	-N(CH$_3$)$_2$
PhSH; NaOH; MeOH/H$_2$O	-SPh
Na; MeOH	-OCH$_3$

Scheme 5

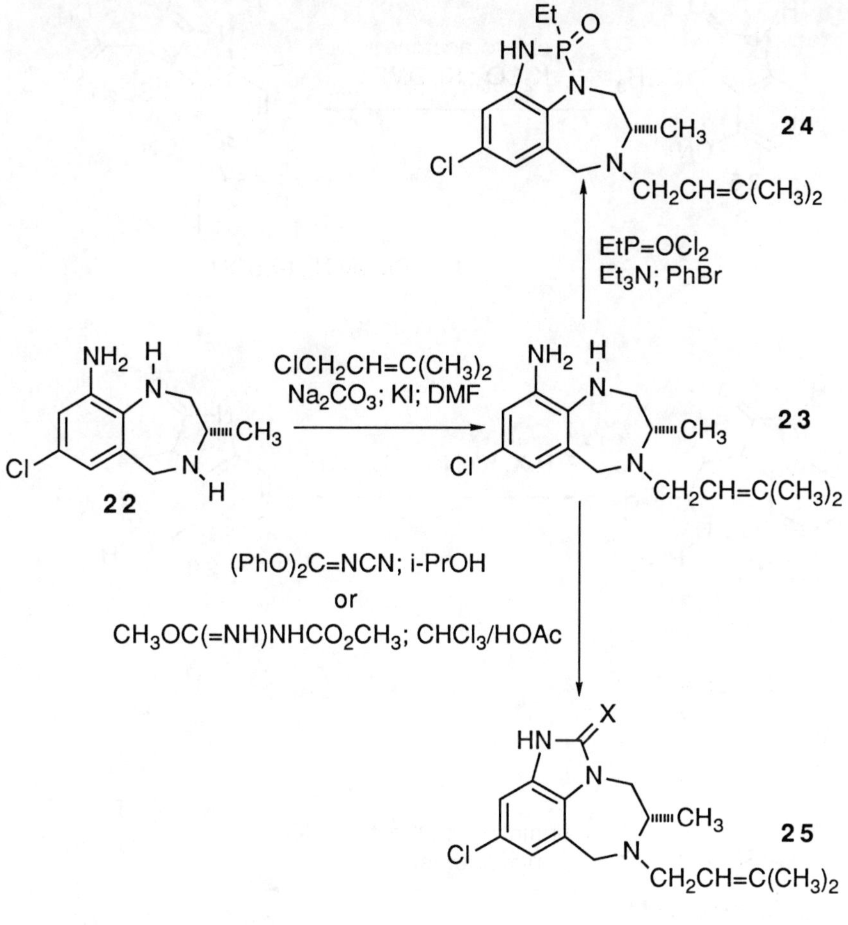

Scheme 6

X = NCN
X = NCO$_2$CH$_3$

In other variations of the cyclic urea ring, the nitrogens were replaced in analogs with carbon or oxygen. The synthesis of one example of this type of analog is shown in Scheme 7. Gassman reported indolone intermediate **26** some years ago as part of a study of indole syntheses.[5] This material was readily alkylated with chloroacetone to give **27** which, when treated with Raney nickel in t-butanol, underwent a series of conversions including desulfurization and nitrile reduction to the primary amine, followed by reductive amination with the internal ketone to yield the desired ring system **28**. This was readily alkylated to give **29**, a TIBO analog without the internal urea nitrogen.

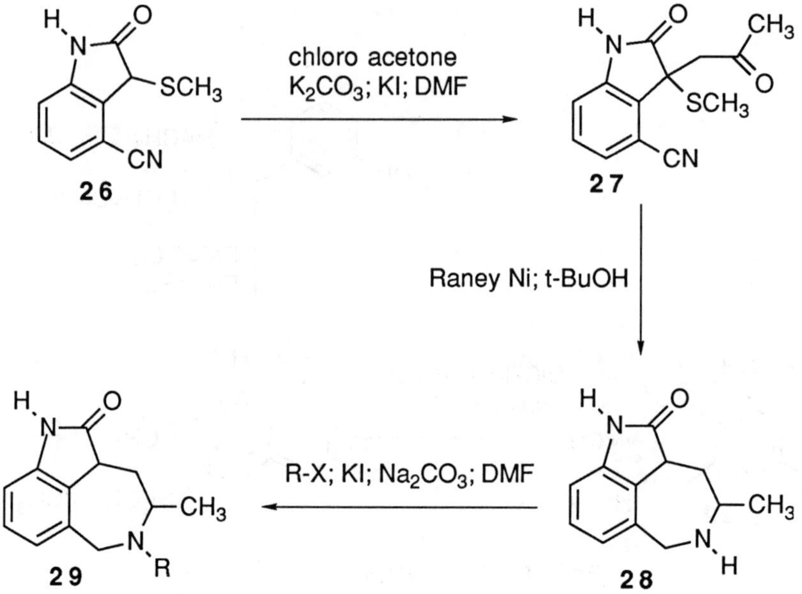

Scheme 7

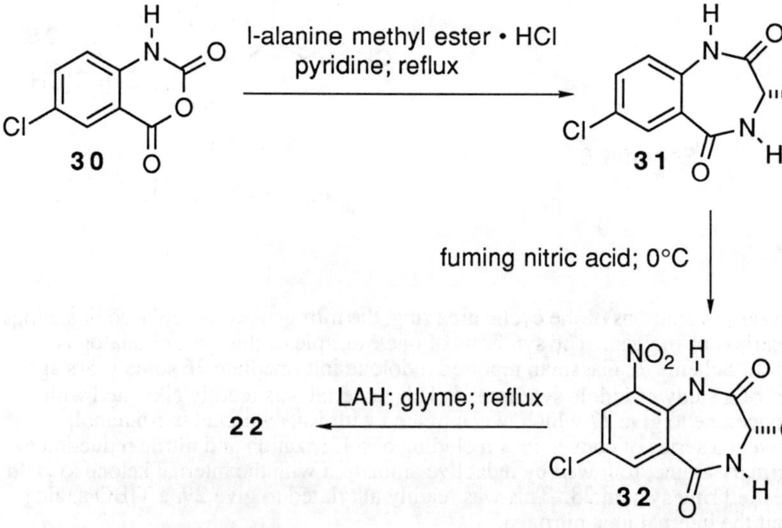

Scheme 8

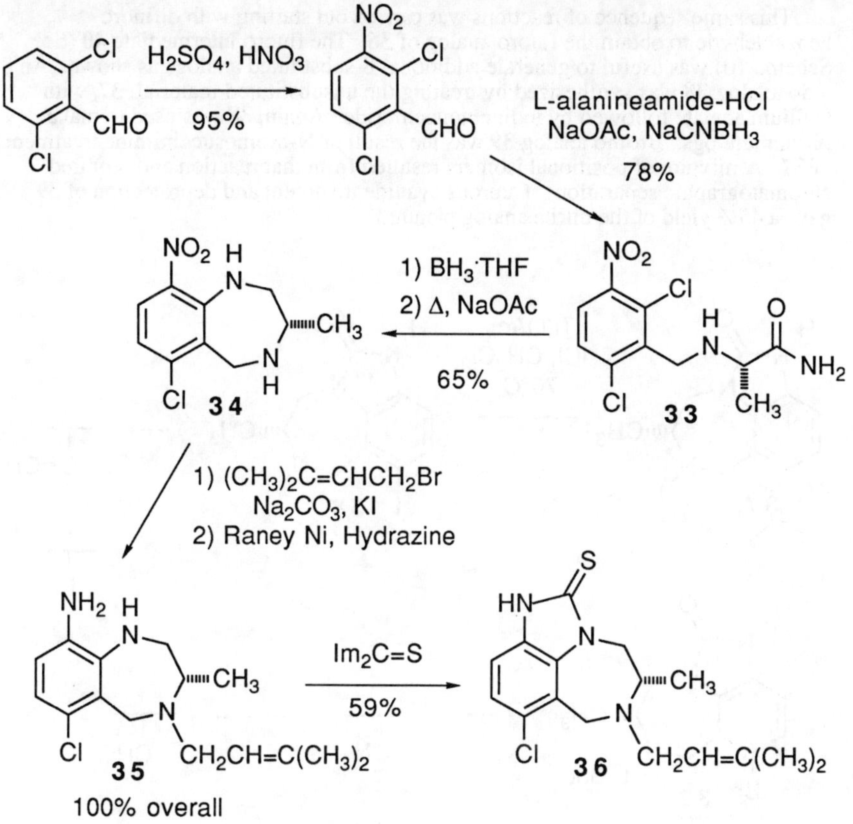

Scheme 9

The final type of structural feature that was explored on the TIBO nucleus involved aromatic ring substitution. A number of syntheses were required to obtain a representative variety of analogs. Schemes 8-10 show a few of the methods used to place groups on the 8 and 9 positions. Chloroisatoic anhydride, **30**, readily reacts with alanine methyl ester in refluxing pyridine to give benzodiazepine derivative **31**.[6] Since the para position was blocked with a chlorine, nitration with fuming nitric acid gave a single product **32** in high yield at the position ortho to the activating anilino amine. Reduction of the lactams and nitro functionalities with LAH gave **22**, a previously described intermediate.

As will be discussed, substituents on the 8 position on the aromatic ring improved activity dramatically. Scheme 9 shows a convenient synthesis of the chloro analog **36**. Nitration of 2,6-dichlorobenzaldehyde with a mixture of nitric and sulfuric acid gave a single product in 95% yield. This aldehyde was reacted with alanine amide under reductive amination conditions to yield **33**. Borane reduction of the amide produced a nucleophilic nitrogen that readily displaced the activated aromatic chloride to yield bicyclic **34**. The secondary amine was alkylated prior to reduction of the nitro group with a mixture of Raney nickel and hydrazine. The urea ring, in this case a thiourea, was generated with 1,1'-thiocarbonyldiimidazole to yield **36**.

This same sequence of reactions was carried out starting with difluoro-benzaldehyde to obtain the fluoro analog of **36**. The fluoro intermediate **40** (See Scheme 10) was useful to generate additional 8-substituted analogs as shown. An 8-iodo analog **38** was synthesized by treating the unsubstituted material, **37**, with thallium acetate followed by iodinemonochloride. Again, **38** was used to make further analogs. Bromo analog **39** was the result of N-bromosuccinimide treatment of **37**. A mixture of positional isomers resulted from that reaction and required chromatographic separation. Cuprous cyanide treatment and deprotection of **39** gave a 45% yield of the nitrile analog pictured.

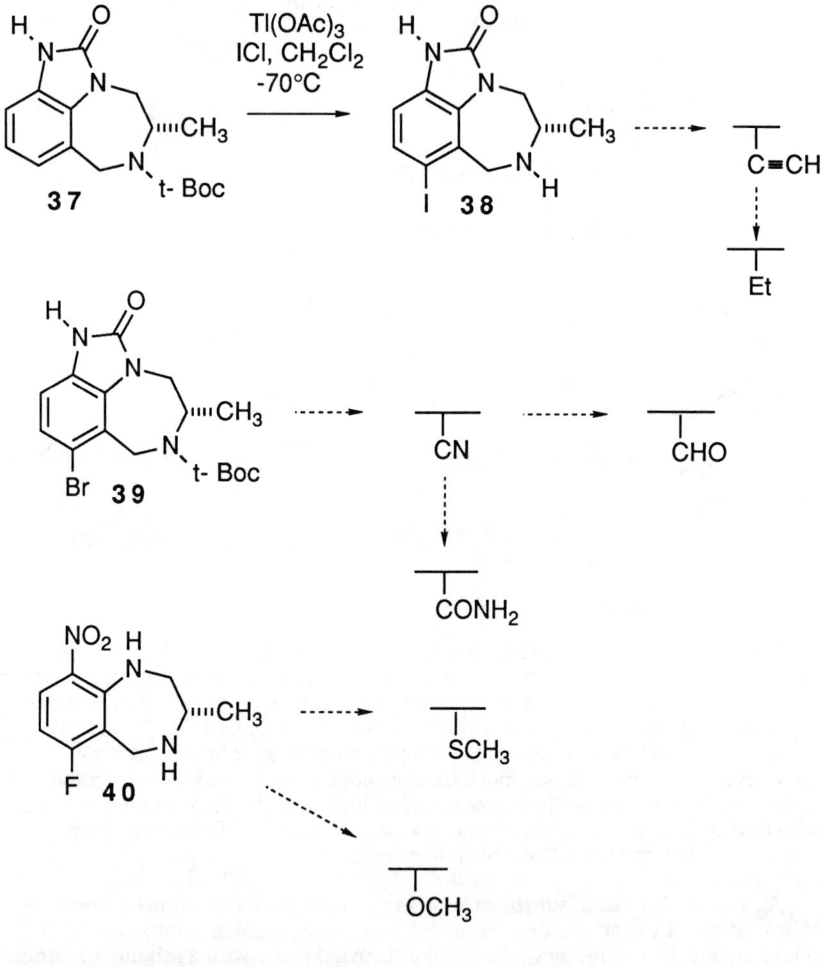

Scheme 10

Only an overview of the chemistry and resultant analogs has been included in the previous schemes. As many as 400-500 analogs were prepared in this project to develop an SAR and ultimately establish a clinical candidate. The "SAR summary" table gives a broad overview of the alterations of the TIBO structure that were carried out and the activity conclusions that were established.

Table 1 **SAR Summary**

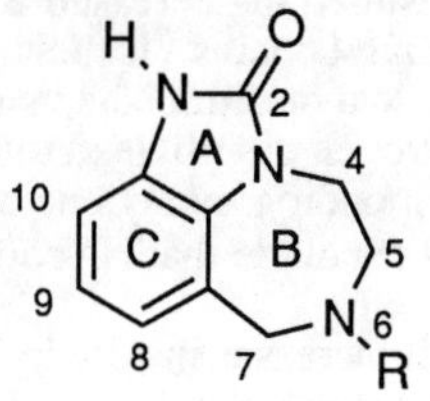

N^6 Substituents:

- Alkyl or alkenyl chains are active,
  4 to 5 carbon length is optimal

- Allyl best, preferably 3,3-disubstituted but not contained in a ring

- Heteroatom in chain results in loss of activity

- N^6 can be non-basic by virtue of an acyl substituent
  and activity remains, although diminished relative to alkyl

A Ring

- A carbonyl (oxygen, sulfur, or selenium) at 2-position required

- All other heterocyclic rings are inactive

B Ring

- Methyl substituents at 5 or 7 positions improve activity,
  larger groups at 5 and 7 are inactive, but retain activity at 4 position.

- Fused rings all inactive

C Ring

- Aromatic Substituents:
    8-Improve activity
    9-Not much effect
    10-Decrease activity

- Heteroaromatics inactive or weak

A large variety of groups (R) were substituted at the N6 position.[2] The best substituents were simple alkyl or alkenyl chains with an an optimum length of 4 to 5 carbons; shorter or longer chains resulted in decreased activity. The best substituents were 3,3-disubstituted allyl chains. If the substituent contained a heteroatom, activity was invariably lost or diminished significantly. Interestingly, a carbonyl attached directly to the nitrogen did not lead to inactive compounds. Although less active than the corresponding methylene derivatives, the analogs were active to the point where one could conclude that N6 could be non-basic.

Ring A change led to significant increases in activity from the original lead compound, but also had a very low tolerability for change.[4] Thus the thiourea analogs were much more active than their urea counterparts and factors of 10-100 difference in activity were usual. Selenourea analogs were as active as the sulfur analogs but were not practical development analogs. Many other variations of the A ring were synthesized and were all inactive. Both nitrogens were separately replaced with carbon or oxygen (for N1) and gave no activity. Several six-membered rings that retained both nitrogens were inactive. The imidazole, triazine, guanidine, and substituted guanidines were all ineffective at inhibiting the HIV-1 virus.

The unsubstituted diazepine ring B as shown in the diagram showed good antiviral activity; however, numerous alkyl substituted analogs have been synthesized and have been shown to affect activity. Methyl groups at either the 5 or 7-positions have improved potency, but larger groups at those positions made the molecules inactive. Substitution at the 4-position had a somewhat different profile. A 4-methyl analog had slightly diminished activity versus the unsubstituted material, but unlike the 5-position larger groups, like n-propyl or isopropyl, they still retained the same level of potency as methyl. The anti HIV activity of the 5-methyl analog displayed an enantiomeric preference. In those cases where both enantiomers were synthesized, the analog with the S configuration was always the more active, or the only active, isomer. In contrast, both the R and S enantiomers of a 4-isopropyl analog were nearly equipotent. Several analogs were synthesized in which 5 or 6 membered rings were fused onto the TIBO structure B-ring in virtually every positional combination, ie. 4-5, 5-6, 6-7, and 7-8. In all cases the compounds were inactive.

Several different aromatic substituents on the 8-position of ring C gave large improvements in potency versus the unsubstituted material. These included halogens, thiomethyl, and methyl. Others like cyano, methoxy, and acetylene were equipotent , while 8-amino, aminoacetyl, dimethylamino, and nitro were inactive. Substituents at the 9-position tended to be have little effect on activity and 10-position substituents decreased activity. Finally, a limited number of heteroaromatic analogs such as pyridine and pyrimidine were synthesized and found to be inactive or very weakly active relative to the carbon counterparts.

The final results of the SAR and lead optimization work are shown below in a comparison of the original lead structure, R14558, and the most recent development candidate, R86183.[7] As indicated, there has been a nearly 14,000 fold improvement in potency in ability to protect the MT-4 cell from death caused by administered HIV-1 virus. This compound is nearly equipotent to AZT, the standard in anti-AIDS therapy. In summary, the molecular changes that have resulted in improved potency include: a thiourea in place of urea, a 5-methyl substituent in the S configuration, an N6 3-methyl-2-butenyl appendage, and an 8-chloro aromatic substituent.

Compound	IC_{50} (μM)	Selectivity (CC_{50}/IC_{50})
R14458	70	10
R86183	0.0052	11,600
AZT	0.0041	5,300

R14458   R86183

As discussed earlier, a broad screening approach was used to find an active lead; consequently, the mechanism of action of this series of compounds was initially unknown. Several studies were carried out to explore the actions of the TIBO series of compounds.[8,9] Some of the results are summarized in the following table.

Table 2

TIBO mechanism of action summary

• Acting after binding and uptake of the virus into the cell, but before synthesis of viral DNA (time of addition study)

• Dose dependent inhibition of reverse transcriptase (RT)

• Inhibition of RT is specific for HIV-1 (same specificity in MT-4 cells)

• Good correlation between inhibition of RT and MT-4 cell assay

• Specifically inhibited RNA-dependent DNA polymerase activity of RT; does not affect other two steps of RT function

• A series of competition experiments using varied template/primers and nucleotide substrate indicates TIBO inhibits HIV-1 by binding at an allosteric site rather than the active site of RT.

• Synergistic with AZT

• HIV-1 viral mutants develop RT variants that are resistant to TIBO

An initial experiment was done in which the time of addition of the inhibitor was postponed relative to the virus addition. Depending on when the inhibitor was acting in the virus life-cycle, it would still be active if acting at a later stage. This study indicated that the compounds were not interfering with binding or uptake of the virus into the cell but were acting prior to synthesis of viral DNA. This time frame was similar to that seen with AZT and other dideoxy nucleotides that inhibit reverse transcriptase, an enzyme that is unique to the retroviruses such as HIV-1.

When tested against the reverse transcriptase, there was a dose dependent inhibition. When plotted, the inhibitory activity of the TIBO series in the MT-4 cell assay and their inhibition of the HIV-1 reverse transcriptase enzyme correlated very well. So the series seemed to be acting at the same point as AZT to block the virus life-cycle. However, when the compounds were tested against HIV-2 infected MT-4 cells, unlike AZT, they were inactive. In fact, they were also ineffective in protecting cells against simian virus and other retroviruses as well, indicating the inhibition of TIBO was specific for HIV-1 virus. This was further exemplified when the isolated HIV-2 reverse transcriptase enzyme was not inhibited by any compounds of the TIBO series. Thus the inhibitory effect of this structural class was, suprisingly, very specific for HIV-1. This property was unique and never seen in any series of compounds reported.

Consequently, a variety of experiments were carried out on HIV-1 reverse transcriptase using different RNA templates/primers and nucleotide substrates. The conclusions from this work were that the molecules were not acting at the active site of the enzyme like AZT, but instead were binding at an allosteric site. Presumably this allosteric binding disrupted the enzyme so that it could no longer carry out its full function.

Reverse transcriptase actually carries out three separate functions: transcription of viral RNA into a DNA chain, destruction of the RNA chain, and transcription of the corresponding DNA chain to form a DNA duplex. The compounds of the TIBO series, once bound to an allosteric site on the enzyme, inhibit the first step in this process, the RNA-dependent DNA polymerase activity.

Another interesting experiment showed that the TIBO compounds were synergistic with AZT in inhibition of HIV-1 caused MT-4 cell death.[7] When given together, the effective dose of each compound is much lower than when each is administered separately. Thus there is the promise that a combination therapy of these two classes of inhibitors might be even more effective and have less side affect potential then a single drug therapy.

Recent experiments have revealed a potential problem with the TIBO series that will require further study, especially in the clinic. After repeated treatment with a TIBO derivative in the MT-4 cell assay, some virus mutants are able to survive and replicate and eventually become prevalent when harvested. Apparently, a mutation in the reverse transcriptase has rendered the strain insensitive to inhibition by these compounds.[10]

A phase I clinical study of one compound from this series was carried out on 22 late stage AIDS patients (mean OKT4 level of 127). It was a dose escalating study of from 10 to 300 mg/day, by i.v. administration, for up to 50 weeks. There were no overt toxicities from the compound; there was a significant lowering (41%) of the p24 antigen that is a marker for virus levels; and after two months, there was a rise in the median OKT4 levels.[11]

Many people have contributed to this project and their names are listed below. The synthetic chemistry and structure activity work was carried out, for the most part, by the US group in Spring House, Pennsylvania. The biological testing and mechanism studies were carried out in the Rega Institute. The impetus for the project, general coordination, development studies and clinical design were done by the group in Beerse, Belgium.

Janssen Research Foundation

Spring House, PA
H. Breslin
C. Diamond
C. Fedde
P. Grous
L. Hecker
K. Hitchens
C. Ho
W. Ho
M. Kukla
T. Kulp
D. Ludovici
M. Miranda
R. Mohrbacher
P. Pitis
J. Rodgers
M. Scott
R. Sherrill
M. Zelesko

Beerse, Belgium
K. Andries
J. Van Gelder
J. Heykants
M. Janssen
P. Janssen
A. Raemaekers
K. Schellekens
P. Stoffels
R. Woestenborghs

Rega Institute

Leuven, Belgium
Z. Debyser
E. De Clercq
J. Desmyter
R. Pauwels
D. Schols

REFERENCES

1. R. Pauwels, E. De Clercq, J. Desmyter, J. Balzarini, P. Goubau, P. Herdewijn, M. Vandeputte; *J. Virol. Methods* **1987**, *16*, 171.

2. M. Kukla, H. Breslin, R. Pauwels, C. Fedde, M. Miranda, M. Scott, R. Sherrill, A. Raeymaekers, J. Van Gelder, K. Andries, M. Janssen, E. De Clercq, P. Janssen; *J. Med. Chem.* **1991**, *34*, 746.

3. M. Suzuki, T. Iwasaki, M. Miyoshi, K. Okumura, K. Matsumoto; *J. Org. Chem.* **1973**, *38*, 3571.

4. M. Kukla, H. Breslin, C. Diamond, P. Grous, C. Ho, M. Miranda, J. Rodgers, R. Sherrill, E. De Clercq, R. Pauwels, K. Andries, L. Moens, M. Janssen, P. Janssen; *J. Med. Chem.* **1991**, *34*, 3187.

5. P. Gassman and T. Van Bergen; *J. Am. Chem. Soc.* **1974**, *96*, 5508.

6. D. Kim; *J. Het. Chem* **1975**, *12*, 1323.

7. R. Pauwels, K. Andries, Z. Debyser, M. Kukla, D. Schols, H. Breslin, R. Woestenborghs, J. Desmyter, M. Janssen, E. De Clercq, and P. Janssen; *Antimicrob. Agents Chemoth.* **1992**, *36*.

8. R. Pauwels, K. Andries, J. Desmyter, D. Schols, M. Kukla, H. Breslin, A. Raeymaekers, J. Van Gelder, R. Woestenborghs, J. Heykants, K. Schellekens, M. Janssen, E. De Clercq, and P. Janssen; *Nature* **1990**, *343*, 470.

9. Z. Debyser, R. Pauwels, K. Andries, J. Desmyter, M. Kukla, P. Janssen, and E. De Clercq; *Proc. Natl. Acad. Sci. USA* **1991**, *88*, 1451.

10. J. Nunberg, W. Schleif, E. Boots, J. O'Brien, J. Quintero, J. Hoffman, E. Emini, and M. Goldman; *J. Virol.* **1991**, *65*, 4887. K. De Vreese, Z. Debyser, R. Pauwels, J. Desmyter, E. De Clercq, and J. Anne; *Virology* **1992**, in press.

11. G. Pialoux, M. Youle, B. Dupont, B. Gazzard, G. Cauwenbergh, P. Stoffels, S. Davies, J. De Saint Martin, and P. Janssen; *Lancet* **1991**, *338*, 140.

19

Discovery and Development of a Non-nucleoside Reverse Transcriptase Inhibitor

Julian Adams and Karl D. Hargrave

DEPARTMENT OF MEDICINAL CHEMISTRY, BOEHRINGER INGELHEIM
PHARMACEUTICALS, INC., 900 RIDGEBURY ROAD, RIDGEFIELD,
CT 06877, USA

1 INTRODUCTION

The HIV viral life cycle offers many opportunities for therapeutic intervention.[1,2] For example, inhibitors of viral entry such as soluble CD4 receptors block the entry of the virion into the target cell by interfering with the interaction of the gp120 glycoprotein on the surface of the virion with the CD4 receptor on the surface of certain target (T) cells. Inhibitors of either the DNA polymerase or ribonuclease (RNase) H functions of reverse transcriptase (RT) block the synthesis of viral DNA in newly infected cells, and inhibition of the integrase enzyme would prevent the incorporation of the double stranded viral DNA into the host cell genome. Modulation of transcriptional regulators such as Tat are reported to prevent the formation of new viral products, and protease inhibitors have been demonstrated to block the processing and packaging of the newly synthesized viral proteins which are used to form new virions. It was our intention at the outset to utilize a mechanistic approach in the design of anti-HIV agents, and the enzyme reverse transcriptase was chosen as the target because of the uniqueness of this enzyme to retroviruses and because of the fact that AZT, the only clinically approved anti-HIV agent at that time, was known to block virus replication by inhibiting reverse transcriptase.

2 DISCOVERY OF THE DIPYRIDODIAZEPINONES

Reverse transcriptase is a heterodimer composed of p66 and p51 kD subunits. The polymerase domain is located in the p66 subunit, with the RNase H domain comprising the 15 kD portion of the C-terminal end of the p66 subunit. A standard assay was established using cloned p66/p51 heterodimer with poly rC as the template, oligo dG as the primer, and ^{3}H-dGTP as the substrate.[3] In part because of the toxicological disadvantages due to the lack of specificity and relatively short half-lives of the known nucleoside RT inhibitors, it was decided to utilize a high capacity screen to test a structurally diverse group of compounds in order to discover a non-nucleoside inhibitor to serve as a lead compound in a rational design program. A weakly potent (6 μM) pyrido[2,3-b]-[1,4]benzodiazepinone (1) was found which was structurally related to the M_1-selective anti-muscarinic drug pirenzepine (2, Figure 1). Evaluation of representative

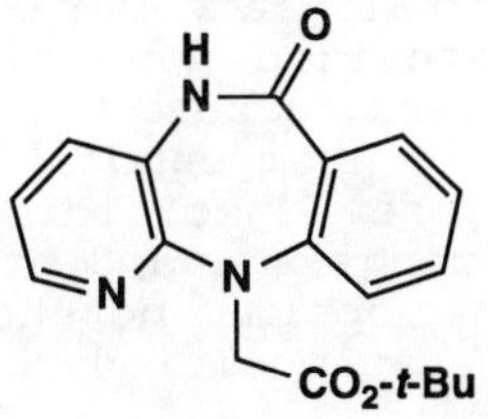

1, IC_{50} = 6 μM

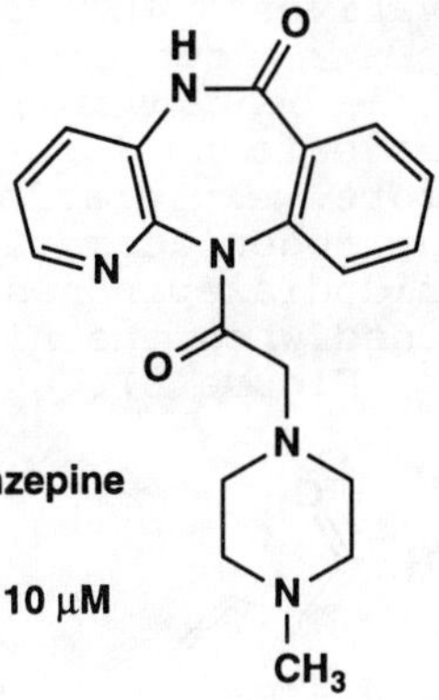

2, Pirenzepine

IC_{50} >> 10 μM

Figure 1   HIV-1 reverse transcriptase high capacity screening

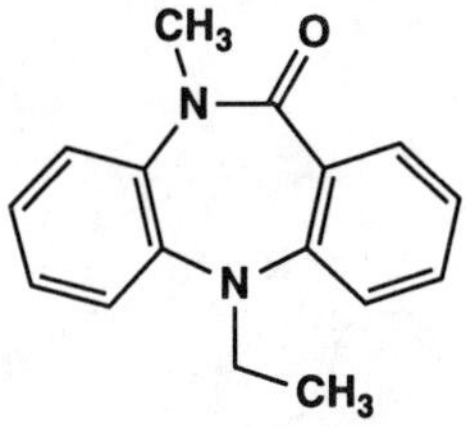

3, IC_{50} = 1500 nM

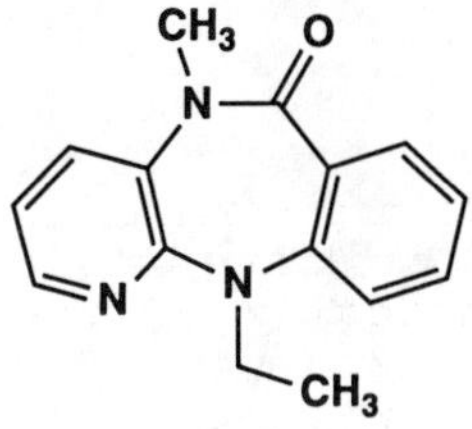

4, IC_{50} = 500 nM

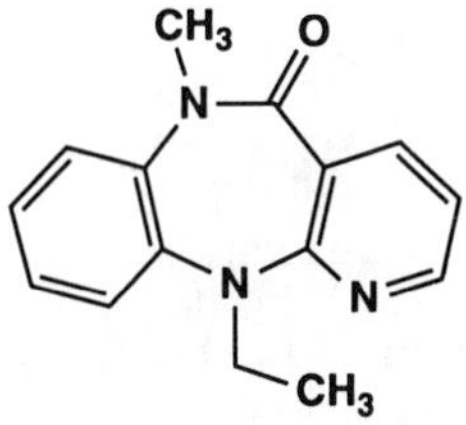

5, IC_{50} = 350 nM

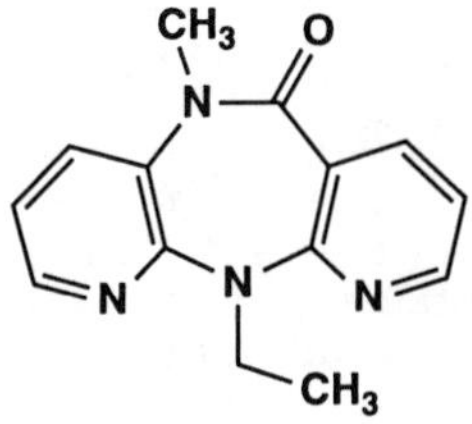

6, BI-RH-397
IC_{50} = 125 nM

Figure 2   Diazepinones

examples of closely related analogs suggests that, in general, the dibenzo[b,e][1,4]diazepinones (3, Figure 2) are less potent than the corresponding pyrido[2,3-b]-[1,4]benzodiazepinones (4) which, in turn, are slightly less potent than the isomeric pyrido[2,3-b][1,5]benzo-diazepinones (5). The preferred analogs are the di-pyrido[3,2-b:2',3'-e][1,4]diazepinones (6) which are equivalent to or slightly more potent than the monopyrido derivatives. In addition, they generally are more sol-uble, less cytotoxic, and are metabolized more slowly than the benzopyrido analogs. For these reasons the syn-thetic efforts focused on the dipyrido compounds.[4] Fur-thermore, exploration of the positional specificity of the nitrogen atoms in the pyridine rings A and C of the dipyridodiazepinones demonstrated that optimum potency is obtained when the nitrogen atoms are at positions 1 and 10 (7, Figure 3).

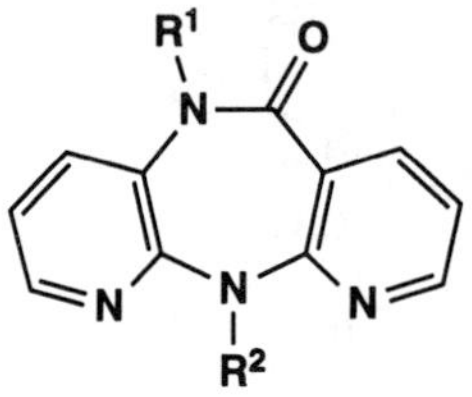

7

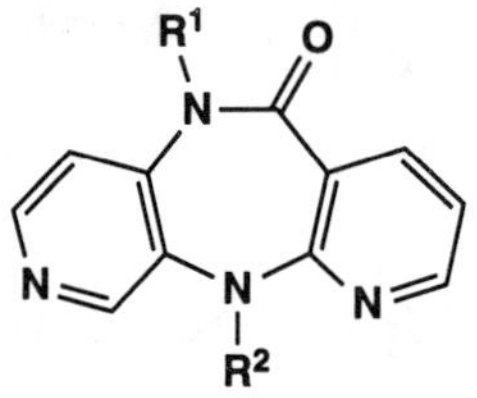

8

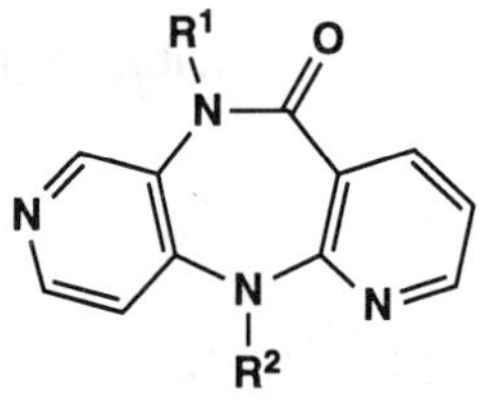

9

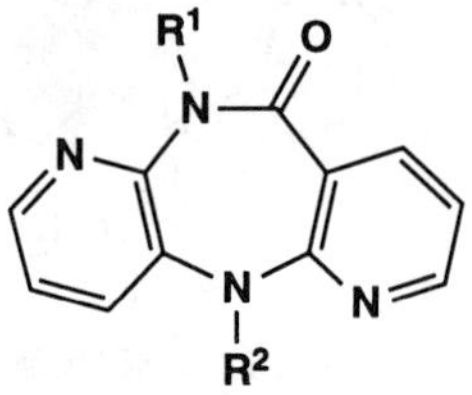

10

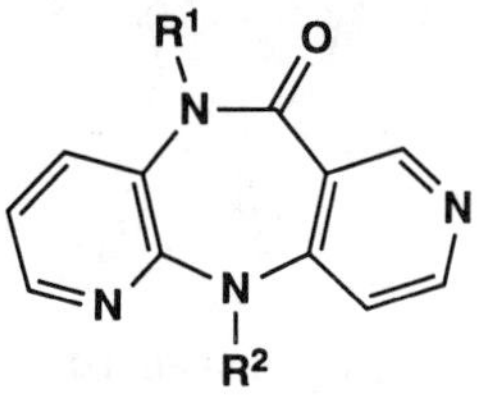

11

Figure 3   Isomeric dipyridodiazepinones

For the ultimate selection of a clinical candidate, the criteria set forth included the requirement that the compound be potent and specific for HIV, that it exhibit reasonable aqueous solubility and bioavailability, that it have a relatively long half-life in vivo, and that it be relatively non-toxic. In addition, since the HIV virus enters the brain it is desirable that inhibitors demonstrate good distribution between the plasma and brain. Lacking predictive pharmacological models, the demonstration of in vivo (animal) efficacy was not a requirement.

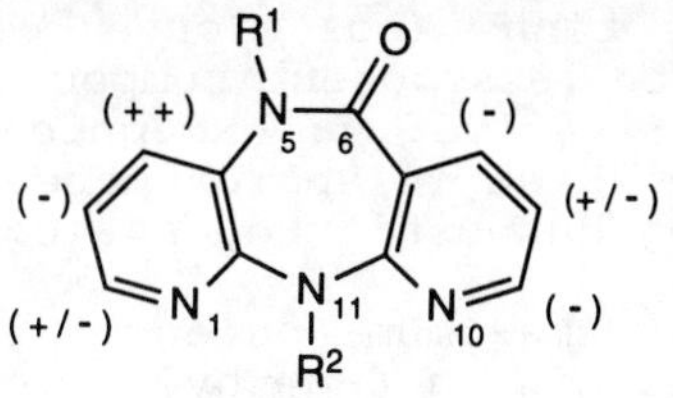

12

<u>Figure 4</u>   Summary SAR of pyridine-ring substituents

Consideration of the structure-activity relationships of dipyrido[3,2-*b*:2',3'-*e*][1,4]diazepinones reveals
the following generalizations: optimum potency is
achieved when there is a small lipophilic substituent on
the nitrogen at position eleven (Figure 4); an ethyl or
cyclopropyl group is generally best, with smaller or
larger lipophilic substituents, or hydrophilic or polar
substituents producing compounds with less potency.
Similarly, a small lipophilic substituent at the lactam
nitrogen (position 5) is generally preferred, with methyl
being optimum except in the cases noted (vide infra).
Exceptions to this observation are clearly seen in A-ring
methylated analogs in which the *N*-5 methyl and *N*-5
unsubstituted compounds are compared.   When the A-ring
methyl substituent is at position 2, then the general-
ization holds that the *N*-5 methyl analog is more potent
than the corresponding *N*-H derivative.   When the A-ring
methyl group is at position 3, the same trend is observed, but the potency differences are not significant.
In the case of the 4-methyl derivative, however, the
reverse is true.   That is, the unsubstituted lactam
nitrogen analog (IC_{50} = 35 nM) is much more potent than
the *N*-methyl analog (IC_{50} = 1.9 µM).   The reason for the
significant drop in potency in the 4,5-dimethyl analog is
not clear, though one factor which may be involved is the
deviation from planarity of the amide bond which results
from the unfavorable steric interaction of the two methyl
groups.   This deviation from planarity (and consequent
reduction of electron delocalization) has been observed
to be 13° in the crystal structure of the *N*-5 methyl
analog of nevirapine.[5]

A more detailed examination of the effects on
potency of substituents on the two pyridine rings
indicates that substitution at position 2 is well
tolerated, especially with small lipophilic moieties. In
contrast, all substitutions at position 3 result in a
reduction in potency.   Although a wide variety of
functional groups have been placed at position 4, a
methyl group is optimum.   In fact, this moiety, in
combination with an unsubstituted lactam nitrogen, is
found in all of the most potent compounds.

In the C ring (positions 7-9), substitution generally leads to less potent compounds compared to the unsubstituted derivative. An exception to this is the 8-amino derivative which is approximately equipotent with the corresponding unsubstituted analog.

In summary, maximum potency is achieved with derivatives containing a 4-methyl moiety in combination with an unsubstituted lactam nitrogen, a small lipophilic group at position N-11, and no substituents on positions 3, 7, 8, and 9.

On the basis of preclinical evaluation of the most promising analogs, the two compounds (Figure 5) selected as potential clinical candidates were the N-11 ethyl and cyclopropyl derivatives [13, BI-RH-414 and 14, BI-RG-587 (nevirapine), respectively]. After extensive comparison studies, the only distinguishing difference between the two compounds was the higher plasma levels of the cyclopropyl derivative in the rat and cynomolgus monkey following oral administration. Subsequent investigations of distribution in the cynomolgus monkey with nevirapine have demonstrated good distribution between the plasma and various other compartments.[6]

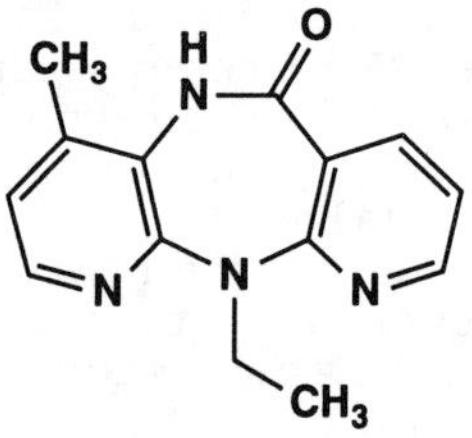

13, BI-RH-414

IC_{50} = 35 nM

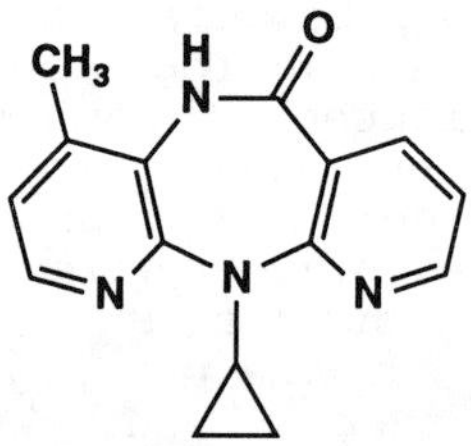

14, BI-RG-587 (Nevirapine)

IC_{50} = 84 nM

Figure 5 Potential clinical candidates

Following the selection of nevirapine as a clinical candidate, numerous synthetic pathways were explored. The synthetic route utilized is shown in Scheme 1 and makes use of the commercially available 2-chloro-4-methyl-3-nitropyridine (15).[4] Reduction of the nitro moiety is effected by Raney nickel hydrogenation, and the resulting 3-amino-2-chloro-4-methylpyridine (16) is condensed with 2-chloronicotinoyl chloride to form the amide (17). Subsequent treatment with cyclopropylamine at elevated temperatures leads to displacement of the chloro substituent almost exclusively at the position *ortho* to the amide carbonyl to give the amine 18, and ring closure to the desired compound (14, nevirapine) occurs upon reaction with sodium hydride. Overall yield of the synthesis from 2-chloro-4-methyl-3-nitropyridine (15) is approximately 70%.

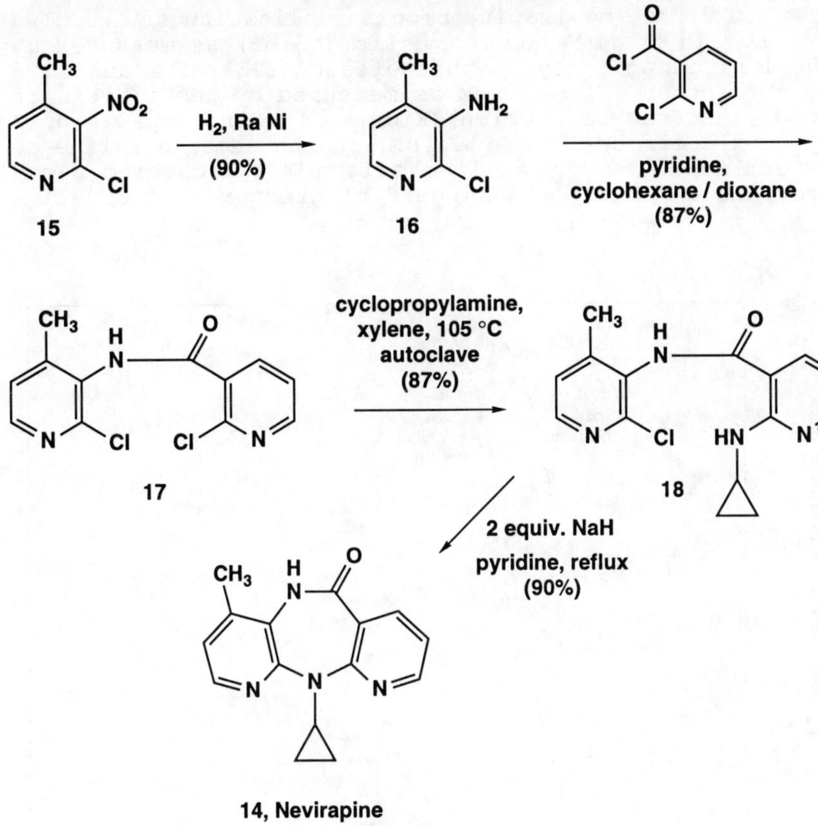

Scheme 1 Synthesis of Nevirapine

3 DEVELOPMENT OF NEVIRAPINE

Nevirapine exhibits a highly reproducible dose response curve when tested in the RT assay and using an RNA template as described above, with an IC_{50} = 84 nM and a K_i = 200 nM. When the assay is performed utilizing DNA as the template, the measured IC_{50} = 196 nM. Partial inhibition of RNase H is also observed with nevirapine, suggesting allosteric inhibition. It is highly specific for HIV-1 RT, being inactive against reverse transcriptases derived from HIV-2, simian immunodeficiency virus (SIV), feline leukemia virus (FLV), murine virus, and avian virus. Significantly, and in contrast to the nucleoside triphosphate analogs which inhibit HIV-1 RT, nevirapine has no effect against the human DNA polymerases α, β, γ, and δ. In addition, when evaluated in a variety of receptor binding assays, nevirapine showed no significant effect.[3]

In vitro, nevirapine blocks replication of the HIV-1_{IIIb} strain in CD4$^+$ human T cells (c8166) as measured by the inhibition of cytopathic effect (CPE) with an IC_{50} = 40 nM, and an IC_{50} = 10 nM as measured by the inhibition of p24 antigen production (Figure 6). In comparison, a cytotoxic effect (CC_{50} = 321 μM) in the same cell line as determined by a tetrazolium salt (MTT) metabolic assay provides a "therapeutic index" of greater than 8000.[3]

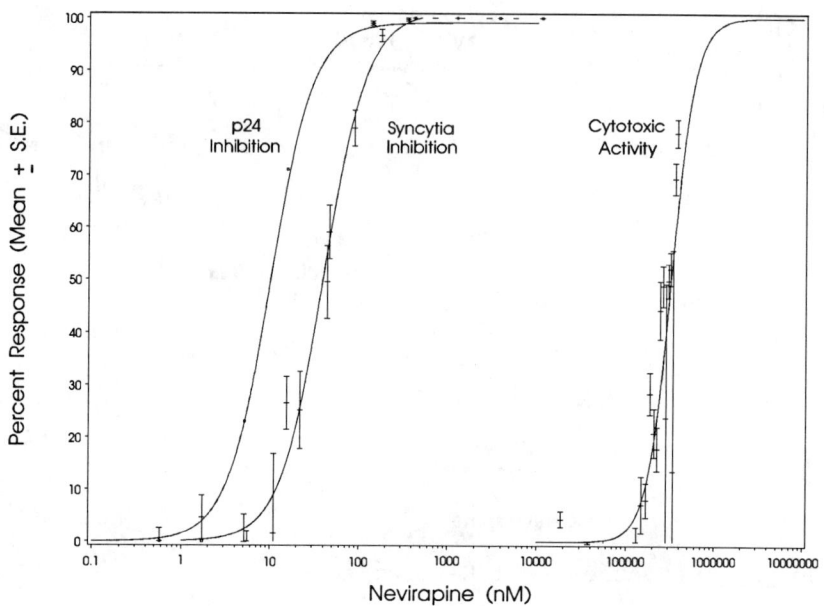

<u>Figure 6</u> Effect of nevirapine on HIV-1 replication and cytotoxic activity in c8166 T cells.[3]   Reprinted by permission.

Early evidence suggesting that this series of compounds exerts its antiviral effect by inhibiting reverse transcriptase was provided by evaluating a group of seventeen analogs with low to high potency against the enzyme. A plot (Figure 7) of the IC_{50} inhibition data for the enzyme and syncytia tests shows the high correlation (Spearman Rank Correlation, r = 0.94) between these two assays.[7]

In an in situ hybridization study with a clinical isolate which had been passaged only in human peripheral blood mononuclear cells (PBMC), many HIV-1 infected cells stained dark purple in the control group, whereas no such infected cells were observed in the group treated with 375 nM nevirapine.[3]

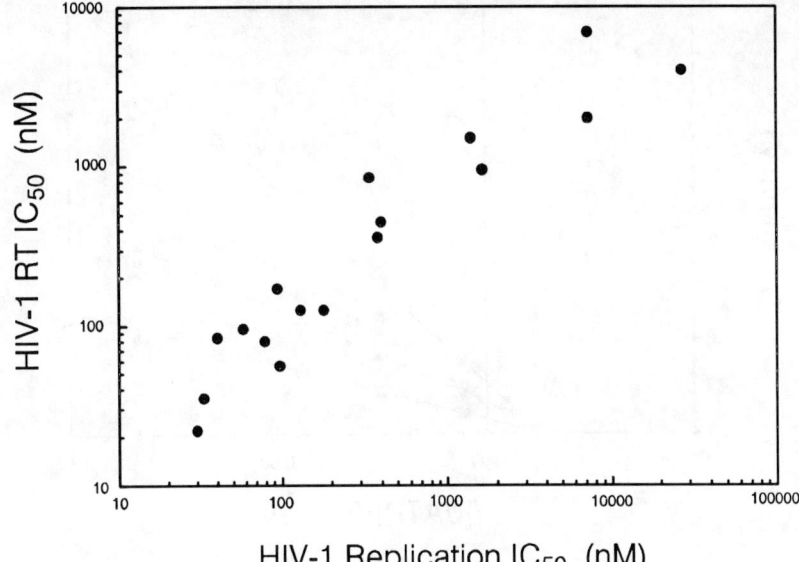

Figure 7 Comparative effect of dipyridodiazepinones on HIV-1 RT and HIV-1 replication

In addition to the cellular inhibition data noted above, a variety of other strains of HIV-1, e.g., HIV-1_{BRU}, HIV-1_{RF} (a Haitian strain), and HIV-1_{UMGL} (a clinical isolate) have been tested in several cell lines including human peripheral blood mononuclear cells (PBMC), CD4$^+$ HeLa cells, and CEM T-cells, and the results were found to give similar 50% inhibition concentrations. As expected based on the lack of activity against HIV-2 RT, nevirapine had no effect against the HIV-2_{ROD} at concentrations up to 100 μM. Importantly, an AZT-resistant strain from a clinical isolate exhibited no cross resistance to nevirapine. In addition, synergy of nevirapine with AZT was observed at all concentrations tested using the isobologram method.[3,8]

Having established the fact that nevirapine is a potent and specific inhibitor of HIV-1 replication, it was desirable to determine the exact mechanism by which it inhibits the reverse transcriptase enzyme. The approaches utilized included kinetics, photoaffinity labeling and mapping, site directed mutagenesis, spectroscopic studies, and X-ray crystallography. Nevirapine was shown to be a noncompetitive inhibitor of RT by a double reciprocal plot using dATP as the variable substrate and a template-primer concentration that was fixed and saturating (Figure 8). The sequence of the template-primer was identical to the natural HIV genomic initiation site. An analogous double reciprocal plot was

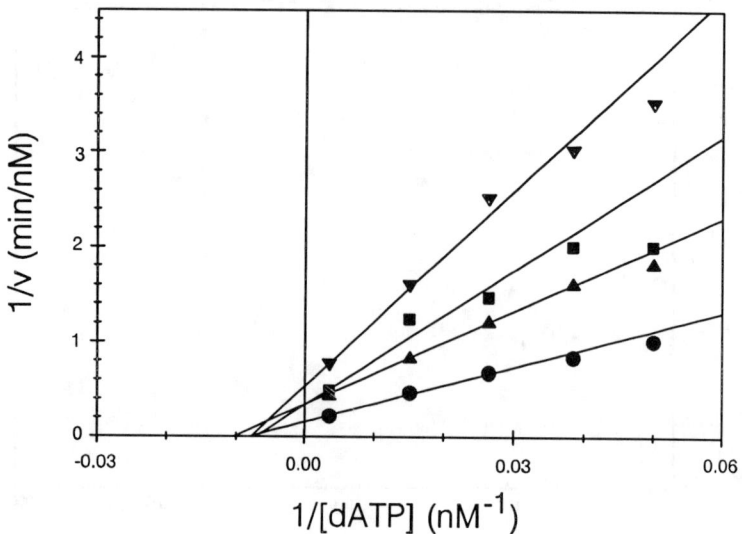

<u>Figure 8</u> A double reciprocal plot of inhibition of HIV-1
RT by nevirapine using a heteropolymeric template-primer
with dATP as the variable substrate. Concentrations of
nevirapine were as follows: • 0 nM, ▲ 60 nM, ■ 125 nM,
▼ 330 nM.[9] Reprinted by permission.

used varying concentrations of this heteropolymeric
template-primer and maintaining the nucleotide
concentration fixed and saturating. The results from
these studies demonstrate that nevirapine is an
allosteric inhibitor and, being noncompetitive with
respect to substrates, binds to both the enzyme:template-
primer and the enzyme:template-primer:nucleotide
complexes of the enzyme.[9]

Having established the noncompetitive (allosteric)
nature of the binding, it was desirable to determine the
exact location of the binding site. To facilitate these
studies, a radiolabeled photoaffinity probe was
synthesized which contained an azido group which, when
irradiated with UV radiation while bound to the enzyme,
would form a reactive nitrene intermediate which in turn
would bind covalently to the enzyme. In the first
experiments with this probe, it was shown that the enzyme
alone was not adversely affected by UV irradiation, and
that reversible inhibition of RT was obtained with the
probe provided that the probe:enzyme mixture was
maintained in the dark. However, when the probe:enzyme
mixture (1:1 stoichiometry) was subjected to UV
irradiation, a time-dependent irreversible inactivation
of the enzyme was observed. In order to establish that
nevirapine and the photoaffinity probe were binding to
the same site on the enzyme, a similar photoinactivation
experiment was carried out using varying concentrations

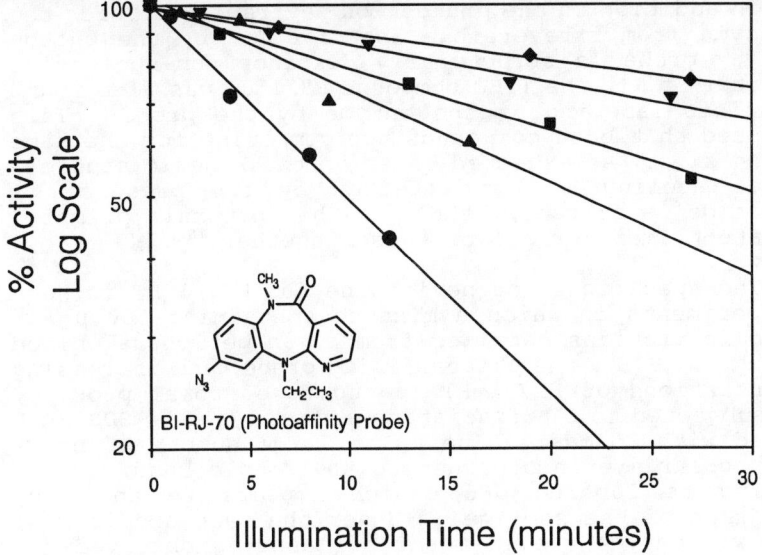

Illumination Time (minutes)

<u>Figure 9</u> Time dependent inhibition of HIV-1 RT by a photoaffinity probe in the presence of nevirapine. Concentrations of nevirapine were as follows: • 0 nM, ▲ 200 nM, ■ 500 nM, ▼ 800 nM, ♦ 1500 nM.

of nevirapine. The findings showed that increasing concentrations of nevirapine provided increasing protection of the enzyme from photoinactivation and thus confirm that the two compounds occupy the same site on the enzyme (Figure 9).[10]

In a separate experiment to determine the stoichiometry and localization of RT labeling by the photoaffinity probe, aliquots were removed at various time points from an irradiated reaction mixture containing a three-fold excess of the probe relative to enzyme. The filtrate was then assayed for enzyme activity, radioactivity, and protein concentration. A plot of the ratio of specific activity of labeled RT to unlabeled RT versus the number of moles of covalently linked probe per mole of RT showed that one mole of the probe was sufficient to inactivate one mole of RT enzyme. When a large excess of the probe was added late in the experiment, some non-specific labeling was observed. An autoradiograph of aliquots which had been analyzed by SDS-polyacrylamide gel electrophoresis (SDS-PAGE) clearly showed that during the first phase of the experiment before the addition of a large excess of the probe, only the p66 subunit of the enzyme was labeled; some labeling of the p51 subunit occurred only in the second phase of the experiment after a large excess of the probe had been added.[10]

In addition to the protection nevirapine affords the RT enzyme from irreversible inactivation by the photo-affinity probe, a structurally distinct non-nucleoside inhibitor of RT, the TIBO compound R82150, was also found to inhibit labeling of the enzyme by the probe. This suggested that both compounds are competing for the same binding site. As expected based on the kinetic studies, enzyme labeling is not affected by the presence of nucleotide substrate, tRNA^{lys} (the natural primer), template-primer, or all of these together.[10]

The specificity of nevirapine for HIV-1 RT is seen in experiments in which a mixture containing 100 µg of cytosolic proteins obtained from human peripheral blood leukocytes (PBL), the photoaffinity probe, and increasing concentrations of HIV-1 RT (up to 5% of total protein) were subjected to UV irradiation and analyzed by SDS-PAGE stained with Coomasie blue. The large number of bands which appear are in contrast to the single band, corresponding to labeled p66, which appears in an autoradiograph of the same gel. Under the same conditions, there was only a small amount of labeling observed for both the p68 and p54 subunits of HIV-2 RT and for BSA.[10]

In order to establish exactly where nevirapine binds on the p66 subunit of HIV-1 RT, the labeled enzyme was subjected to tryptic digestion, isolation of the labeled fragments by high-performance liquid chromatography, and these fragments subjected to peptide sequencing.[11] Comparison of the chromatographic traces of radioactivity and of absorbance at 335 nM were used to indicate which of the more than 65 fragments contained the probe. The labeled peptide fragment was sequenced, and was found to correspond to residues 174-199 (Figure 10). Analysis of the radioactivity of the isolated amino acids during the sequencing indicated that the primary sites labeled were the tyrosine residues at positions 181 and 188. This

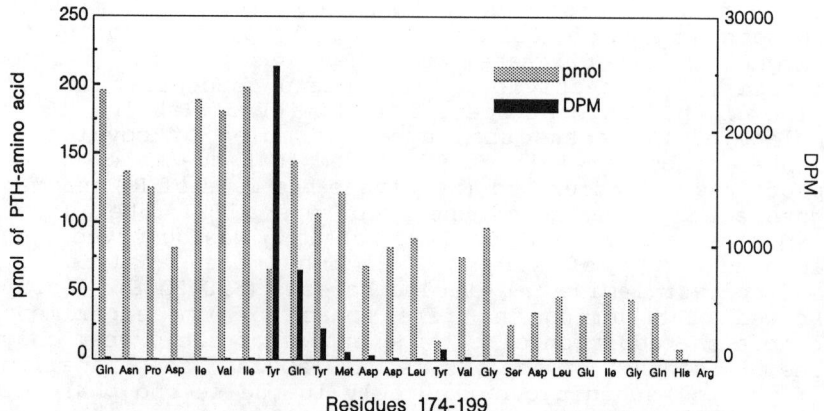

Residues 174-199

Figure 10 Sequence of HIV-1 RT tryptic fragment labeled with photoaffinity probe.[11] Reprinted by permission.

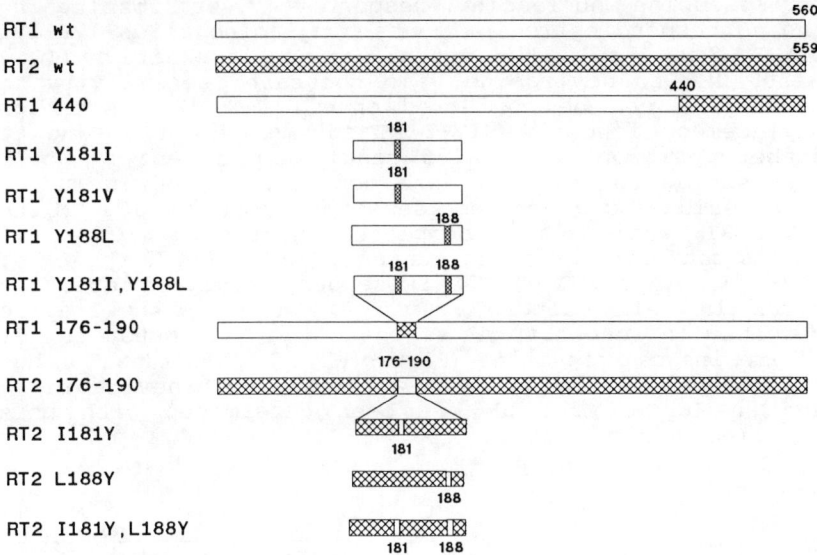

<u>Figure 11</u>   Schematic representation of chimeric con-
structs between HIV-1 and HIV-2 RTs. Open boxes indicate
the HIV-1 RT coding sequence. Shaded boxes indicate the
coding sequences of HIV-2$_{ROD}$ RT.   Position numbers
indicate amino acid residues at which the substitution
was made – e.g., "RT1 Y181I" is HIV-1 RT in which
tyrosine (Y) was replaced with isoleucine (I).   Other
abbreviations are V for valine and L for leucine.   RT1
176-190 has amino acids 176-190, inclusive, from HIV-2
substituted into HIV-1 RT.   The amino acid sequence of
peptide 176-190 from HIV-1 RT is PDIVIYQYMDDLYVG and the
sequence from HIV-2 RT is KDVIIIQYMDDILIA.[12]   Reprinted
by permission.

region of amino acids 181-188 of the p66 subunit is
therefore believed to be involved in the binding of
nevirapine as well as the TIBO compound and other
reported non-nucleoside HIV-1 RT inhibitors.

Support for the results from the mapping work was
provided by site directed mutagenesis studies.[12]
Replacement of the tyrosine at position 181 by isoleucine
which is found in the corresponding position in HIV-2 RT
results in an active enzyme which has lost most of its
sensitivity to nevirapine (Figure 11).   For comparison,
the IC$_{50}$ of nevirapine using the crude lysate containing
this mutant enzyme is 73 μM compared to 0.25 μM for crude
lysate containing the wild-type (i.e., [181]tyrosine)
enzyme.   Similarly, point mutation of the tyrosine at
position 188 with the leucine found at the corresponding
position in HIV-2 RT results in a complete loss of
sensitivity to nevirapine (IC$_{50}$ > 200 μM).   Not unexpectedly, a double mutation at both positions 181 and 188

to isoleucine and leucine, respectively, or substituting the corresponding region containing amino acids 176-190 of HIV-2 RT resulted in mutant enzymes insensitive ($IC_{50}s$ > 200 μM) to nevirapine. In contrast, sensitivity to nevirapine can be conferred onto HIV-2 RT. Although replacement into the HIV-2 RT enzyme by a tyrosine at either position 181 or 188 results in an enzyme still insensitive to nevirapine, a double mutation including both residues confers some sensitivity (IC_{50} = 121 μM) to HIV-2 RT, while almost complete sensitivity (IC_{50} = 0.8 μM) is conferred by substitution of the 176-190 region of HIV-1 RT into HIV-2 RT. Subsequent studies passaging virus in cell culture in the presence of nevirapine results in the rapid emergence of a mutation of [181]tyrosine to [181]cysteine (Figure 12).[13] This mutation has also been observed in clinical trials with nevirapine,[14] and an IC_{50} of 2.6 μM has been determined with this purified mutant enzyme.[15]

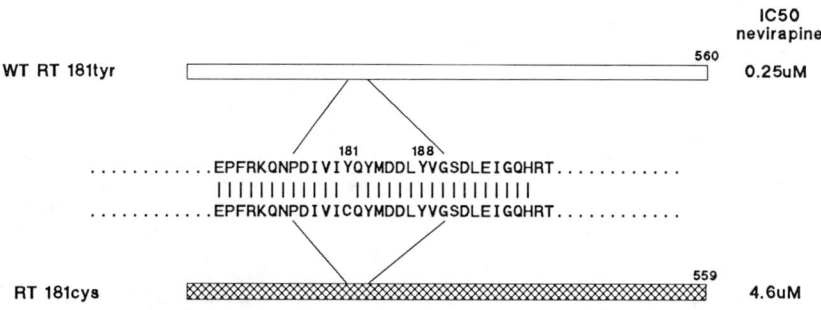

Figure 12   HIV-1 RT point mutation at [181]tyrosine

On the basis of the photoaffinity probe, mapping, and site-directed mutagenesis studies described here, the binding site for nevirapine was demonstrated to be located somewhere near the polymerase catalytic site on the p66 subunit of reverse transcriptase. Nevertheless, in the design of more potent inhibitors of the enzyme, it is desirable to know its detailed 3-dimensional structure. Although all attempts at obtaining crystals suitable for X-ray structural determination had been unsuccessful in various laboratories, it was felt that co-crystallization of the enzyme with an inhibitor such as nevirapine may prove successful. After exploring a large number of crystallization conditions which included purified HIV-1 reverse transcriptase and nevirapine, data from crystals which were obtained of the RT-nevirapine complex were collected and the structure was solved at 3.5 Å resolution.[16] The results establish that nevirapine and, by implication, the other non-nucleoside RT inhibitors, is close to the non-mutable aspartic acid residues at positions 110, 185 and 186, which are believed to be located at the polymerase catalytic site on the p66 subunit. In addition, the proximity of the A-ring of nevirapine to the tyrosine residues at positions 181 and 188 supports the possibility of a similar binding orientation (vis-à-vis the amide linkage) of the photo-affinity probe. It has been suggested[16] that nevirapine may inhibit the enzyme by reducing the conformational flexibility which the enzyme needs for catalysis and/or it may alter the conformation of residue(s) critical for catalysis. This is supported by spectroscopic studies[17] which indicate that the native enzyme undergoes a confor-mational change upon binding of the nevirapine, and this inhibitor-induced conformational change may be sufficient to block the polymerase activity.

4   CONCLUSION

Nevirapine entered Phase I clinical trials in early 1991. In these initial single dose studies in patients, the drug was administered orally at doses ranging from 2.5 to 400 mg. The results indicate that the drug is well tolerated and safe up to the maximum dose tested. Furthermore, excellent bioavailability was demonstrated, and plasma concentrations of nevirapine increased with increasing dose, though maximum plasma concentration and area under the plasma level curve tended toward non-linearity at the highest doses. Maximum plasma levels are achieved approximately one hour after drug administration, and nevirapine has a half-life of > 24 hours which suggested once-a-day dosing in the subsequent multiple-dose trials. Although the lower dose mono-therapy (multiple-dose) trials were discontinued due to the relatively rapid appearance of virus with reduced sensitivity to nevirapine, preliminary results from the 400 mg multiple-dose monotherapy trials and the nevirapine-AZT combination multiple-dose trials are sufficiently encouraging to warrant continuing clinical studies to determine the therapeutic efficacy of nevirapine in the treatment of HIV-1 infection.

REFERENCES

1.   W. C. Greene, <u>N. Eng. J. Med.</u>, 1991, <u>324</u>, 308.
2.   W. A. Haseltine, <u>FASEB J.</u>, 1991, <u>5</u>, 2349.
3.   V. J. Merluzzi, K. D. Hargrave, M. Labadia, K. Grozinger, M. Skoog, J. Wu, C.-k. Shih, K. Eckner, S. Hattox, J. Adams, A. S. Rosenthal, R. Faanes, R. J. Eckner, R. A. Koup, and J. L. Sullivan, <u>Science</u>, 1990, <u>250</u>, 1411.
4.   K. D. Hargrave, K. D., J. R. Proudfoot, K. G. Grozinger, E. Cullen, S. R. Kapadia, U. R. Patel, V. U. Fuchs, S. C. Mauldin, J. Vitous, M. L. Behnke, J. M. Klunder, K. Pal, J. W. Skiles, D. W. McNeil, J. M. Rose, G. Chow, M. T. Skoog, J. C. Wu, G. Schmidt, W. W. Engel, W. G. Eberlein, T. D. Saboe, S. J. Campbell, A. S. Rosenthal, and J.

Adams, J. Med. Chem., 1991, 34, 2231.

5. P. W. Mui, S. P. Jacober, K. D. Hargrave, and J. Adams, unpublished results.

6. S. E. Hattox, P. S. Riska, S. H. Norris, J. N. Johnstone, R. L. St. George, H. Silverstein, and J. W. Streett, unpublished results.

7. K. D. Hargrave, R. A. Koup, R. J. Eckner, V. J. Merluzzi, F. E. Brewster, T. Roy, J. P. Sabo, and J. L. Sullivan, unpublished results.

8. D. Richman, A. S. Rosenthal, M. Skoog, R. J. Eckner, T.-c. Chou, J. P. Sabo, and V. J. Merluzzi, Antimicrob. Agents Chemother., 1991, 35, 305.

9. E. B. Kopp, J. J. Miglietta, A. G. Shrutkowski, C.-k. Shih, P. M. Grob, and M. T. Skoog, Nucleic Acids Res., 1991, 19, 3035.

10. J. C. Wu, T. C. Warren, J. Adams, J. Proudfoot, J. Skiles, P. Raghavan, C. Perry, I. Potocki, P. R. Farina, and P. M. Grob, Biochem., 1991, 30, 2022.

11. K. A. Cohen, J. Hopkins, R. H. Ingraham, C. Pargellis, J. C. Wu, D. E. H. Palladino, P. Kinkade, T. C. Warren, S. Rogers, J. Adams, P. R. Farina, and P. M. Grob, J. Biol. Chem., 1991, 266, 14670.

12. C.-k. Shih, J. M. Rose, G. L. Hansen, J. C. Wu, A. Bacolla, and J. A. Griffin, Proc. Natl. Acad. Sci. USA, 1991, 88, 9878.

13. J. W. Mellors, G. E. Dutschman, G.-j. Im, E. Tramontano, S. R. Winkler, and Y.-c. Cheng, Molecular Pharmacol., 1992, 41, 446.

14. D. Richman, unpublished results.

15. E. David and P. M. Grob, unpublished results.

16. L. A. Kohlstaedt, J. Wang, J. M. Friedman, P. A. Rice, and T. A. Steitz, Science, 1992, 256, 1783.

17. C. A. Grygon, D. J. Greenwood, R. C. Cousins, P. McGoff, and J. Stevenson, unpublished results.

20
Structure-based Inhibitors of HIV Protease

D. J. Kempf, D. W. Norbeck, L. Codacovi, X. C. Wang,
W. F. Kohlbrenner, N. E. Wideburg, A. Saldivar, A. Craig-
Kennard, S. Vasavanonda, J. J. Clement, and J. Erickson
PHARMACEUTICAL PRODUCTS DIVISION, ABBOTT LABORATORIES,
ABBOTT PARK, IL 60064, USA

1   INTRODUCTION

The alarming spread of the human immunodeficiency virus
(HIV) types 1 and 2 has prompted unprecedented efforts
aimed at the discovery of chemotherapeutic agents which
retard the progression of or delay the onset of acquired
immunodeficiency syndrome (AIDS). As a result of an
increased understanding of the replication cycle of HIV, a
variety of potential molecular targets for therapeutic
intervention have been identified.[1-6] These include, among
others, the envelope glycoprotein, the regulatory proteins
and the virally encoded enzymes, reverse transcriptase,
integrase and protease. Site-directed mutagenesis of these
targets has produced HIV phenotypes which either are unable
to replicate *in vitro* or show drastically attenuated virulence. Chemical agents which block the action of these gene
products are therefore potential candidates for clinical
therapy.

The HIV protease represents a particularly attractive
target for the development of HIV-specific inhibitors.
Encoded by the *gag-pol* gene, HIV protease plays an essential role in the late stages of the viral life cycle by
proteolytically processing the *gag* and *gag-pol* polyproteins
to release the structural proteins of the viral capsid as
well as the three viral enzymes.[7-11] Blocking the action of
HIV protease through site-directed mutagenesis of residues
in the active site[12-14] or through the use of chemical inhibitors[15-18] results in production of morphologically distinct viral particles which are incapable of maturing to an
infectious form. As a result, inhibitors of the protease
have been shown to effectively block the spread of both
HIV-1 and HIV-2 *in vitro*, and several compounds have
emerged as promising candidates for clinical evaluation.[19,20]

The discovery of potent inhibitors of HIV protease has
progressed rapidly due in part to the identification of the
protease as a member of the aspartic proteinase family.[21,22]
Strategies for inhibitor design based on transition-state

analogues which had previously been employed for the other
aspartic proteinases renin and pepsin[23] were successfully
applied to HIV-1 protease. As a result, potent inhibitors
containing hydroxyethylene,[24-28] hydroxyethylamine,[19,29] di-
hydroxyethylene,[30] phosphinate,[24,31] and other transition
state isosteres[32-34] have been described. In addition, the
unique characteristics of the retroviral proteases in general, and of HIV-1 protease in particular, have inspired
the design of novel inhibitors based on a knowledge of the
three-dimensional structure of the enzyme. In particular,
initial indications of the homodimeric, C_2-symmetric structure of active HIV protease[22,35,36] were subsequently confirmed by X-ray crystallography.[37-39] In response, we[40,41]
and others[42,43] have described C_2-symmetry-based inhibitors
which complement the symmetry characteristics of the HIV
protease active site. Structure-activity studies to
improve the physical properties of lead compounds led to
symmetric and pseudo-symmetric inhibitors with therapeutic
potential.[20]

2   DESIGN OF SYMMETRY-BASED INHIBITORS

The conceptual process which we employed[41] for the design of
C_2-symmetric and pseudo-C_2-symmetric inhibitors is shown
schematically in Figure 1. The initial step entailed the
definition of a two-fold axis of rotation onto the tetrahedral intermediate for cleavage of an asymmetric substrate. Placement of the axis either through or near to
the tetrahedral carbon undergoing catalytic cleavage was
based on mechanistic considerations wherein the two hydro-

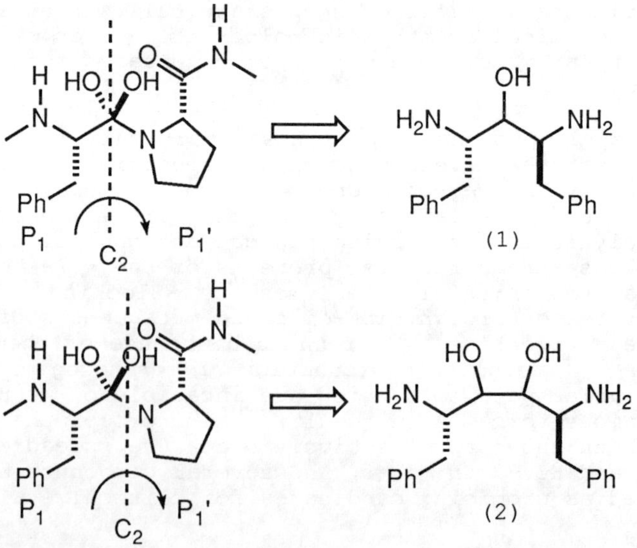

Figure 1. Design of C_2-symmetric inhibitors of HIV
protease. Reprinted with permission from *J. Med. Chem.*,
1990, **33**, 2687. Copyright 1990, American Chemical Society.

gen-bonded catalytic aspartate residues, both of which must be located near the C_2-axis, interact with the amide carbonyl to stabilize the transition state for formation of the tetrahedral intermediate. Further support for this placement was obtained by modeling the structure of an inhibitor of the fungal aspartic proteinase *Rhizopus* pepsin into the active site of Rous sarcoma virus protease.[40] In the second step, the P' portion of the substrate was deleted, leaving the N-terminal half of the dipeptide residue. The arbitrary decision to delete the P' region was based on consideration of the known substrate sequences for HIV-1 protease, which show greater homology in the P_1 site than in the P_1' position, as well as the considerable precedence with inhibitors of renin, wherein compounds truncated in the P' region retain activity. The third step of the design process entailed use of a C_2-rotational operation on the remaining P region of the substrate to generate C_2-symmetric or pseudo-C_2-symmetric inhibitor core units (1) or (2), depending of the placement of the axis in the first step. Diamine (1) is inherently pseudo-C_2-symmetric due to the tetrahedral nature of its central carbon. Diaminodiol (2) exists as three distinct diastereomers, two of which ($3R,4R$ and $3S,4S$) are C_2-symmetric, the other of which ($3R$, $4S$) is pseudosymmetric.

C_2-Symmetric Diaminoalcohols and Diaminodiols

The synthesis of diaminoalcohol (1) and the three diastereomers of diaminodiol (2) have been described.[41,44] Acylation of (1) and (2) with peptide residues led to potent, selective inhibitors of HIV-1 protease.[41] Table 1 provides an overview of the structure-activity relationships for the inhibition of recombinant HIV-1 protease in a fluorogenic inhibition assay[45] for derivatiaves of (1). The parent diamine (1) was only a weak inhibitor. However, diacylated derivatives of (1) showed increased activity, particularly compounds (4) and (5), which contain lipophilic groups to interact with the symmetry-related P_2 subsites. Acylation of (1) with Cbz-protected amino acids produced potent inhibitors of the protease. Compound (6), with one such substitution, showed submicromolar inhibition. Attachment of Boc to the remaining amino group produced the nanomolar inhibitor (7). Substitution of Asn for Val at the P_2 position (compound (8)) was tolerated; however, extension with longer peptide chains, as in compound (9), provided no benefit. Finally, attachment of Cbz-Val to both ends of (1) provided the potent lead compound A-74704, (10).[40,41] Other amino acid protecting groups were less effective than Cbz, as shown by compounds (11) and (12); however, compound (13), with unprotected Val at one end, retained most of the potency of (10). Changes in the symmetry-related P_2 positions are shown with compounds (14)-(16). Substitution with Ile gave (14), which was nearly equipotent to (10). Interestingly, although (8) showed the same potency as (7), the bis Cbz-Asn substituted inhibitor (15) was substantially less potent than (10).

Table 1. Inhibition of HIV-1 protease by derivatives of
         diaminoalcohol (1)

No	A	B	IC$_{50}$ (nM)
(1)	H	H	>50,000[a]
(3)	Ac	Ac	>10,000[b,c]
(4)	Boc	Boc	3,000
(5)	H-Val	H-Val	590
(6)	Cbz-Val	H	305
(7)	Cbz-Val	Boc	17
(8)	Cbz-Asn	Boc	18
(9)	Cbz-Leu-Asn	Boc	21
(10)	Cbz-Val	Cbz-Val	3[b]
(11)	Ac-Val	Ac-Val	12[b]
(12)	Cbz-Val	Boc-Val	90
(13)	Cbz-Val	H-Val	10
(14)	Cbz-Ile	Cbz-Ile	5[b]
(15)	Cbz-Asn	Cbz-Asn	90[b]
(16)	Cbz-Leu	Cbz-Leu	>100[b,d]

[a]19% inhibition at 50 μM. [b]Reference 41. [c]12%
inhibition at 10 μM. [d]15% inhibition at 100 nM.

Substitution with Leu was even less favorable (compound
16).

    Attachment of Cbz-Val to diaminodiol (2) provided the
diastereomeric inhibitors (17)-(19) with subnanomolar
inhibition constants for HIV-1 protease.[41] Unexpectedly,
3R,4R-diol (17), 3S,4S-diol (18) and 3R,4S-diol (19) were
all of similar potency, in contrast to observations of

(17)   (18)

(19)

(20)

others with inhibitors based on transition-state analogues.[19,24,25,29,46,47] The implications of this lack of
stereochemical dependence on the modes of binding of the
symmetry-based inhibitors have been addressed elsewhere.[48]
The high lipophilicity and low aqueous solubility of (10)
and (17)-(19) prompted the design of analogues in which the
Cbz phenyl groups were replaced by pyridinyl groups. Investigation of the antiviral and pharmacokinetic properties
of the resulting inhibitors led to the identification of A-
77003 (20), which showed good aqueous solubility while
retaining high potency *in vitro*.[20]

Structure-Activity Studies on Symmetry-Based Inhibitors

In view of the superior activity of the diol-based
inhibitors (17)-(19) over monohydroxy inhibitor (10), we
sought to ascertain the factors responsible for the significant potency difference. Structurally, (17)-(19) are
derived by insertion of -CH(OH)- into (10). It was therefore of interest to determine whether the increase in
potency resulted from the presence of the second hydroxyl
group of (17)-(19) or from the additional length of the
inhibitor provided by the insertion of a carbon atom.
Structure-activity relationships observed with analogues of
(17)-(19) indicated that the contribution of the two hydroxyl groups to the total binding energy of the inhibitor
may not be equal. For example, the K_i values for the
unsymmetrically substituted 3R,4S inhibitors (21) and (22)
differed by a factor of five, indicating that the bound
structures of (21) and (22) were not symmetrically dis-
posed. In order to determine the relative contribution of

(21), K_i = 3.9 nM

(22), K_i = 0.75 nM

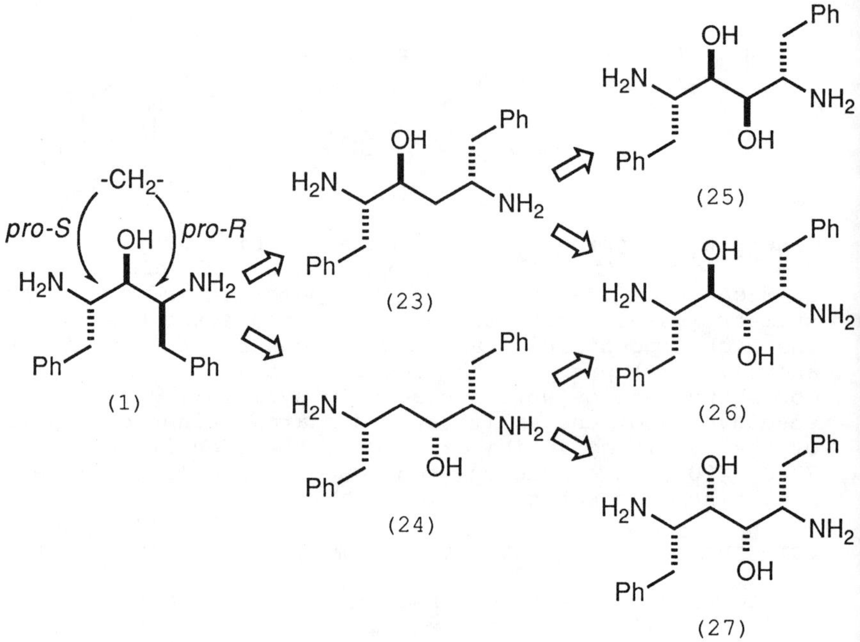

Figure 2. Design of symmetry-based core diamines (23) and (24).

the two hydroxyl groups, we therefore prepared inhibitors in which the second hydroxyl group of the core unit was modified. In particular, we were interested in diamines (23) and (24), which represent conceptual intermediates between diamine (1) and the three diastereomeric diamino-diols (25), (26) and (27). As shown in Figure 2, insertion of a methylene group into the *pro-R* carbon-carbon bond of (1) generates 3S-diaminoalcohol (23). Hydroxylation of the methylene group gives rise to either R,R-diol (25) or R,S-diol (26). On the other hand, insertion of methylene into the *pro-S* carbon-carbon bond of pseudo-C$_2$-symmetric (1) leads to the 3R-isomer (24), with subsequent hydroxylation giving either S,S-diol (27) or, again, R,S-diol (26).

Our initial synthesis of diamine (23) is shown in Scheme 1. Diol (28)[44] was converted to cyclic sulfate (29) by the two-step protocol of Sharpless.[49] The cyclic sulfate was opened by treatment with sodium thiophenoxide at 150°C to provide thioether (30), which was directly reduced with Raney nickel to give the protected diamine (31). Deprotection with trifluoroacetic acid provided (23). The electrophilicity of cyclic sulfate (29) also allowed access to other core units of interest. Opening of (29) with azide anion led to the azidoalcohol (32), which upon reduction provided aminoalcohol (33).

A more efficient synthesis of diamines (23) and (24) is shown in Scheme 2. Treatment of the Cbz protected R,R-

Scheme 1

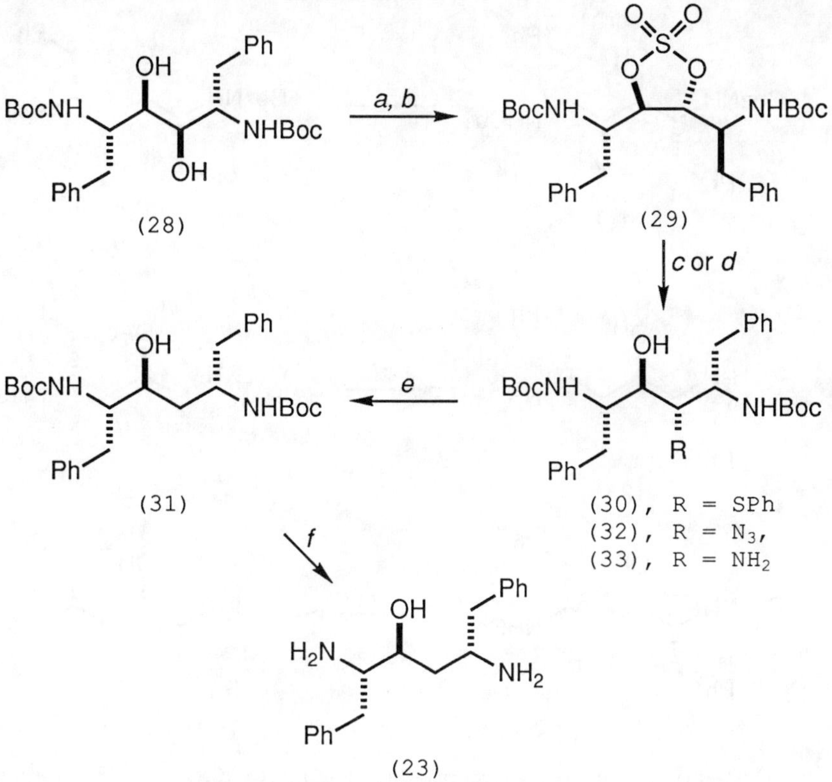

(28)

a, b

(29)

c or d

(31)

e

(30), R = SPh
(32), R = N$_3$,
(33), R = NH$_2$

f

(23)

[a]SOCl$_2$, 4-methylmorpholine, CH$_2$Cl$_2$; [b]RuCl$_3$, NaIO$_4$, CH$_3$CN, CCl$_4$, H$_2$O; [c]PhSH, NaOCH$_3$, DMF, 150°C; [d]LiN$_3$, DMF, 150°C; [e]RaNi, EtOH, reflux; [f]CF$_3$CO$_2$H, CH$_2$Cl$_2$

diaminodiol (34)[44] with α-acetoxyisobutyryl bromide[50] in the presence of lithium bromide provided the acetoxybromide (35). The presence of excess lithium bromide was essential in order to prevent intramolecular displacement of the activated hydroxyl group with the neighboring Cbz carbonyl oxygen.[44] Reductive debromination of (35) with tri-*n*-butyl tin hydride gave the acetate (36) which was hydrolyzed to diamine (23). In a similar fashion, protected S,S-diaminodiol (37)[44] was converted to (24).

Table 2 lists the inhibition constants of the di-Boc derivatives of various diamine core units against recombinant HIV-1 protease. Diols (28), (38) and (39)[41] were all substantially more potent than the alcohol (4). Unexpectedly, the deoxygenated inhibitor (31) was twice as active as the most potent diol, (39). Conversely, the amino-alcohol (33) was 40-fold less active than (39). The activity of the intermediate azidoalcohol (32), which was apparently too large to efficiently bind into the active site, showed an additional decline.

Scheme 2

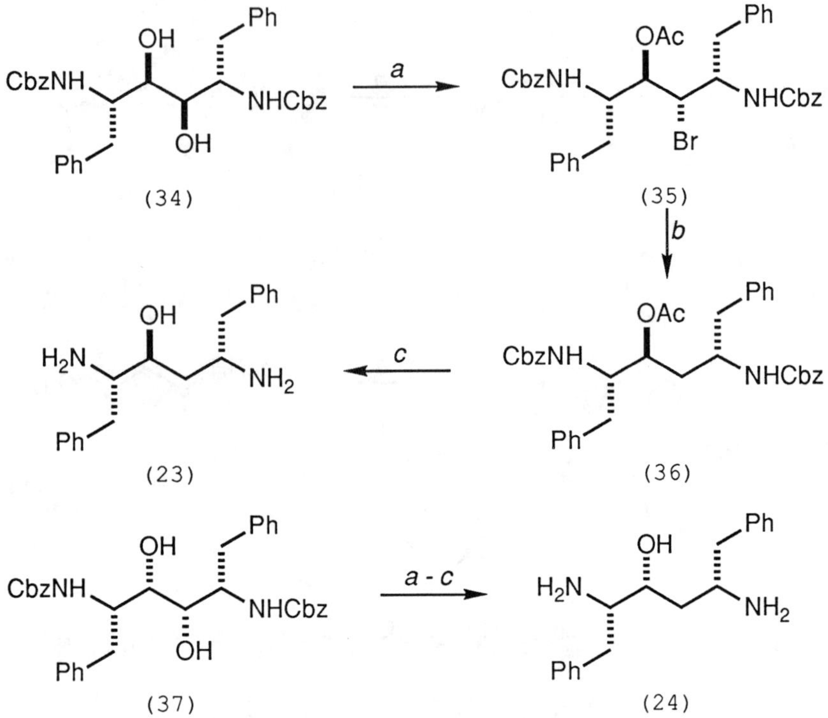

(34)   (35)

(23)   (36)

(37)   (24)

^{a}α-Acetoxyisobutyryl bromide, LiBr, CH$_3$CN, rt, 86%; ^{b}(n-Bu)$_3$SnH, AIBN, THF, reflux; ^{c}Ba(OH)$_2$·8H$_2$O, H$_2$O, dioxane, reflux, 57% (2 steps).

A similar trend in activities was observed with peptide derivatives of the core diamines, as shown in Table 3. All

Table 2. Inhibition of HIV-1 protease by di-Boc-diamines

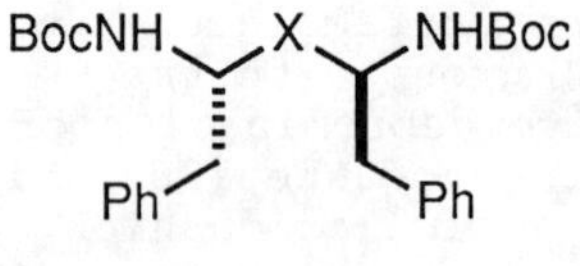

No	X	IC$_{50}$ (nM)
(4)	CH(OH)	3,000
(28)	R,R-CH(OH)CH(OH)	40[a]
(38)	S,S-CH(OH)CH(OH)	280[a]
(39)	R,S-CH(OH)CH(OH)	12[a]
(31)	S-CH(OH)CH$_2$	6
(32)	R,S-CH(OH)CH(N$_3$)	3,200
(33)	S,S-CH(OH)CH(NH$_2$)	480

[a]Reference 41.

Table 3. Inhibition of HIV-1 protease by peptide
derivatives of core diamines

No	X	K_i (nM)
(40)	CH(OH)	43[a,b]
(41)	R,R-CH(OH)CH(OH)	1.66[c]
(42)	S,S-CH(OH)CH(OH)	0.18[c]
(20)	R,S-CH(OH)CH(OH)	0.15[c]
(43)	S-CH(OH)CH$_2$	0.09[c]
(44)	R-CH(OH)CH$_2$	0.82[c]
(51)	CH$_2$CH(OH)CH$_2$	260[a]

[a]IC$_{50}$ value; [b]Reference 20; [c]Reference 48.

of the diastereomeric diols (41), (42) and (20) were substantially more potent than the monohydroxy inhibitor (40).
Again, the activity of compound (43), derived from the
deoxygenated diamine (23), was twofold higher than any of
the diols.[48] However, the diastereomeric deoxygenated
inhibitor (44), based on diamine (24), was less potent than
the corresponding diol, (42).

The high activity of inhibitors (31) and (43) compared
to (28) and (41), respectively, indicates that the presence
of the second hydroxyl group of the R,R diol-based inhibitors is actually detrimental to activity. Thus, the key
feature responsible for the greater potency of the diol
inhibitors over the monohydroxy inhibitors derived from
diamine (1) is apparently the spacing between the functional groups and side chains of the end pieces, dictated
by the length of the core unit. Consistent with this
result is the observation that in the X-ray crystal structure of A-74704, (10), bound to HIV-1 protease, the hydrogen bonds formed between the amide NH groups of the core
unit of (10) and Gly$_{27}$ (and Gly$_{127}$) of the enzyme are
significantly longer than normal.[40] Lengthening the
distance between the nitrogen atoms of the core diamine
would presumably strengthen those hydrogen-bonding interactions. Upon changing the stereochemistry of the hydroxyl
groups, the above trend is reversed, and the activity of
S,S diol (42) is higher than that of the corresponding
deoxygenated inhibitor (44). Both hydroxyl groups therefore contribute to the high activity of (42), in contrast
to R,R diol (41). The implications of this difference for
the modes of binding of the diastereomeric diols have been
discussed in detail elsewhere.[48]
    In view of the high potency of inhibitors derived from
diamine (23), we prepared the homologated core unit (45).

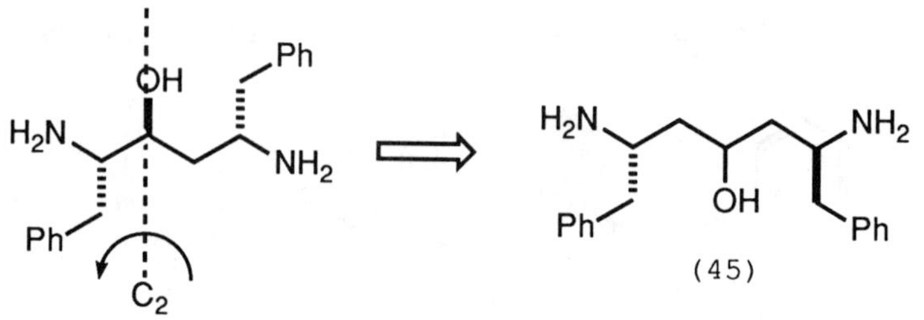

Figure 3.   Design of pseudo-C_2-symmetric diaminoalcohol (45)

Conceptually, (45) was derived from (23) as shown in Figure 3, using the protocol described above for the design of (1) and (2).   Alternately, (45) can be envisioned to result from insertion of two methylene groups into diamine (1). We reasoned that if the carbon bearing the hydroxyl group in inhibitors based on (23) lies close to the center of symmetry of HIV protease, duplication of the P' portion containing the methylene spacer might result in further potency enhancement.   We were recognizant of the fact, however, that because of the prochiral nature of the central carbon of (45), insertion of a methylene group into (23) to give (45) is not stereochemically equivalent to insertion of a methylene group into (1) to give (23) (Figure 2).   Moreover, if the hydroxyl-bearing carbon of (23) does not align closely to the C_2 axis, the length of the resulting core diamine (45) could exceed that which allows optimal interactions with the enzyme subsites.

The synthesis of (45) is detailed in Scheme 3.   Following a route partially analogous to that reported for different symmetry-based inhibitors,[43]   acyloxazolidinone (46) was asymmetrically alkylated[51] with 3-iodo-2-(iodomethyl)propene to give the C_2-symmetric product (47). Hydrolysis of (47) and Curtius rearrangement of the resulting diacid (48) provided protected diamine (49). Oxidative cleavage of (49) followed by borohydride reduction led to alcohol (50), which was deprotected to diamine (45).

Diacylation of (45) led to the inhibitor (51).   As shown in Table 3, the potency of (51) against HIV-1 protease (IC_{50} = 260 nM) was sixfold lower than monohydroxy inhibitor (40) and nearly 3,000 times lower than that of (43), the corresponding inhibitor derived from diamine (23).   The activity of (51) was also substantially lower than that of inhibitors based on the retro-inverted

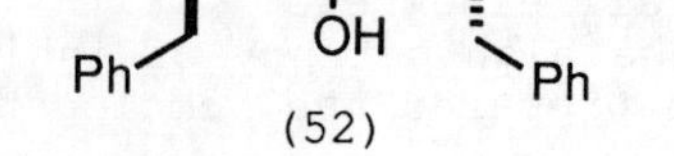

(52)

Scheme 3

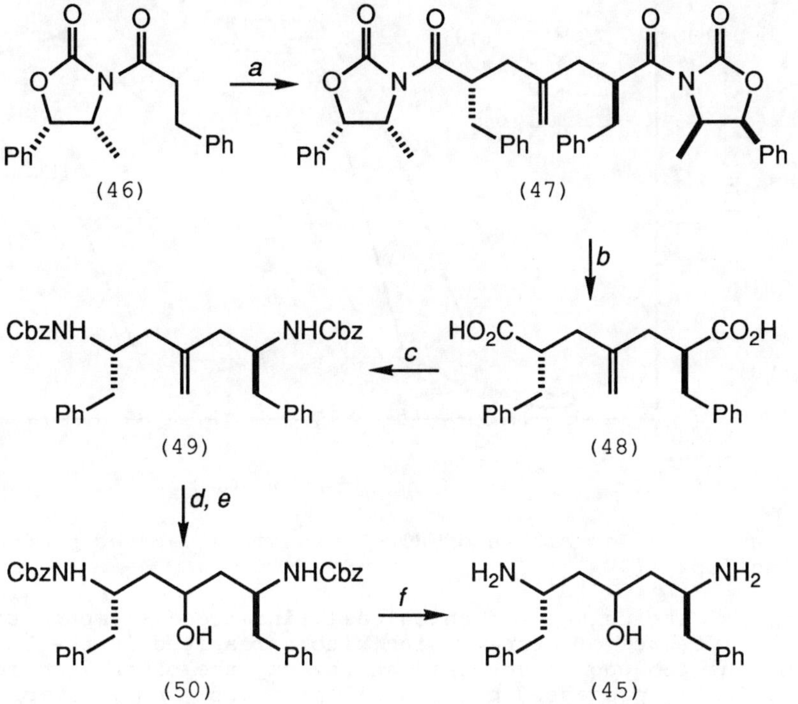

[a]LDA, THF; 3-iodo-2-(iodomethyl)propene; [b]LiOH, H2O2, THF/H2O; [c]diphenylphosphoryl azide, triethylamine, PhCH3, reflux; [d]OsO4, NaIO4, dioxane/H2O; [e]NaBH4, CH3OH; [f]Ba(OH)2·8H2O, H2O, dioxane, reflux.

analogue (52).[42,43] The reason for this difference is presumably that diacid (52) contains the same number of atoms between the hydroxyl-bearing carbon and the carboxyl group as the P' region of both peptide substrates and potent, unsymmetric inhibitors.[42] Diamine (45), in contrast, does not directly resemble the amino terminus of the substrate from which it was derived, and apparently forces the functional groups involved in critical interactions too far from the center of the active site. The fact that there is only a sixfold difference in activity between (51) and (40), which contains two fewer carbons, may indicate that, between the two respective inhibitors, different functional groups are responsible for a greater contribution to the total energy of binding.

3   ANTIVIRAL ACTIVITY OF SYMMETRY-BASED INHIBITORS

The ability of potent HIV protease inhibitors to block the spread of HIV-1 and HIV-2 *in vitro* is by now well documented.[15,19,20,25,52-54] Consistent with the role of HIV protease in the processing of the *gag* and *gag-pol* polypro-

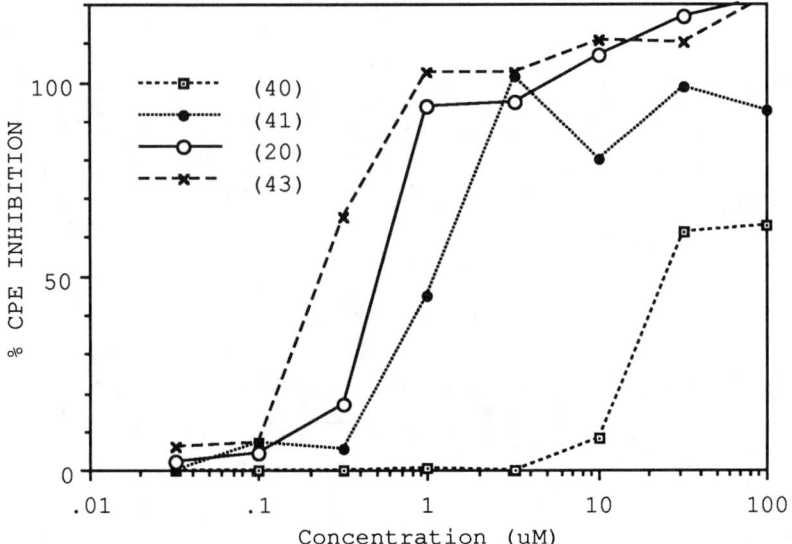

Figure 4.   Inhibition of HIV-1 by symmetry-based protease inhbitors

teins, the culture of chronically infected T-lymphocytes in the presence of protease inhibitors resulted in the buildup of the p55 *gag* polyprotein at the expense of the proteo-lytically processed p24 and p17 proteins.[55]   Moreover, electron micrographs of infected cells in the presence of protease inhibitors revealed the production of morpho-logically immature, non-infectious virions.[15,16,18]

   The antiviral properties of the symmetry-based inhibitor A-77003 (20) have been described in detail.[20] Briefly, A-77003 showed broad-spectrum activity against acute infection by a variety of laboratory and primary strains of HIV-1 and HIV-2 in both transformed and primary human lymphocytes.   Antiviral efficacy, as measured by inhibition of either p24 production or cytopathic effect, ranged from 0.03 to 0.3 µM, depending predominantly upon the assay method.   Equivalent activity was observed between pre- and post-AZT-resistant HIV.[20]

   Within a given series of inhibitors of similar polar-ity, we have observed a general correlation between inhi-bitory potency against purified HIV protease and anti-HIV activity *in vitro*.   Figure 4 shows the inhibition of the cytopathic effect of HIV-1$_{3B}$ in MT4 cells by inhibitors (20), (40), (41) and (43).   In accordance with its K_i of 90 picomolar for protease inhibition, deoxygenated inhibitor (43) showed the greatest efficacy against HIV *in vitro*. The activity of inhibitor (41) (K_i 1.7 nM) paralleled that of (43) at *ca.* 10-fold higher concentrations.   The moderate inhibitor (40) showed only weak antiviral activity in this assay.   None of the inhibitors showed evidence for cellular toxicity at the highest concentration level (100 µM). The consistent correlation of inhibitory potencies against purified HIV protease and anti-HIV activity *in vitro* provides evidence that the antiviral effects of these compounds result from specific inhibition of HIV protease in infected cells.

4   CONCLUSION

In view of its essential biological role in the replication of HIV as well as its unique structural characteristics, HIV protease constitutes an attractive target for chemotherapeutic intervention of HIV disease. Our studies in this area have shown that the use of structural information about the enzyme active site can result in the rational design of unique classes of potent inhibitors. With increasing understanding of the subtle features which contribute to the binding of these peptide-based inhibitors, the availability of three dimensional structures of both native[37-39] and complexed HIV protease[27,40,42,47,56-59] will undoubtedly lead to the design of second generation inhibitors which do not resemble peptide substrates.[60]

It is now well established through their activity *in vitro* that inhibitors of HIV protease show high potential to be effective clinical therapeutics for AIDS. The peptide-like character of most inhibitors, however, presents a formidable drug delivery challenge due to poor oral bioavailability and rapid clearance.[61] Although intravenous administration of protease inhibitors[20] may be useful clinically, ultimately, orally bioavailable agents are needed for widespread treatment of chronic HIV infection. Issues such as CNS penetration by the inhibitors and the development of viral resistance also present serious development challenges. Increased knowledge of the structure-activity relationships of protease inhibitors in these areas is needed for the future design of superior agents.

REFERENCES

1. H. Mitsuya, R. Yarchoan and S. Broder, Science, 1990, 249, 1533.

2. J.J. McGowan, N. Sarver, H.S. Allaudeen and M.M. Johnston, 'Peptides: Chemistry, Structure and Biology. Proceedings of the Eleventh American Peptide Symposium', J.E. Rivier and G.R. Marshall, Ed., ESCOM, Leiden, 1990, Vol. p. 815.

3. D.W. Norbeck, Ann. Rep. Med. Chem., 1990, 25, 149.

4. E. Arnold and G.F. Arnold, Advances in Virus Research, 1991, 39, 1.

5. C.A. Rosen, AIDS Res. Human Retroviruses, 1992, 8, 175.

6. M. Stevenson, M. Bukrinsky and S. Haggerty, <u>AIDS Res.
    Human Retroviruses</u>, 1992, <u>8</u>, 107.

7. B.M. Dunn and J. Kay, <u>Antiviral Chem. Chemotherapy</u>,
    1990, <u>1</u>, 3.

8. C. Debouck and B.W. Metcalf, <u>Drug Dev. Res.</u>, 1990, <u>21</u>,
    1.

9. A.G. Tomasselli, W.J. Howe, T.K. Sawyer, A. Wlodawer
    and R.L. Heinrikson, <u>Chimica oggi</u>, 1991, <u>1991</u>, 6.

10. J.R. Huff, <u>J. Med. Chem.</u>, 1991, <u>34</u>, 2305.

11. C. Debouck, <u>AIDS Res. Human Retroviruses</u>, 1992, <u>8</u>, 153.

12. N.E. Kohl, E.A. Emini, W.A. Schleif, L.J. Davis, J.C.
    Heimbach, R.A.F. Dixon, E.M. Scolnick and I.S. Sigal,
    <u>Proc. Natl. Acad. Sci. U.S.A.</u>, 1988, <u>85</u>, 4686.

13. S.F. Le Grice, J. Mills and J. Mous, <u>EMBO J.</u>, 1988, <u>7</u>,
    2547.

14. C. Peng, B.K. Ho, T.W. Chang and N.T. Chang, <u>J. Virol.</u>,
    1989, <u>63</u>, 2550.

15. P. Ashorn, T.J. McQuade, S. Thaisrivongs, A.G.
    Tomasselli, W.G. Tarpley and B. Moss, <u>Proc. Natl. Acad.
    Sci. USA</u>, 1990, <u>87</u>, 7472.

16. C. Baboonian, A. Dalgleish, L. Bountiff, J. Gross, S.
    Oroszlan, G. Rickett, C. Smith-Burchnell, P. Troke and
    J. Merson, <u>Biochem. Biophys. Res. Comm.</u>, 1991, <u>179</u>, 17.

17. H. Nitschko, H. Schatzi, H.R. Gelderblom, M. Oswald and
    K. von der Helm, <u>Biomed. Biochim. Acta</u>, 1991, <u>50</u>, 655.

18. H. Schatzl, H.R. Gelderblom, H. Nitschko and K. von der
    Helm, <u>Arch. Virol.</u>, 1991, <u>120</u>, 71.

19. N.A. Roberts, J.A. Martin, D. Kinchington, A.V.
    Broadhurst, J.C. Craig, I.B. Duncan, S.A. Galpin, B.K.
    Handa, J. Kay, A. Krohn, R.W. Lambert, J.H. Merrett,
    J.S. Mills, K.E.B. Parkes, S. Redshaw, A.J. Ritchie,
    D.L. Taylor, G.J. Thomas and P.J. Machin, <u>Science</u>,
    1990, <u>248</u>, 358.

20. D.J. Kempf, K.C. Marsh, D.A. Paul, M.F. Knigge, D.W.
    Norbeck, W.E. Kohlbrenner, L. Codacovi, S. Vasavanonda,
    P. Bryant, X.C. Wang, N.E. Wideburg, J.J. Clement, J.J.
    Plattner and J. Erickson, <u>Antimicrob. Agents
    Chemother.</u>, 1991, <u>35</u>, 2209.

21. H. Toh, M. Ono, K. Saigo and T. Miyata, <u>Nature</u>, 1985,
    <u>315</u>, 691.

22. L.H. Pearl and W.R. Taylor, <u>Nature</u>, 1987, <u>329</u>, 351.

23. W.J. Greenlee, Med. Res. Rev., 1990, 10, 173.

24. G.B. Dreyer, B.W. Metcalf, T.A. Tomaszek Jr., T.J.
    Carr, A.C. Chandler III, L. Hyland, S.A. Fakhoury, V.W.
    Magaard, M.L. Moore, J.E. Strickler, C. Debouck and
    T.D. Meek, Proc. Natl. Acad. Sci. USA, 1989, 86, 9752.

25. J.P. Vacca, J.P. Guare, S.J. deSolms, W.M. Sanders,
    E.A. Giuliani, S.D. Young, P.L. Darke, J. Zugay, I.S.
    Sigal, W.A. Schleif, J.C. Quintero, E.A. Emini, P.S.
    Anderson and J.R. Huff, J. Med. Chem., 1991, 34, 1225.

26. T.A. Lyle, C.M. Wiscount, J.P. Guare, W.J. Thompson,
    P.S. Anderson, P.L. Darke, J.A. Zugay, E.A. Emini, W.A.
    Schleif, J.C. Quintero, R.A.F. Dixon, I.S. Sigal and
    J.R. Huff, J. Med. Chem., 1991, 34, 1228.

27. W.J. Thompson, P.M.D. Fitzgerald, M.K. Holloway, E.A.
    Emini, P.L. Darke, B.M. McKeever, W.A. Schleif, J.C.
    Quintero, J.A. Zugay, T.J. Tucker, J.E. Schwering, C.F.
    Homnick, J. Numberg, J.P. Springer and J.R. Huff, J.
    Med. Chem., 1992, 35, 1685.

28. S.D. Young, L.S. Payne, W.J. Thompson, M. Gaffin, T.A.
    Lyle, S.F. Britcher, S.L. Graham, T.H. Schultz, A.A.
    Deana, P.L. Darke, J. Zugay, W.A. Schleif, J.C.
    Quintero, E.A. Emini, P.S. Anderson and J.R. Huff, J.
    Med. Chem., 1992, 35, 1702.

29. D.H. Rich, J. Green, M.V. Toth, G.R. Marshall and
    S.B.H. Kent, J. Med. Chem., 1990, 33, 1285.

30. S. Thaisrivongs, A.G. Tomaselli, J.B. Moon, J. Hui,
    T.J. McQuade, S.R. Turner, J.W. Strohbach, W.J. Howe,
    W.G. Tarpley and R.L. Heinrikson, J. Med. Chem., 1991,
    34, 2344.

31. D. Grobelny, E.M. Wondrak, R.E. Galardy and S.
    Oroszlan, Biochem. Biophys. Res. Commun., 1990, 169,
    1111.

32. T. Mimoto, J. Imai, S. Tanaka, N. Hattofi, O.
    Takahashi, S. Kisanuki, Y. Nagano, M. Shintani, H.
    Hayashi, H. Sakikawa, K. Akaji and Y. Kiso, Chem.
    Pharm. Bull., 1991, 39, 2465.

33. B. Raju and M.S. Deshpande, Biochem. Biophys. Res.
    Commun., 1991, 180, 181.

34. E. Matayoshi, G. Wang, G. Krafft, D. Kempf, L. Codacovi
    and J. Erickson, 'AIDS: Anti-HIV Agents, Therapies, and
    Vaccines; Annals of the New York Academy of Sciences',
    V. St. Georgiev and J.J. McGowan, Ed., The New York
    Academy of Sciences, New York, 1990, Vol. 616, p. 566.

35. T.D. Meek, B.D. Dayton, B.W. Metcalf, G.B. Dreyer, J.E.
    Strickler, J.G. Gorniak, M. Rosenberg, M.L. Moore, V.W.

Magaard and C. Debouck, Proc. Natl. Acad. Sci. USA,
1989, 86, 1841.

36. I. Katoh, Y. Ikawa and Y. Yoshinaka, J. Virology, 1989,
63, 2226.

37. M.A. Navia, P.M.D. Fitzgerald, B.M. McKeever, C.-T.
Leu, J.C. Heimbach, W.K. Herber, I.S. Sigal, P.L. Darke
and J.P. Springer, Nature, 1989, 337, 615.

38. A. Wlodawer, M. Miller, M. Jaskolski, B.K.
Sathyanarayana, E. Baldwin, I.T. Weber, L.M. Selk, L.
Clawson, J. Schneider and S.B.H. Kent, Science, 1989,
245, 616.

39. R. Lapatto, T. Blundell, A. Hemmings, J. Overington, A.
Wilderspin, S. Wood, J.R. Merson, P.J. Whittle, D.E.
Danley, K.F. Geoghegan, S.J. Hawrylik, S.E. Lee, K.G.
Scheld and P.M. Hobart, Nature, 1989, 342, 299.

40. J. Erickson, D.J. Neidhart, J. VanDrie, D.J. Kempf,
X.C. Wang, D.W. Norbeck, J.J. Plattner, J.W.
Rittenhouse, M. Turon, N. Wideburg, W.E. Kohlbrenner,
R. Simmer, R. Helfrich, D.A. Paul and M. Knigge,
Science, 1990, 249, 527.

41. D.J. Kempf, D.W. Norbeck, L. Codacovi, X.C. Wang, W.E.
Kohlbrenner, N.E. Wideburg, D.A. Paul, M.F. Knigge, S.
Vasavanonda, A. Craig-Kennard, A. Saldivar, W.
Rosenbrook Jr., J.J. Clement, J.J. Plattner and J.
Erickson, J. Med. Chem., 1990, 33, 2687.

42. R. Bone, J.P. Vacca, P.S. Anderson and M.K. Holloway,
J. Am. Chem. Soc., 1991, 113, 9382.

43. R. Babine, 32nd Annual Medicinal Chemistry Symposium,
Buffalo, NY, June 3-5, 1991.

44. D.J. Kempf, T.J. Sowin, E.M. Doherty, S.M. Hannick, L.
Codacovi, R.F. Henry, B.E. Green, S.G. Spanton and D.W.
Norbeck, J. Org. Chem., 1992, in press.

45. E.D. Matayoshi, G.T. Wang, G.A. Krafft and J. Erickson,
Science, 1990, 247, 954.

46. D.H. Rich, C.-Q. Sun, J.V.N.V. Prasad, A. Pathiasseril,
M.V. Toth, G.R. Marshall, M. Clare, R.A. Mueller and K.
Houseman, J. Med. Chem., 1991, 34, 1222.

47. A. Krohn, S. Redshaw, J.C. Ritchie, B.J. Graves and
M.H. Hatada, J. Med. Chem., 1991, 34, 3340.

48. W.E. Kohlbrenner, A. Craig-Kennard, D.J. Kempf, L.
Codacovi, H.L. Sham and D.W. Norbeck, submitted for
publication.

49. Y. Gao and K.B. Sharpless, *J. Amer. Chem. Soc.*, 1988, 110, 7538.

50. S. Greenberg and J.G. Moffatt, *J. Am. Chem. Soc.*, 1973, 95, 4016.

51. D.A. Evans, M.D. Ennis and D.J. Mathre, *J. Am. Chem. Soc.*, 1982, 104, 1737.

52. T.J. McQuade, A.G. Tomasselli, L. Liu, V. Karacostas, B. Moss, T.K. Sawyer, R.L. Heinrikson and W.G. Tarpley, *Science*, 1990, 247, 454.

53. H.A. Overton, D.J. McMillan, S.J. Gridley, J. Brenner, S. Redshaw and J.S. Mills, *Virology*, 1990, 179, 508.

54. J.C. Craig, I.B. Duncan, D. Hockley, C. Grief, N.A. Roberts and J.S. Mills, *Antiviral Res.*, 1991, 16, 295.

55. T.D. Meek, D.M. Lambert, G.B. Dreyer, T.J. Carr, T.A. Tomaszek Jr., M.L. Moore, J.E. Strickler, C. Debouck, L.J. Hyland, T.J. Matthews, B.W. Metcalf and S.R. Petteway, *Nature*, 1990, 343, 90.

56. M. Miller, J. Schneider, B.K. Sathyanarayana, M.V. Toth, G.R. Marshall, L. Clawson, L. Selk, S.B.H. Kent and A. Wlodawer, *Science*, 1989, 246, 1149.

57. A.L. Swain, M.M. Miller, J. Green, D.H. Rich, J. Schneider, S.B.H. Kent and A. Wlodawer, *Proc. Natl. Acad. Sci. U.S.A.*, 1990, 87, 8805.

58. P.M.D. Fitzgerald, B.M. McKeever, J.F. VanMiddlesworth, J.P. Springer, J.C. Heimbach, C.-T. Leu, W.K. Herber, R.A.F. Dixon and P.L. Darke, *J. Biol. Chem.*, 1990, 265, 14209.

59. M. Jaskolski, A.G. Tomasselli, T.K. Sawyer, D.G. Staples, R.L. Heinrikson, J. Schneider, S.B.H. Kent and A. Wlodawer, *Biochemistry*, 1991, 30, 1600.

60. R.L. DesJarlais, G.L. Seibel, I.D. Kuntz, P.S. Furth, J.C. Alvarez, P.R. Ortiz de Montellano, D.L. DeCamp, L.M. Babe and C.S. Craik, *Proc. Natl. Acad. Sci. USA*, 1990, 87, 6644.

61. J.J. Plattner and D.W. Norbeck, 'Drug Discovery Technologies', R. Clark and W.H. Moos, Ed., Ellis Horwood Ltd., Chichester, 1990, Vol. p. 92.

21

Tat Inhibitors: A New Class of Anti-HIV Agents

S. Tam,* J. Borgese, A. Cislo, C. Cook, J. Earley, M. Holly,
M. Holman, M.-C. Hsu, D. M. Huryn, D. D. Keith,
K.-C. Luk, D. Moore, A. Richou, A. Schutt, B. Sluboski,
D. D. Richman,[1] and D. Volsky[2]

ROCHE RESEARCH CENTER, HOFFMANN-LA ROCHE INC., NUTLEY, NJ, USA
[1]VA MEDICAL CENTER, UNIVERSITY OF CALIFORNIA, SAN DIEGO, CA, USA
[2]SAINT LUKE'S-ROOSEVELT HOSPITAL, COLUMBIA UNIVERSITY, NEW
YORK, NY, USA

The human immunodeficiency virus type-1 (HIV-1), the
etiologic agent of the acquired immunodeficiency
syndrome (AIDS), differs from most of the other known
retroviruses in the possession of several regulatory
genes in its genome. One of these genes, *tat*, codes for
a protein Tat, which is a strong positive regulator of
the viral gene expression and is required for HIV-1
replication.[1] Tat is a small, basic protein (86-101
amino acids long, depending on the viral strain).
Through the interaction with its target sequence, TAR,
which is downstream of the transcription start site
(mapped at +19 to +44 nucleotides from the +1 start
site), Tat can greatly activate the transcription of
HIV-1 gene and subsequently lead to elevated production
of viral proteins and progeny virions.[2] Complementation
of HIV-1 defective in *tat* gene with recombinant Tat
protein increased viral production many thousandfold.[3]
TAR-RNA forms a stem loop structure,[4] and *in vitro*
studies have shown that Tat binds to a bulge in the
stem.[5] Several cellular factors have also been shown to
interact with the TAR-RNA,[6] but the exact mechanism of
how Tat works together with the cellular transcription
factors in the activation of HIV-1 gene expression
remains a subject of several intensive studies.

Due to its importance for productive HIV-1 replication
and the fact that there is no known analogous human
protein, Tat is an attractive target for the development
of anti-HIV agents. However, since the molecular
mechanism of Tat action was not understood, an *in vitro*
functional assay for Tat was not possible at the onset
of our program in 1987. For this reason, we designed a
cell-based assay as a measure of the transcriptional
transactivation activity of Tat. The detail of this
assay was published recently.[7] Briefly, in this assay,
an indicator gene encoding secreted alkaline phosphatase
(SeAP) is placed under the control of the HIV-1 LTR
promoter in a plasmid construct pBC12/HIV/SeAP. The *tat*

gene is similarly placed under the control of the HIV-1 LTR promoter in a second plasmid pBC12/HIV/tat. Cotransfection of COS cells with the two plasmids leads to secretion of alkaline phosphatase into the culture media. The activity of this enzyme is then quantitated by a colorimetric assay. Inhibition of the production of alkaline phosphatase is correlated with inhibition of the HIV-1-Tat activity. As an indication of cytotoxicity, a control plasmid, pBC12/RSV/SeAP, containing the SeAP gene driven by the Rous sarcoma virus LTR, which is not responsive to HIV-Tat, is used to replace the plasmid pBC12/HIV/SeAP in the assay. Inhibition of alkaline phosphatase activity in the HIV-1 but not the RSV promoter construct indicates selective activity against HIV-1-Tat function, whereas inhibition of alkaline phosphatase activity in both constructs indicates cytotoxicity. Using this assay, we were able to identify from the Roche compound repository a number of inhibitors with potent activity selective against the HIV system. One of these compounds, Ro 5-3335, has been reported by us recently.[7]

Ro 5-3335

This lead compound, Ro 5-3335, was subsequently found to have a broad spectrum of activity against different strains of HIV-1 (3B, Bru, BAL/85, NIT, and RF), in several cell lines (CEM, Jurkat, H9, JM, C8166, and HT4-6C), and in primary peripheral blood lymphocytes and macrophages.[7] Similar potencies were found (IC_{50} = 0.1 - 1 µM, IC_{90} = 1 - 3 µM) with different endpoint measurements (p24 ELISA, plaque reduction, syncytium inhibition, antibody staining of infected cells, and viral RNA quantitation). The compound was also active against HIV-2 (strains ROD and UC-1) and AZT-resistant clinical isolates.

In addition, Ro 5-3335 has been shown to effectively inhibit viral replication from proviral DNA as well as in chronically infected cells.[7] These data show that Ro 5-3335 indeed possesses properties expected of a Tat-inhibitor, namely the ability to arrest viral replication at the latency stage. Such a property is not shared by other types of anti-HIV drugs (e.g. inhibitors of reverse transcriptase or protease). An HIV-Tat inhibitor could therefore have the potential to prevent disease progression in asymptomatic carriers.

Because of our experience in benzodiazepine chemistry at Roche, the discovery of Ro 5-3335 as a Tat inhibitor came as a pleasant surprise. However, after a computer sub-structure search of our in-house compound repository and subsequent testing of the hit compounds, it was realized that the Tat inhibitory activity is not a general property of benzodiazepines or pyrroles. Simple methylation of either or both of the pyrrole and the amide NHs (see structures **1-3**) led to a complete loss of Tat inhibitory activity, indicating that the biological activity resides in the entire Ro 5-3335 molecule (Figure 1).

Figure 1. N-Methyl derivatives of Ro 5-3335

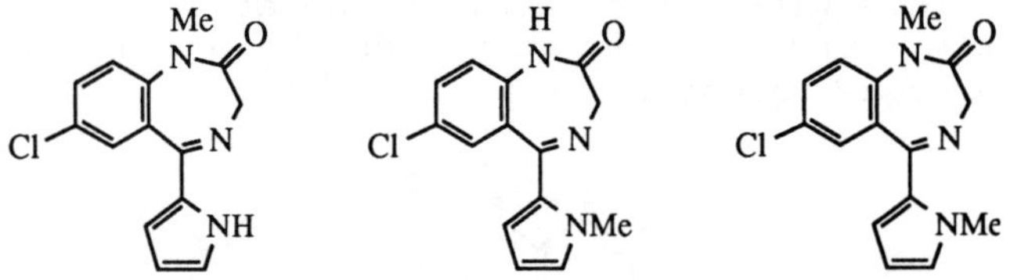

In order to perform a systematic study of the structure-activity relationship, six regions of the molecule of Ro 5-3335 were targeted for structural modifications. These included changes at the N1-C2 amide group, substitution at the C3 carbon, modification of the N4-C5 imine group, replacement of the pyrrole moiety at C5 with different heterocycles, new substitution patterns on the benzene ring and complete replacement of the benzodiazepine skeleton. In this program, about 400 analogs were made and tested in the Tat assay.

Isosteric replacement of the amide oxygen with sulphur (**4**), or conversion of the amide to an imidate (**5**), resulted in a loss in Tat inhibitory activity. However, good activity was observed in the amidine series (**6-9**), and the activity was clearly related to the size of the substituent(s) on the amine function.   Small substituents such as hydrogen (**6**), or hydrogen and methyl (**7**), retained high activity, whereas bulky or aromatic substituents on the amino group, such as butyl (**8**), and benzyl (**9**), were detrimental to the activity (Figure 2).

Other changes at the amide function, such as removal of the carbonyl group, **10**, or replacement with an olefin, **11**, caused a large reduction in anti-Tat activity. Fusion of a five membered ring onto this region of the molecule (12–15) also led to a complete loss of activity (Figure 3).

Figure 2. Amide analogs of Ro 5-3335

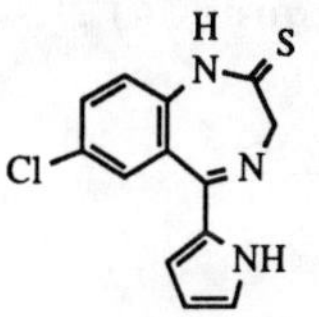

4

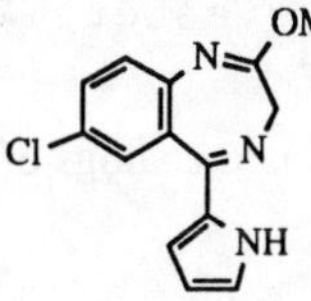

5

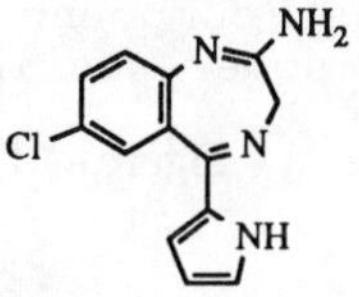

6

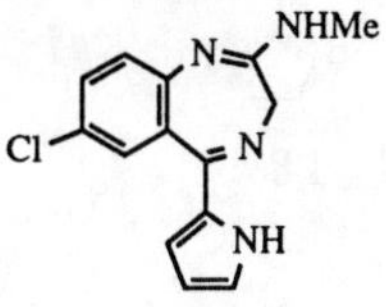

7

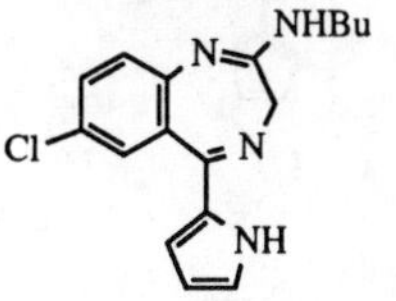

8

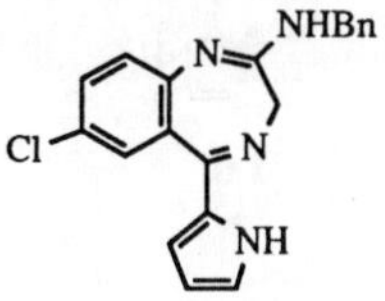

9

Figure 3. New analogs at N1-C2 position of Ro 5-3335

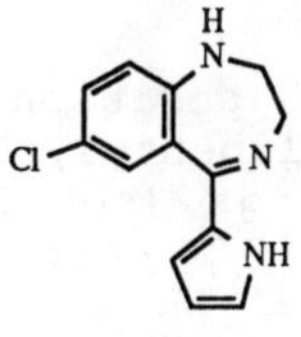

10

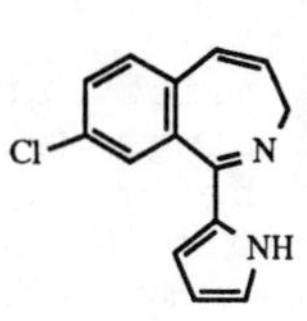

11

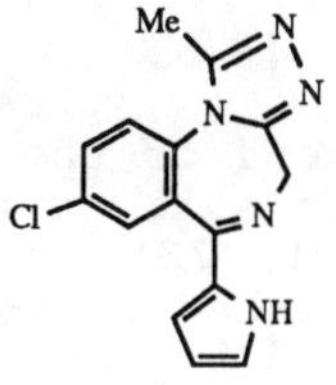

12

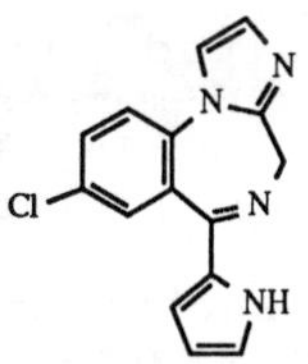

13

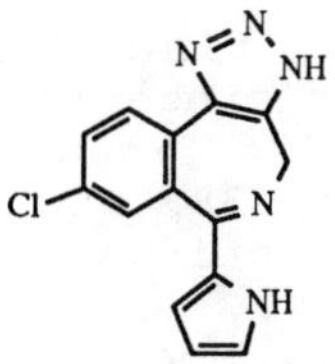

14

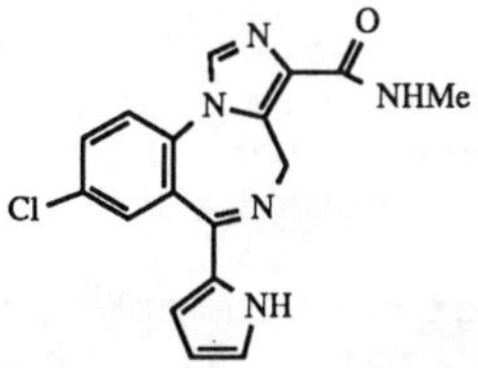

15

Incorporation of heteroatoms (OH, OAc, OMe or NH_2) at the C3 position of Ro 5-3335 was detrimental to the anti-Tat activity. However, it is interesting to note that of the two 3-methyl-substituted derivatives, the R-isomer, 16, showed high anti-Tat activity, while its enantiomer, 17, was inactive (Figure 4). This structure-activity relationship is opposite to that observed in the CNS activity of diazepam derivatives.[8]

Parallel to this finding, the gem-dimethyl derivative, **18**, was inactive.   As observed in the amide analog series discussed above, a poor tolerance for bulky substituents at the C3 position was also observed.   Thus the 3-isopropyl and the 3-benzyl, **19** and **20**, were not active.   Out of a large series of derivatives prepared, the best compound was the ester, **21** (Figure 4).   Again the S-isomer was inactive.

Figure 4. C3-Substituted analogs of Ro 5-3335

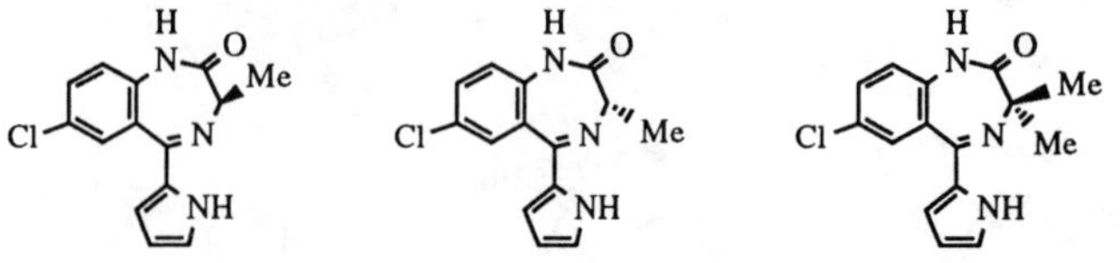

16   **17**   **18**

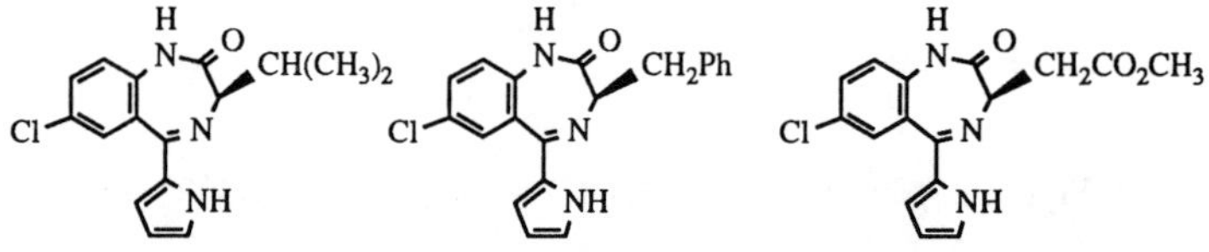

19   **20**   **21**

Replacement of the pyrrole moiety at the C5 position of Ro 5-3335 with carbocyclic rings (e.g. cyclopentyl, **22**, or phenyl, **23**) or various heterocyclic rings (see **24**–**33**) abolished anti-Tat activity (Figure 5).   In contrast to Ro 5-3335, none of these compounds have an NH as part of the heterocyclic substituent.   However, analogs such as the indole, **34**, the benzimidazole, **35**, or the imidazole, **36**, were also inactive.

The most interesting findings resulted from substitutions with other nitrogen-containing 5-membered heterocycles (Figure 6).   The 3-substituted pyrrole, **37**, the 4-substituted pyrazole, **38**, and the 1,2,3-triazole, **39**, all exhibited high anti-Tat activity.   The 5-substituted pyrazole, **40**, and the 5-substituted imidazole, **41**, showed only moderate anti-Tat activity whereas the 5-substituted 1,2,4-triazole, **42**, was inactive.   Based on single crystal X-ray studies of these compounds, we were not able to detect any conformational change that can be used to explain the differences in activity.   On the other hand, the electrostatic potential maps of these compounds were clearly different and suggest that the electron distribution in these molecules may be important for anti-Tat activity.

Figure 5. New analogs at C-5 position of Ro 5-3335

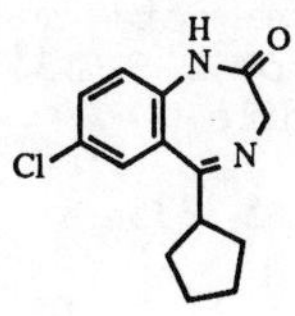

22 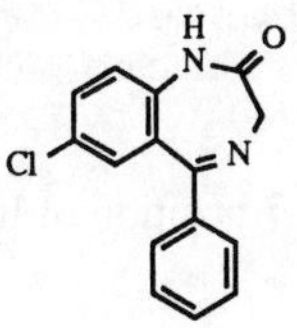   23 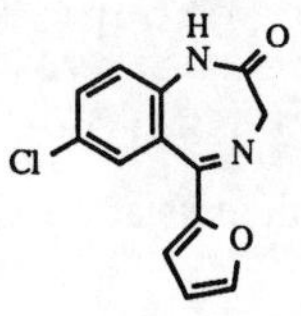   24 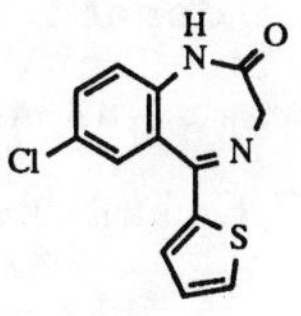   25 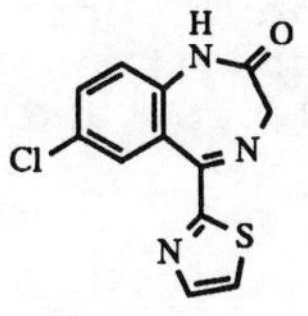   26

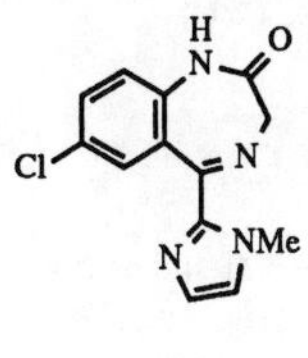

27 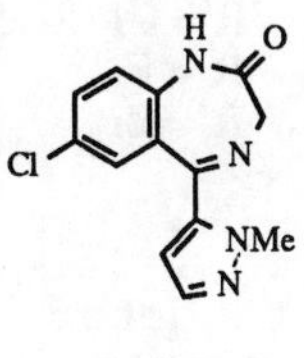   28 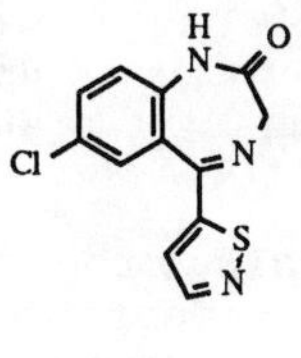   29 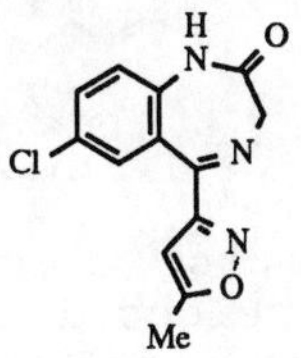   30 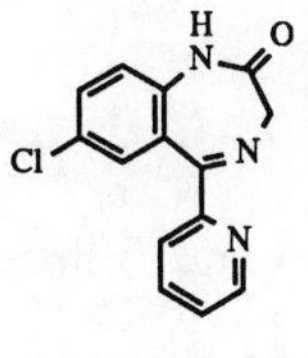   31

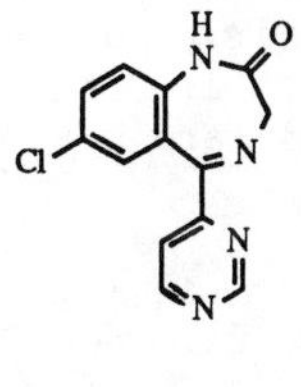

32 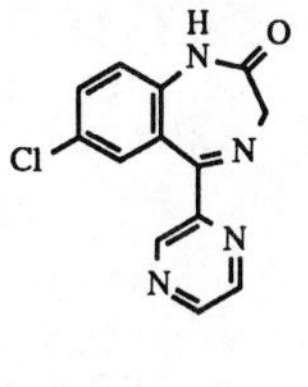   33 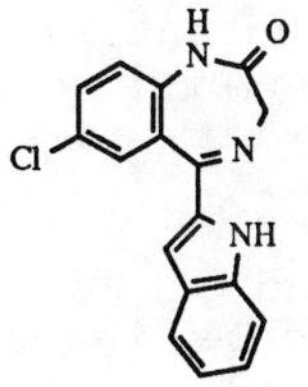   34 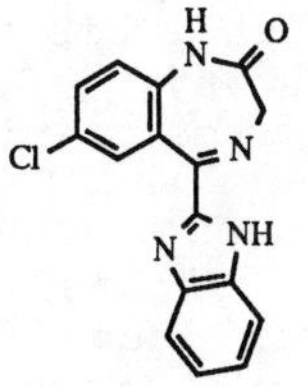   35 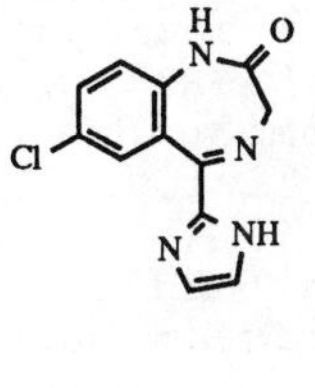   36

Figure 6. Pyrrole analogs of Ro 5-3335

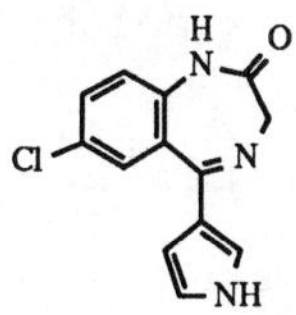

37 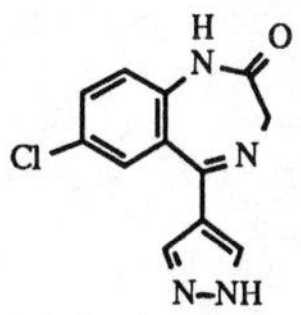   38 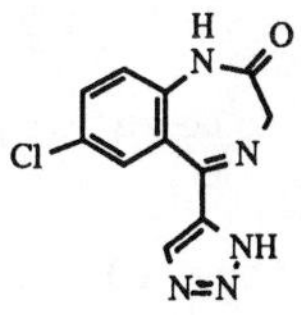   39

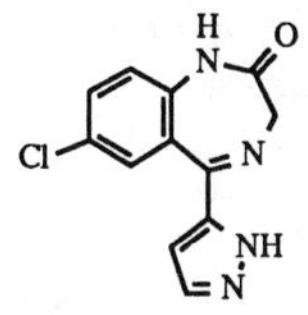

40 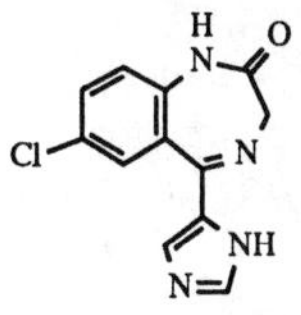   41 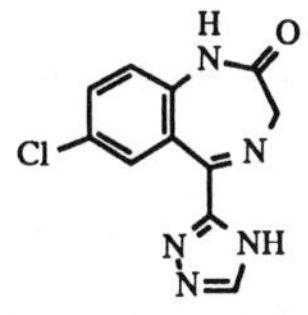   42

For substitutions in the benzene ring, it was clear from our results that only substitution at C7 is allowed. Substitutions at positions 6, 8, or 9 all led to loss of activity. Among a whole range of electron withdrawing and electron donating groups tested, the optimal substituents at C7 were methyl, chloro (Ro 5-3335), fluoro, and bromo (**43-46**, respectively in Figure 7).

Figure 7. Optimal substituents at the C-7 position of Ro 5-3335

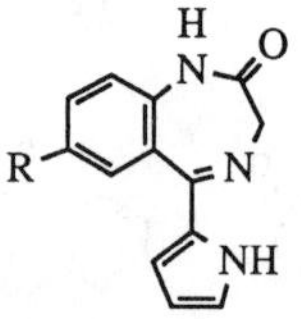

43   R = Me
44   R = Cl (Ro 5-3335)
45   R = F
46   R = Br

During the course of this program, a large number of analogs were selected for testing to determine anti-HIV activity in cell culture. The results of our anti-Tat assay compared very well with the results of antiviral assays performed in Dr. D. Volsky's and Dr. D. Richman's laboratories using different assay protocols. All of the analogs that exhibited anti-Tat activity also showed anti-HIV activity. Conversely none of the inactive compounds in the anti-Tat assay exhibited anti-HIV activity.

Ro 5-3335 was synthesized at Roche in 1959 for the CNS program but was found to be inactive. Its affinity to the benzodiazepine receptor in the rat CNS was less than 1% of that of diazepam. This compound was 85% bioavailable with an half-life of 2 hours in dog. With a 1mg/kg dose, a 0.8 μM plasma concentration was achieved. However, when tested in rats, nephrotoxicity was observed, and there was also tissue discoloration. For this reason, a major criteria in our choice of a clinical candidate was based on animal toxicology data.

As an exploratory screen for toxicity, all of our potential development candidates were tested in rats for 5 days with three different daily dosages. Ro 24-7429 (**7** in Figure 2) was chosen for further study due to its low cytotoxicity and a favorable animal toxicological profile. This compound had highly selective anti-Tat activity, good anti-HIV activities in acutely and in chronically infected cell lines as well as in AZT resisitant cell line. In addition, Ro 24-7429 lacked affinity for the benzodiazepine receptor and was virtually inactive on free behavior parameters in mice when adminstered at doses up to 300 mg/kg, p.o. Studies in dogs further revealed a favorable pharmacokinetic profile. Roche management approved the development of Ro 24-7429 in June 1990. An extensive preclinical evaluation program led to an IND for Ro 24-7429 in March 1991.

References

(1)   (a) S.K. Arya, G. Guo, S.F. Josephs and F. Wong-Staal, *Science* 1985, *229*, 69. (b) J.G. Sodroski, C.A. Rosen, F. Wong-Staal, S.Z. Salahuddin, M. Popovic, S.K. Arya and R.C. Gallo, *Ibid.* 1985, *227*, 171.

(2)   (a) A. Jakobovits, D.H. Smith, E.B. Jakobovits and D.J. Capon, *Mol. Cell. Biol.* 1988, *8*, 2555. (b) J. Hauber and B.R. Cullen, *J. Virol.* 1988, *62*, 673. (c) M.J. Selby, E.S. Bain, P.A. Luciev and B.M. Peterlin, *Genes & Dev.* 1989, *3*, 547.

(3)   M.B. Feinberg, D. Baltimore and A.D. Frankel, *Proc. Natl. Acad. Sci. U.S.A.* 1991, *88*, 4045.

(4)   S. Feng and E.C. Holland, *Nature* 1988, *334*, 165.

(5)   (a) C. Dingwall, I. Ernberg, M.J. Gait, S.M. Green, S. Heaphy, J. Karn, A.D. Lowe, M. Singh, M.A. Skinner and R. Valerio, *Proc. Natl. Acad. Sci. U.S.A.* 1989, *86*, 6925. (b) C. Dingwall, I. Ernberg, M.J. Gait, S.M. Green, S. Heaphy, J. Karn, A.D. Lowe, M. Singh and M.A. Skinner, *EMBO J.* 1990, *9*, 4145. (c) S. Roy, U. Delling, C.H. Chen, C.A. Rosen and N. Sonenberg, *Genes & Dev.* 1990, *4*, 1365. (d) K.M. Weeks, C. Ampe, S.C. Schultz, T.A. Steitz and D.M. Crothers, *Science* 1990, *249*, 1281. (e) M.G. Cordingley, R.L. LaFemina, P.L. Callahan, J.H. Condra, V.V. Sardana, D.J. Graham, T.M. Nguyen, K. LeGrow, L. Gotlib, A.J. Schlabach and R.J. Colanno, *Proc. Natl. Acad. Sci. U.S.A.* 1990, *87*, 8985.

(6)   (a) A. Gatignol, A. Kuman, A. Rabson and K.T. Jeang, *Proc. Natl. Acad. Sci. U.S.A.* 1989, *86*, 7828. (b) R. Gaynor, M. Soultanakis, M. Kuwabara, J. Garcia and D.S. Sigman, *Proc. Natl. Acad. Sci. U.S.A.* 1989, *86*, 4858. (c) R.A. Marciniak, B.J. Calnan, A.D. Frankel and P.A. Sharp, *Cell* 1990, *63*, 791. (d) F. Wu, J. Garcia, D. Sigman and R. Gaynor, *Genes & Del.* 1991, *5*, 2128. (e) C.T. Sheline, L.H. Milocco and K.A. Jones, *Genes & Del.* 1991, *5*, 2508.

(7)   M.-C. Hsu, A.D. Schutt, M. Holly, W.S. Lee, M. Sherman, D.D. Richman, M.J. Potash and D.J. Volsky, *Science* 1991, *254*, 1799.

(8)   J. Blount, R.I. Fryer, N.W. Gilman and L.J. Todaro, *Mol. Pharmacol.* 1983, *24*, 425.

22
Atovaquone – A Novel Agent for the Treatment of Malaria, PCP and Toxoplasmosis

A. T. Hudson

THE WELLCOME RESEARCH LABORATORIES, LANGLEY COURT, SOUTH
EDEN PARK ROAD, BECKENHAM, KENT BR3 3BS, UK

1   INTRODUCTION

Despite significant advances in the chemotherapy of diseases caused by invading organisms, treatment of tropical diseases, such as malaria, and many of the AIDS-associated opportunistic infections, remains problematic. Malaria has increased dramatically over the last 2 decades due to the spread of drug-resistant disease,[1] whilst the AIDS epidemic has resulted in escalating numbers of patients dying from AIDS-associated pathogens, such as *Pneumocystis carinii* and *Toxoplasma gondii*.[2] This has created a need for new therapeutics for both treatment and prophylaxis of these diseases. This review will focus on the novel drug, atovaquone, which is showing great promise for use against malaria, *P. carinii* pneumonia (PCP) and toxoplasmosis.

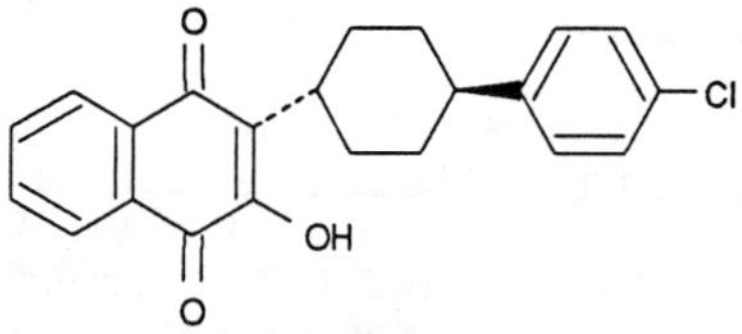

Figure 1.   Atovaquone

2   MALARIA

The research which led to atovaquone was initiated primarily in an attempt to produce a new antimalarial.[3] Malaria is endemic in much of the tropics and sub-tropics placing at risk some 40% of the world's population. More than 100 million clinical cases of the disease are thought to occur annually, resulting in 1 to 2 million deaths. These figures are very crude estimates and could be even higher. What is certain is that malaria is on the increase due to the spread of drug-resistant parasites and insecticide-resistant mosquitoes which carry the disease.[4]

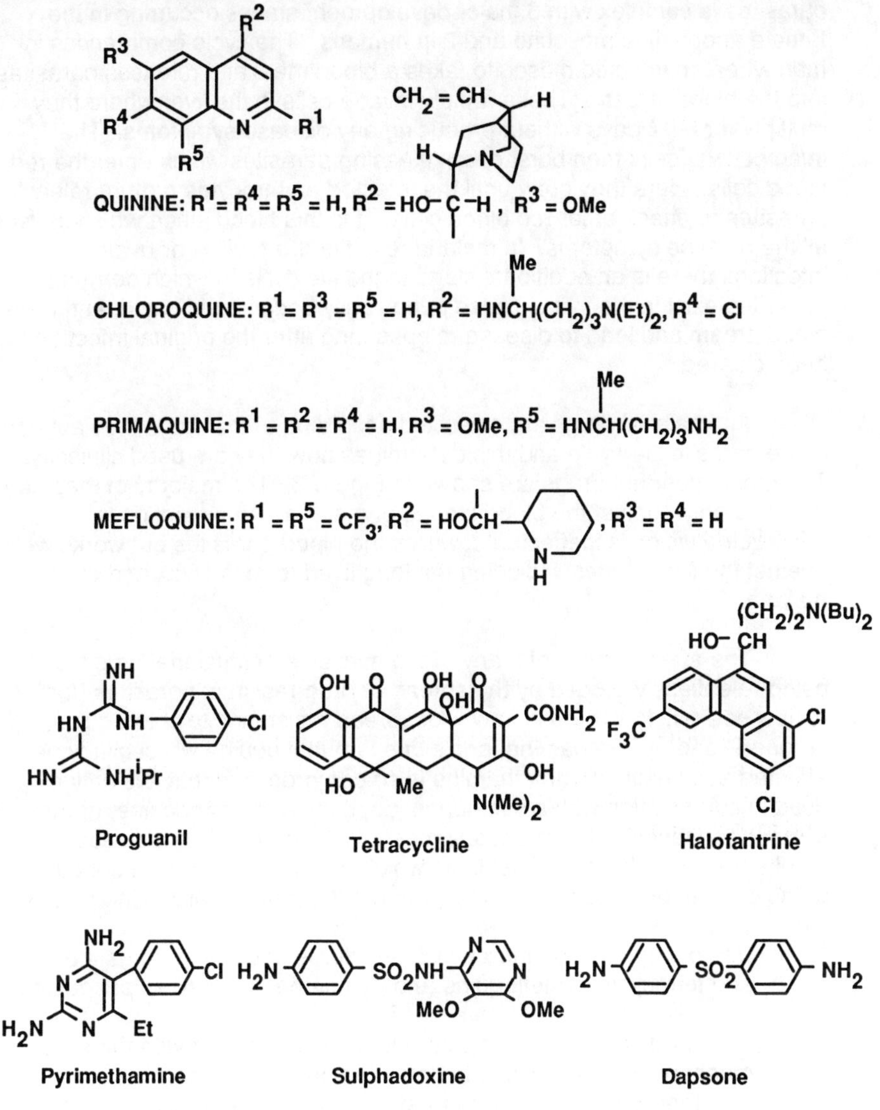

QUININE: $R^1 = R^4 = R^5 = H$, $R^2 = HO-\overset{\displaystyle }{\underset{\displaystyle }{C}}-H$, $R^3 = OMe$

CHLOROQUINE: $R^1 = R^3 = R^5 = H$, $R^2 = HN\overset{\displaystyle Me}{CH}(CH_2)_3N(Et)_2$, $R^4 = Cl$

PRIMAQUINE: $R^1 = R^2 = R^4 = H$, $R^3 = OMe$, $R^5 = HN\overset{\displaystyle Me}{CH}(CH_2)_3NH_2$

MEFLOQUINE: $R^1 = R^5 = CF_3$, $R^2 = HOCH-$, $R^3 = R^4 = H$

Proguanil

Tetracycline

Halofantrine

Pyrimethamine

Sulphadoxine

Dapsone

Figure 2. Major Antimalarial drugs

Malaria is caused by mosquito-transmitted infection with 4 species of protozoa - *Plasmodium falciparum, P. vivax, P. ovale* and *P. malariae. Falciparum* and *vivax* account for 95% of disease incidence, with *vivax* being the more common parasite but *falciparum* inducing the more severe disease. In non-immune patients *falciparum* infections will rapidly cause death unless treated promptly. The other parasites usually don't induce the same serious disease pathology although *vivax* and *ovale* are problematic because they cause a recurring form of malaria. The life cycle of the parasites is complex with 3 major development stages occurring in the female anopheline mosquito and 2 in humans. The cycle commences in man when an infected mosquito takes a blood meal and releases parasites into the blood stream. These rapidly invade cells in the liver where they multiply for 8-12 days without producing any disease symptoms. The infected liver cells then burst open releasing parasites which enter the red blood cells. Here they grow until the infected erythrocytes rupture releasing parasites to attack other red blood cells. It is this blood stage which causes all the disease symptoms. In malaria resulting from *vivax* or *ovale* infections there is an additional stage in the life cycle in which dormant parasites exist in the liver. These periodically become activated, enter the bloodstream and lead to disease relapse long after the original infection has been cleared.

Antimalarial drugs are classified according to the stage they act upon in the parasite life cycle and this determines how they are used clinically. The most important drugs are shown in Figure 2. The majority of these are effective only against the blood stages of malaria. An exception is primaquine which is ineffectual towards the blood parasites but works well against the liver stages, including the long lived form in *vivax* and *ovale* malaria.

The effectiveness of many of the available antimalarial drugs is being relentlessly eroded by the spread of drug-resistant parasites (for review see ref. 1). There is now widespread resistance to chloroquine, proguanil and pyrimethamine, and signs that mefloquine will similarly be affected eventually. Pyrimethamine is so compromised that it is only ever used in combination with a sulphonamide, such as sulphadoxine, or the sulphone, dapsone. Unfortunately most of the drug-resistant disease results from *P. falciparum* infections which give rise to the most serious pathology. In order to combat this problem 3 strategies are being pursued:-

1.   Use of combination therapy with mixtures of existing drugs *e.g.* quinine and tetracycline, mefloquine and pyrimethamine and sulphadoxine.

2.   Re-sensitise parasites to chloroquine by interfering with the efflux mechanisms responsible for lowering drug concentrations in resistant *plasmodia e.g.* by treatment with desipramine (Fig. 3).

3.   Identify novel antimalarials unrelated structurally and functionally to the current drugs.

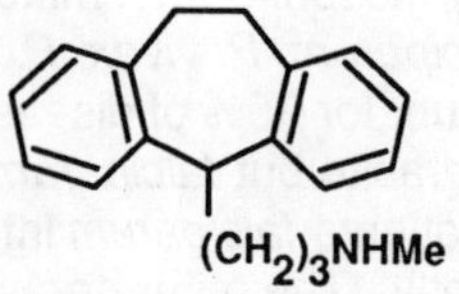

$(CH_2)_3NHMe$

Figure 3.   Desipramine

3   ANTIMALARIAL HYDROXYNAPHTHOQUINONES

In seeking to pursue the latter strategy described above, it was decided to re-evaluate the potential of a series of compounds, the hydroxynaphtho-quinones, whose antimalarial properties have been extensively investigated over the last 50 years. The activity of these quinones towards *plasmodia* was originally discovered in the early 1940's as a result of the American effort to find a replacement for quinine. Large numbers of structurally diverse compounds were screened against *P. lophurae* in ducks and this showed several hydroxynaphthoquinones to be effective in this model. Over 300 members of this series were synthesised by Fieser and his associates in an extensive study of structure-activity relationships.[5] This revealed the 2-OH group to be essential for activity and showed that direct attachment of a cycloalkyl group to the 3-position was especially favourable for activity. Unfortunately, although many of these compounds were very effective *in vivo* against rodent and avian malaria, they were devoid of activity in man when given orally (for review see ref. 6). This was attributed to major pharmacokinetic problems resulting from poor absorption and rapid metabolic deactivation. For example quinones containing cyclohexyl groups were oxidised to secondary alcohols with only weak antimalarial activity[7] - see Scheme 1.

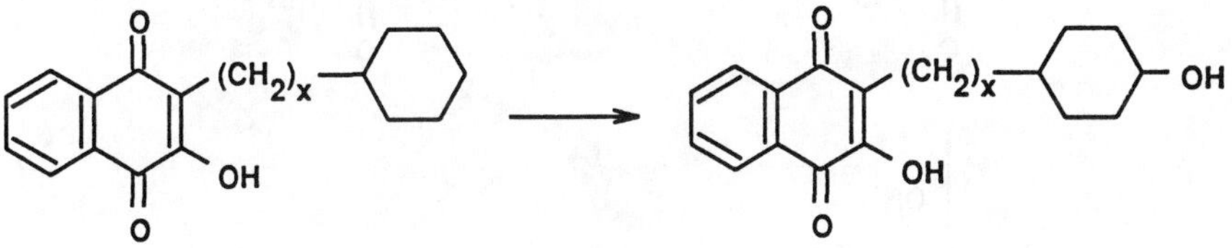

Scheme 1

When the above, and subsequent studies,[8] were carried out on this series, the only screens available were rodent and avian models of malaria. In more recent times both *in vitro*[9] and *in vivo*[10] test systems have been developed using the human parasite *P. falciparum*. These offered the opportunity to conduct more meaningful structure-activity studies on the hydroxynaphthoquinones.

Strategy used to identify Atovaquone

In order to achieve the objective of identifying a quinone with potent activity against *P. falciparum* and good metabolic stability in man, attention was focused on the synthesis of analogues of 2-cyclohexyl-3-hydroxy-1,4-naphthoquinone (Scheme 1, x=O). Derivatives were prepared with the metabolically labile 4'-position of the cyclohexyl ring substituted with a variety of groups. The strategy was to screen these compounds against *P. falciparum in vitro* and then to determine the metabolic stability of the most active using human liver microsome preparations. Quinones with the appropriate potency and resistance to metabolism were then evaluated for lack of cross-resistance to other antimalarials and tested against *P. falciparum*-infected monkeys.

The target compounds were all prepared - see Scheme 2 - as previously described[11] using standard methodology.[12] Structure-activity relationships on these compounds have already been reported[11,13] but Table 1 summarises some of the more important data.

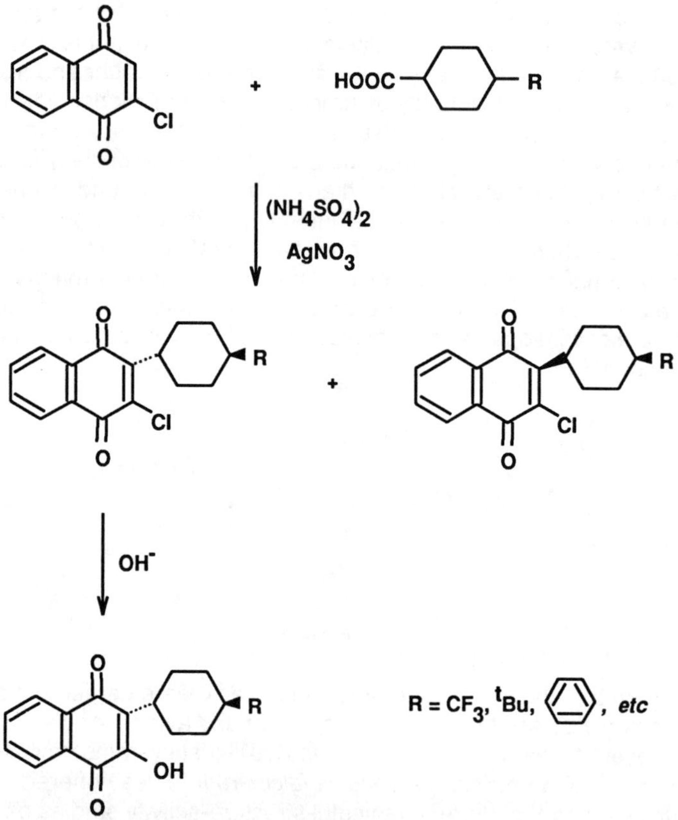

$R = CF_3$, ^tBu, phenyl, *etc*

Scheme 2

The most active quinones were the t-butyl, phenyl and 4-chlorophenyl derivatives of the parent cyclohexylquinone. All had similar potency and were subsequently tested for metabolic stability using human, and other mammalian, hepatic microsome preparations. This revealed striking differences between the compounds - see Table 2.[3,13]

Table 1. Activity of Hydroxynaphthoquinones *v P. falciparum*
in vitro

Ra	EC$_{50}$(M)
H	7.8 X 10^{-8}
OH	>1 X 10^{-4}
tBu(t)	7 X 10^{-10}
(t)	6 X 10^{-10}
Cl (t)	7 X 10^{-10}
Cl (c)	2.6 X 10^{-9}
CHLOROQUINE	7.3 X 10^{-8}

ATOVAQUONE (for the Cl (t) row)

a t=trans isomer, c=cis isomer

Both the t-butyl and phenyl analogues were readily metabolised by the human microsomes to a primary alcohol and phenol, respectively - see Scheme 3. In contrast atovaquone was completely inert in all the microsome preparations. It was of interest that despite being very sensitive to human microsomes, the phenyl derivative was inert to the murine preparation. This explains why this compound had similar activity to atovaquone when tested against *P. yoelii* in mice (ED$_{50}$'s 0.05-0.1mg/kg)[3] and underscores the danger of using this screen to predict the activity of the quinones in man.

Table 2.   <u>Comparative microsomal NADPH-dependent metabolism of</u>

<u>Hydroxynaphthoquinones</u>

Species	R		
	tBu	 ⬡ (phenyl)	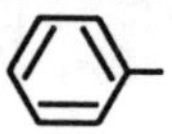 ⬡—Cl (Atovaquone)
MAN	+++	+++	-
AOTUS	++	+	-
CYNOMOLGUS	++	-	-
DOG	+	NT	-
MOUSE	+	-	-

NT = not tested

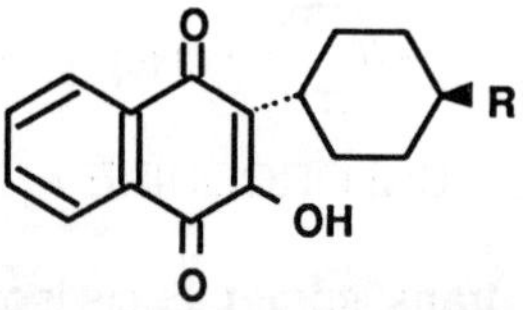

Prior to the availability of the microsome assay the t-butyl derivative had been given to healthy volunteers and shown to be rapidly metabolised to the same alcohol formed *in vitro* with the microsomes.[13] This gave considerable confidence in the predictive value of this assay and suggested that atovaquone should be a good candidate for clinical evaluation. This view strengthened as the compound was further examined in a variety of test systems. The data showed atovaquone to be curative at 1mg/kg orally in *P. falciparum*-infected *Aotus* monkeys and to be fully active *in vitro* against drug-resistant strains of this parasite.[3] In addition the compound was a potent inhibitor of development of liver stages of rodent malaria both *in vitro* and *in vivo*.[14]

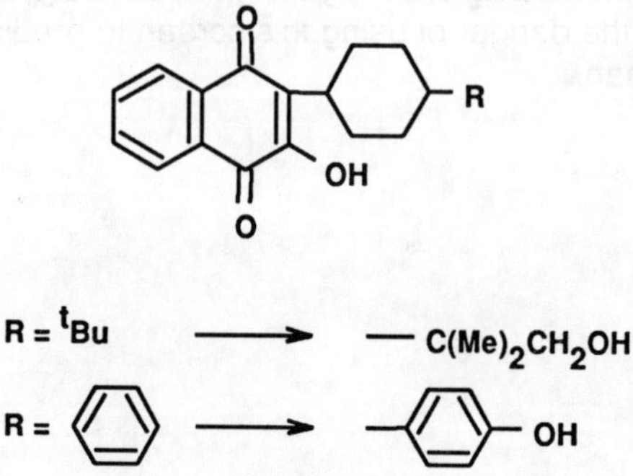

$R = {}^{t}Bu \longrightarrow$ — $C(Me)_2CH_2OH$

$R = $ ⬡ $\longrightarrow$ ⬡— OH

Scheme 3

Clinical Activity of Atovaquone

Clinical evaluation of atovaquone in healthy volunteers showed it to be well tolerated with a plasma elimination half life of around 70 hours. No evidence of metabolic degradation was observed, as predicted from the microsome assay. Plasma levels considered sufficient for antimalarial activity - 0.1-0.3µg/ml (2.7-8.1 x 10^{-7}M)- were readily achieved with a single oral dose of 225mg.[3] In later studies in HIV positive patients, repeat dosing of 3g a day for 12 days resulted in plasma concentrations of 10-50µg/ml.[15]

As a result of the above data, atovaquone was tested in Thailand in patients with *falciparum* malaria. In this region of the world drug-resistant disease is prevalent and single agent therapy cannot be expected to be fully effective. The drug was administered orally in 750mg doses every 8 hours over a period of 4 to 7 days. In all cases there was an early resolution of clinical symptoms and the parasites were removed from the bloodstream.[16] However, during the 28 day follow up period 25% of the patients recrudesced. Two of these were re-treated with drug but failed to respond leading to the suspicion that their disease resulted from parasites which had acquired resistance to atovaquone. Further evidence for this was provided by the observation that parasite isolates from re-lapsed patients had reduced sensitivity to the drug by up to 2 orders of magnitude.[16]

The overall cure rate obtained with atovaquone in the above trials was comparable to that achievable with the best single agent therapy in this region of the world. Nevertheless it was decided that atovaquone should not be developed as a single entity antimalarial. Instead, because combination therapy is now routine for combating resistant disease in S.E. Asia, trials were initiated in which the drug was combined either with tetracycline or proguanil. These drugs were chosen because of their extensive use in combination therapy with other antimalarials and because in *in vitro* studies over a wide dose range they were compatible with atovaquone against *P. falciparum*.[17] The clinical trial results[16] vindicated the selection of these drugs as partners for atovaquone. Both combinations were 100% curative in patients with *falciparum* malaria. Further clinical trials to optimise dose schedules are now on-going in order to choose which combination is to be progressed further.

Mode of Action of Atovaquone

The results which have been obtained clinically with atovaquone show it to be a potent, fast acting antimalarial, with an outstanding safety profile in both healthy volunteers and patients. The reason why this compound, and others in the series, should be so selectively toxic for malaria parasites has been investigated in some detail. As far back as 1946[18] it was observed that various hydroxynaphthoquinones could inhibit the oxygen-uptake of *P. lophurae in vitro*. Some decades later evidence was provided that these compounds could block mitochondrial electron transport processes by competing with the biological electron carrier, ubiquinone.[19] In most eukaryotic organisms, ubiquinone plays a fundamental role in electron transport whereby various substrates are oxidised and the electrons passed through dehydrogenase complexes to oxygen. In mammalian cells, ubiquinone-linked dehydrogenases using NADH or succinate as substrates are involved in energy generation *via* the synthesis of ATP whilst oxidation of dihydroorotate to orotate is central to pyrimidine biosynthesis,[20] - see Scheme 4.

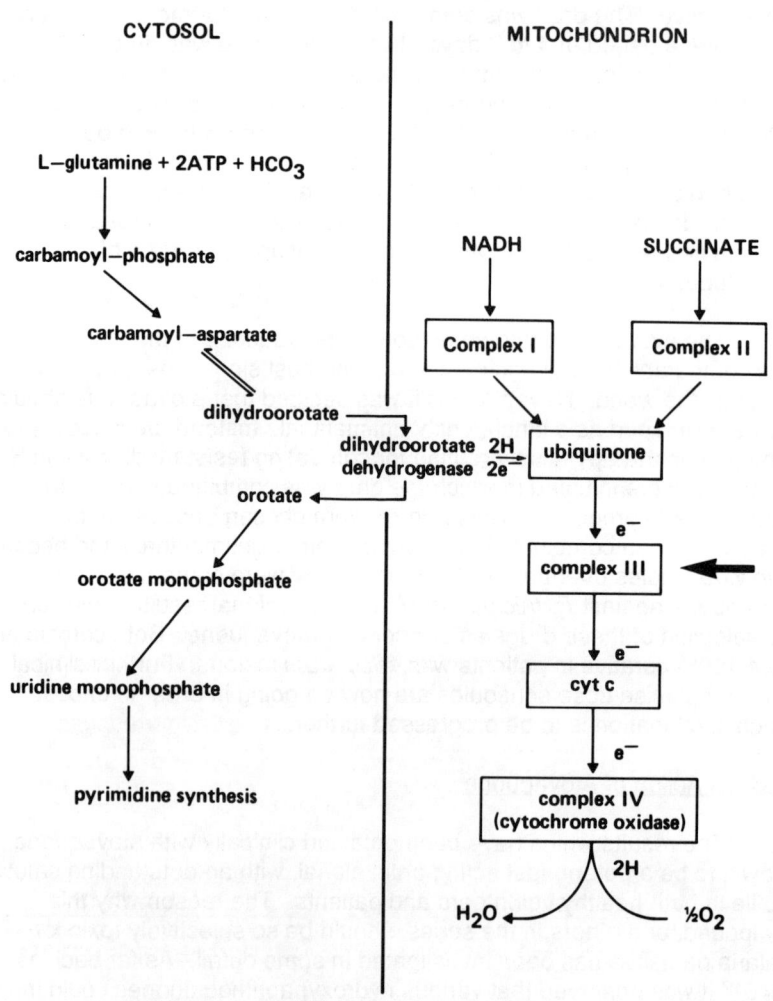

Scheme 4

In earlier studies[21] it was shown that the atovaquone analogue 2-(4'-t-butylcyclohexyl)-3-hydroxy-1,4-naphthoquinone was a potent inhibitor of pyrimidine biosynthesis in *P. falciparum* cells. Since malaria parasites, unlike man, are unable to salvage pre-formed pyrimidines (other than orotate), this inhibition blocks plasmodial nucleic acid synthesis with fatal results. The compound was shown to have no effect on ATP levels, which is reasonable as *plasmodia* primarily generate their energy requirements by anaerobic catabolism of glucose.[22] These studies were repeated with atovaquone[23] and again the compound was found to be a potent inhibitor of parasite pyrimidine biosynthesis. Concentrations as low as 3nM caused a significant accumulation of the pyrimidine intermediate carbamoyl-aspartate. Addition of orotate (which is taken up by *plasmodia*) reversed sensitivity of *P.falciparum* cultures to the drug. In contrast dihydroorotate could not overcome the effect of atovaquone. These results provide good evidence that inhibition of dihydroorotate dehydrogenase is the major mode of action of this drug.

Table 3   Inhibition of Mitochondrial Cytochrome c Reductase by Atovaquone

Respiratory Substrate	$EC_{50}(M)$	
	P. falciparum	Rat Liver
α-glycerophosphate	8.7×10^{-10}	4.7×10^{-7}
succinate	9.6×10^{-10}	4.8×10^{-7}
dihydroorotate	9.5×10^{-10}	5.1×10^{-7}

In an extension of the above work mitochondria were isolated from *P. falciparum* and used to study the effects of atovaquone on the parasite respiratory system.[24] The results obtained -Table 3- showed the compound to be a very potent inhibitor of the reduction of cytochrome c irrespective of substrate. In contrast mammalian mitochondria from rat liver were considerably less sensitive. Because the degree of inhibition was not substrate variable the binding site for the drug must be at a point on the respiratory chain common to electron flow for all the substrates. By assaying spectral changes associated with the redox status of cytochrome c, the primary site of action of atovaquone was located on the ubiquinol-cytochrome c reductase region of the respiratory chain - the so-called Complex III. There was a 2000 fold difference between the response to the drug of Complex III from mammalian and parasite preparations - see Table 4.

Table 4   Inhibition of respiratory chain complexes by Atovaquone

Assay	$EC_{50}(M)$	
	P. falciparum	Rat Liver
succinate-ubiquinone reductase	5.4×10^{-5}	6.1×10^{-5}
α-glycerophosphate-ubiquinone reductase	4.8×10^{-5}	8.2×10^{-5}
cytochrome oxidase	$>10^{-4}$	$>10^{-4}$
ubiquinol-cytochrome c reductase	1.7×10^{-9}	3.2×10^{-6}

The specificity of atovaquone can thus be rationalised from this data - selective inhibition of plasmodial Complex III results in selective and lethal blockade of pyrimidine biosynthesis which the parasites are unable to relieve by taking up preformed extracellular pyrimidines.

Elucidating the mode of action of atovaquone towards *plasmodia* allows rational judgements to be made as to what other organisms could be sensitive to the drug. In particular it was thought of interest to test the compound against *P. carinii* and *T. gondii*. These parasites are sensitive to inhibitors of dihydrofolate reductase, an enzyme involved in the regulation of pyrimidine biosynthesis.[25] Both *P. carinii* and *T. gondii* have a cosmopolitan distribution and many adults harbour these organisms asymptomatically. However in immuno-compromised individuals both can cause severe disease resulting in death. Clinical problems with these infections have increased dramatically as a direct result of the AIDS epidemic.[2]

4   *P. CARINII* PNEUMONIA

In immuno-compromised patients *P. carinii* concentrates in lung tissue resulting in a diffuse bilateral pneumonitis commonly termed PCP (see ref 26 for review). This disease is usually fatal unless treated and until the advent of AIDS it was a clinical rarity. The AIDS virus destroys natural immunity and allows *P. carinii* and other opportunistic invaders to flourish unchecked. It was the increasing incidence of PCP in America which first lead to the AIDS problem being identified. More than 70% of AIDS patients suffer from PCP and it is the leading cause of death in this patient class. Fortunately several therapies for PCP were available prior to the AIDS epidemic, namely pentamidine and the combination of the dihydrofolate reductase inhibitor trimethoprim plus sulphamethoxazole (Fig. 4).

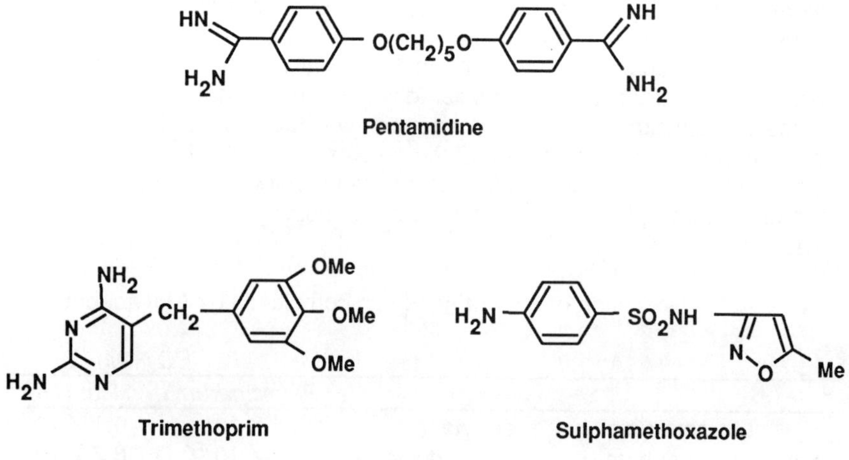

Pentamidine

Trimethoprim   Sulphamethoxazole

Figure 4.   Major drugs for PCP treatment

Unfortunately both are considerably more toxic in AIDS patients than in other immunocompromised groups with a 60% frequency of serious adverse reactions reported.[27] Consequently there is an urgent requirement for new drugs which will be as effective but with fewer side effects.

Activity of Atovaquone

Atovaquone was evaluated against *P. carinii* in both *in vitro* and *in vivo* test systems. *In vitro*[28] the compound had an IC_{50} of 5μM, which was considerably less active than against malaria parasites. In the immunocompromised rat model of PCP a dose of 100mg/kg was totally effective in both the treatment and prevention of pneumonitis.[27] These results prompted the clinical evaluation of the drug in AIDS patients with mild to moderate disease. In open trials atovaquone was found to be safe, well tolerated and effective, achieving an 79% success rate.[29] Further evaluation in AIDS patients is now on-going comparing the drug in double blind trials to the established therapies.

5   TOXOPLASMOSIS

Another AIDS-associated pathogen that atovaquone is being evaluated against is *T. gondii*. This parasite, which is closely related to *plasmodia*, exists in man in 2 forms - tachyzoites and bradyzoites. In the latter state, cysts form around the parasites presenting a barrier to drug access. These cysts occur in brain and other organs and because of their refactory response to drug treatment are an important source of re-infection. Toxoplasma encephalitis is a major problem in AIDS patients with very severe disease pathology which can result in coma, blindness and eventually death. Current therapy of toxoplasmosis relies on the dihydrofolate reductase inhibitor pyrimethamine combined either with sulphadiazine or clindamycin.[30] Again, as with PCP therapy, adverse reactions in AIDS patients pose severe problems in the management of this disease.

Activity of Atovaquone

Atovaquone was tested *in vitro* against tachyzoites from various clinical strains of *T. gondii*. Good activity was obtained but was highly strain-dependent, MIC's varying over a range $4.8 \times 10^{-6}M$ to $4.8 \times 10^{-9}M$.[30] The compound was also tested against the tissue cyst form of *T. gondii* and found to be active, albeit at the high concentration of $1 \times 10^{-4}M$.[31] This is an important result because it is this stage in the life cycle of the parasite which causes toxoplasmic encephalitis and which seems to be resistant to current therapies. When atovaquone was tested against different strains of *T. gondii* in mice, it was active against both tachyzoite and bradyzoite forms with doses ranging from 10-100mg/kg/day given over a 1-2 month period.[30]

The above data encouraged the clinical evaluation of atovaquone in AIDS patients with toxoplasmosis who had failed or were intolerant of all previous therapy. The results obtained with daily oral dosing of 4 x750mg of drug for 6 weeks and beyond were extremely encouraging.[32] The disease was stabilised or improved in 75% of the patients as measured by clinical symptoms and magnetic resonance imaging of brain lesions. Again, the drug was well tolerated with minimal side effects. Further clinical trials with the compound for this indication are now underway both as single agent therapy and in combination with pyrimethamine.

6   SUMMARY

The path to atovaquone has been long and tortuous, driven by changing patterns of disease incidence, and guided by test systems which have evolved to meet the need for more meaningful assays. Although problems with atovaquone still remain to be solved, it now seems likely that after half a century of research, a hydroxynaphthoquinone will finally make a major contribution to the management of human disease.

REFERENCES

1.   A.T. Hudson. Current Opinion in Therapeutic Patents. 1992, 2(3), 227.
2.   H. Masur. J. Infec. Dis., 1990, 161, 858.
3.   A.T. Hudson, M. Dickens, C.D. Ginger, W.E. Gutteridge, T. Holdich, D.B.A. Hutchinson, M. Pudney, A.W. Randall and V.S. Latter. Drugs Exptl. Clin. Res., 1991, 17, 427.
4.   S.C. Oaks Jr., V.S. Mitchell, G.W. Pearson and C.C.J. Carpenter. 'Report of the committee for the study on malaria prevention and control: conclusions and recommendations.' National Academy Press. Washington D.C, 1991, Chapter 1, p1.
5.   L.F. Fieser and A.P. Richardson. J. Am. Chem. Soc., 1948, 70, 3156.
6.   A.T. Hudson in 'Handbook of Experimental Pharmacology. Volume 68/2. Antimalarial drugs'. Eds. W. Peters and W.H.G. Richards. Springer-Verlag. Berlin. 1984, p343.
7.   L.F. Fieser, F.C. Chang, W.G. Caulsen, C. Heidelberger and A.M. Seligman. J. Pharmacol. Exp. Ther., 1948, 94, 85.
8.   L.F. Fieser, J.P. Schirmer, S. Archer, R.R. Lorenz and P.I. Pfaffenbach. J. Med. Chem., 1967, 10, 513.
9.   W. Trager and J.B. Jensen. Science. 1976, 193, 673.
10.   L.H. Schmidt. Am. J. Trop. Med. Hyg., 1978, 27, 718.
11.   A.T. Hudson, M.J. Pether, A.W. Randall, M. Fry, V.S. Latter and N. McHardy. Eur. J. Med. Chem., 1986, 21(4), 271.
12.   N. Jacobsen and K. Torssell. Liebigs Ann. Chem., 1972, 763, 135.
13.   A.T. Hudson in 'Topics in Medicinal Chemistry.' Ed. P.R. Leeming. Special Publication no. 65. Proceedings of the 4th SCI-RSC Medicinal Chemistry Symposium. Royal Society of Chemistry. London. 1988, p266.

14. C.S. Davies, R.E. Sinden and M. Pudney. Abstracts of the XIIth International Congress for Tropical Medicine and Malaria. Amsterdam. 1988, FrP-2-6.
15. W. Hughes, W. Kennedy, J. Shenep, P. Flynn, W. Hetherington, G. Fullen, D. Lancaster, D. Stein, S. Palte, D. Rosenbaum, S. Liao, R. Blum and M. Rogers. Abstracts of the 30th Interscience Conference on Antimicrobial Agents and Chemotherapy. American Society for Microbiology. Atlanta. 1990, Abs. 861.
16. D.B.A. Hutchinson, S. Looareesuwan and J. Farquhar. Data presented at the British Society for Parasitology, 4th Malaria Meeting. London, 1992.
17. M. Pudney. Unpublished data.
18. W.B. Wendel. Fed. Proc., 1946, 5, 406.
19. T.H. Porter and K. Folkers. Angew. Chem. Int. Ed., 1974, 13, 559.
20. M.E. Jones. Ann. Rev. Biochem., 1980, 49, 253.
21. D.J. Hammond, J.R. Burchell and M. Pudney. Mol. Biochem. Parasitol., 1985, 14, 97.
22. I.W. Sherman. Microbiol. Rev., 1979, 43(4), 453.
23. A. Chapman, W.E. Gutteridge and M. Pudney. Unpublished data.
24. M. Fry and M. Pudney. Biochem. Pharmacol., 1992, 43(7), 1545.
25. R. Ferone in 'Handbook of Experimental Pharmacology. Vol 68/2. Antimalarial drugs.' Eds. W. Peters and W.H.G. Richards. Springer-Verlag. Berlin. 1984, p215.
26. A.T. Hudson. Current Opinion in Therapeutic Patents. 1991, 1(3), 289.
27. W.T. Hughes, V.L. Gray, W.E. Gutteridge, V.S. Latter and M. Pudney. Antimicrob. Ag. Chemother., 1990, 34(2), 225.
28. M.D. Rogers and S.W. Lafon. Abstracts of the 30th Interscience Conference on Antimicrobial Agents and Chemotherapy. American Society for Microbiology. Atlanta. 1990, Abs 588.
29. J. Falloon, J. Kovacs, W. Hughes, D. O'Neill, M. Polis, R.T. Davey Jr., M. Rogers, S. Lafon, I. Feuerstein, D. Lancaster, M. Land, C. Tuazon, M. Dohn, S. Greenberg, H.C. Lane and H. Masur. N. Engl. J. Med., 1991, 325(22), 1534.
30. F.G. Araujo, J. Huskinson and J.R. Remington. Antimicrob. Ag. Chemother., 1991, 35(2), 293.
31. J. Huskinson-Mark, F.G. Araujo and J.R. Remington. J. Inf. Dis., 1991, 164, 170.
32. H. Masur, D. O'Neill, I. Feuerstein, B. Baird, R. Walker, R.T. Davey, M.A. Polis, J. Falloon, G. Galetto, E. Seidel, S. Milwee, H.C. Lane, and J.A. Kovacs. Abstracts of the VIIth International Conference on AIDS. Istituto Superiore Di Sanita. Florence. 1991, Abs. W.B. 31 (vol. 2).

23

Heteroatom Substituted 2′,3′-Dideoxynucleoside Analogues as Anti-HIV Agents

T. S. Mansour, B. Belleau, K. Bednarski, T. Breining, W. L. Brown, A. Cimpoia, M. DiMarco, D. M. Dixit, C. A. Evans, H. Jin, J. L. Kraus, D. Lafleur, N. Lee, N. Nguyen-Ba, M. A. Siddiqui, H. L. A. Tse, and B. Zacharie

BIOCHEM PHARMA INC., 531 BLVD. DES PRAIRIES, LAVAL, QUÉBEC, H7V 1B7, CANADA

1 Introduction

Over ten years after the description of the acquired immunodeficiency syndrome (AIDS), the spread of its etiological agents the human immunodeficiency viruses HIV-1 and HIV-2 remains unabated. In 1985, NIH scientists found that 3'-azido-3'-deoxythymine **AZT** inhibited the replication of HIV-1 and HIV-2 *in vitro*[1]. Subsequent reports demonstrated that nucleosides lacking the 2'- and 3'-hydroxyl groups (2',3'-dideoxy nucleosides **ddN**) were potent inhibitors of the HIV viruses[2]. To date, three nucleoside analogues **AZT**, **ddI**, and **ddC** are approved for the treatment of HIV-1 infections in humans. At least three other nucleoside analogues[3,4] are at various stages of clinical evaluation (Figure 1).

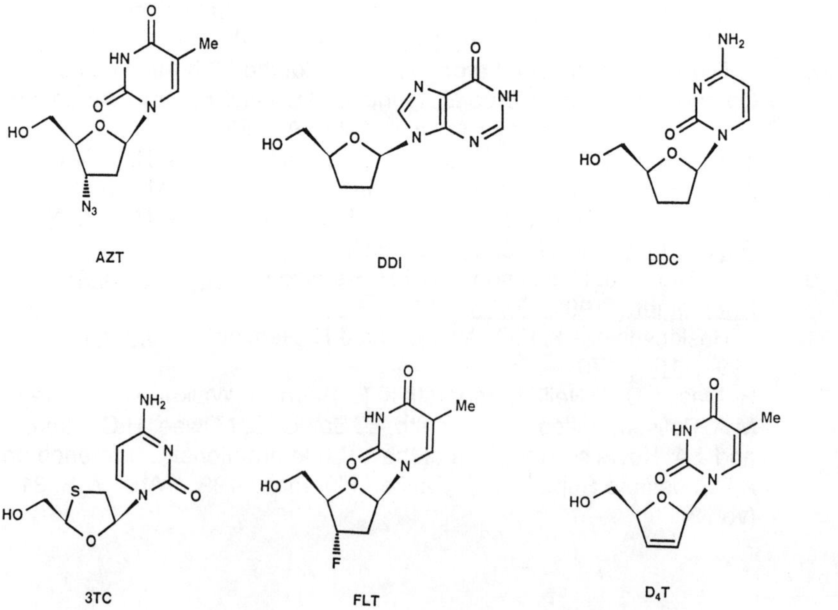

Figure 1. Anti-HIV nucleoside analogues in clinical use or under clinical evaluation

Toxicity effects are the major drawbacks associated with the clinically approved nucleoside analogues **AZT**, **ddI** and **ddC**. **AZT** may induce bone marrow suppression and neutropenia in patients; the side effects of **ddI** include pancreatitis and peripheral neuropathy and **ddC** causes severe peripheral neuropathy. Futhermore, the drug resistance problem has begun to emerge as a significant issue in AIDS therapy. AZT-resistant isolates of HIV-1 have been isolated from patients receiving **AZT** and the treated patients continue to progress to full-blown AIDS. Resistance to **ddI** and **ddC** appears to develop less easily than **AZT**[5]. The potency of **3TC** (currently in Phase II/III) coupled with its lack of toxicity *in vitro* makes it an excellent alternative candidate for the primary treatment of AIDS.

2 Mode of Action

The virally encoded reverse transcriptase (RT) is the biochemical target for **ddN** analogues. **AZT**, for example, is anabolically phosphorylated to **AZT** monophosphate (**AZTMP**) by the action of thymidine kinase. **AZTMP** then serves as a substrate for thymidylate kinase to produce **AZT** diphosphate (**AZTDP**) which is further phosphorylated to the active triphosphate species by nucleotide diphosphate kinase (Figure 2). **AZTTP** competes with other phosphorylated pyrimidines for incorporation into the HIV DNA by the viral enzyme reverse transcriptase and when incorporated into a growing strand causes chain termination[5,6]. The ability of the major intracellular metabolite **AZTMP** to inhibit the RNase H activity of RT may confer additional anti-HIV activity[7].

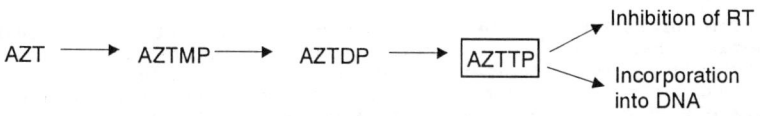

Figure 2. Mechanism of action of nucleoside analogues exemplified by AZT.

3 Heteroatom Substituted 2',3'-Dideoxynucleoside Analogues.

Numerous analogues containing 2'- or 3'-heteroatom substituted **ddN** were synthesized as potential anti-HIV agents of interest, such as the 2'-ß-fluoro derivatives of **ddA**, **ddI** and **ddC** which confer metabolic stability of the glycosidic linkage, particularly for the purine analogues, while retaining the anti-HIV activity of the parent compound[8]. Replacement of the azido group of **AZT** by a wide variety of substituents in particular electron withdrawing functionalities[9] such as cyano, isocyano, isothiocyano, isoseleno and triazolyl[10] have resulted in loss of activity. The critical role of the 3'-azido group for enzyme interaction remains obscure.

Dideoxynucleoside analogues having a heteroatom in the 2'- or 3'- position have been described. Derivatives where the ring oxygen atom and the 3'-carbon atom are interchanged (iso-dideoxynucleoside analogues) displayed moderate anti-HIV activity (base=adenine and guanine)[11]. The corresponding thia[12] (base=uracil, thymine, cytosine, adenine, guanine), aza analogues[13] (base=thymine, adenine, guanine) and the isometric 2'-oxa analogues[14] (base=uracil, cytosine, guanine) are inactive. However, the adenine derivative (2'-oxa analogues)[15] has been reported to have anti-HIV activity. 1'-Aza pyrimidine analogues[16] including those with 3'-fluoro, hydroxy and azido substituents[17] are also inactive (Figure 3).

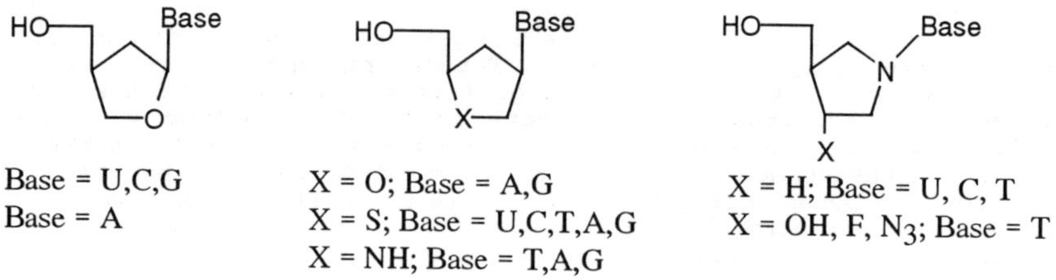
HO
Base
O
Base = U,C,G
Base = A
HO
Base
X
X = O; Base = A,G
X = S; Base = U,C,T,A,G
X = NH; Base = T,A,G
HO
Base
N
X
X = H; Base = U, C, T
X = OH, F, N3; Base = T

Figure 3. Various heteroatom containing nucleoside analogues.

An original contribution in the **ddN** field concerns analogues that contain a heteroatom (oxygen, sulfur) at the 3'-position while still retaining the oxygen atom in the sugar ring. These classes of heteroatom substituted **ddN** analogues were reported by Belleau et al.[18] as potential inhibitors in which an oxygen or a sulfur isostere of the natural sugar hydroxy methylene group may mimic the interaction with its putative binding site on the enzyme. Nucleosides of thio sugars containing a 3'-heteroatom are also of significant biological interest[33].

The presence of the ring heteroatom presents considerable synthetic challenge in controlling the relative and absolute stereochemistry of the acetal centres. This stereochemical issue proved to be of paramount biological consequence in allowing the achievement of selective inhibition of HIV RT by **3TC** in the absence of competitive inhibitors of mamalian DNA polymerase[29] (Figure 4).

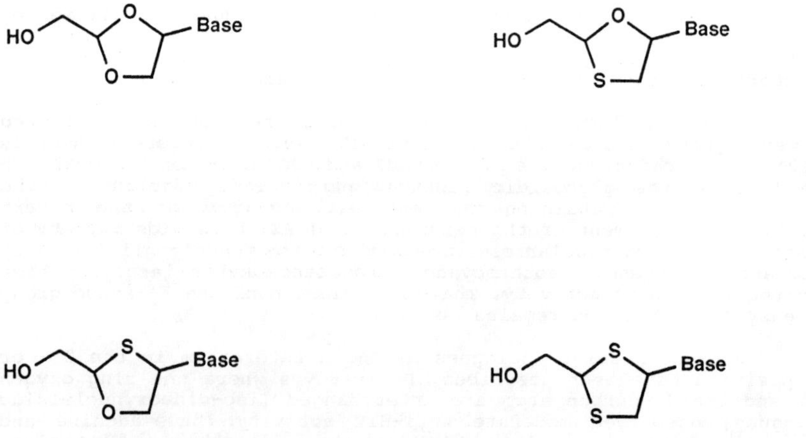
HO
O
Base
O
HO
O
Base
S
HO
S
Base
O
HO
S
Base
S

Figure 4. Various 1,3-heteroatom substituted nucleoside analogues.

4 1,3-Oxathiolane Nucleoside Analogues
   Synthesis

The initial route to the synthesis of 2,5-disubstituted 1,3-oxathiolane nucleoside analogues focussed on the coupling of an appropriate oxathiolane precursor with silylated bases under Vorbruggen conditions. The 5-ethoxy oxathiolane precursor **3** was prepared by cyclocondensation of 2-thiolacetaldehyde diethyl acetal **1** with a glyoxylate derivative followed by reduction[19]. Alternatively, the use of benzoyloxyacetaldehyde instead of glyoxylate gave the desired oxathiolane **3** directly[20] (Scheme 1).

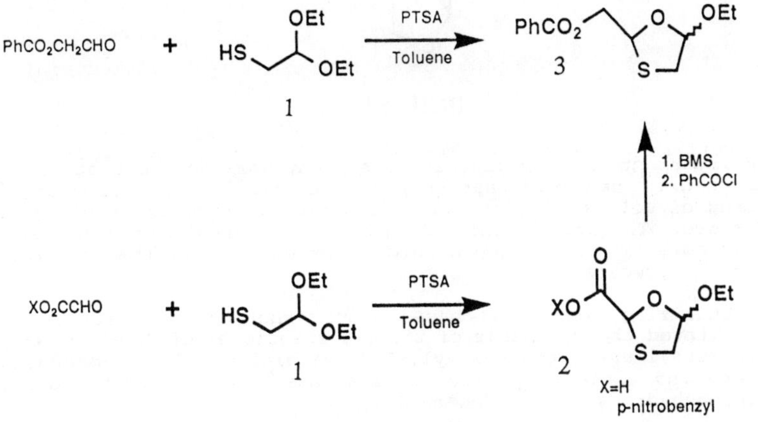

Scheme 1. Synthesis of <u>cis</u> and <u>trans</u> 2-benzoyloxymethyl-5-ethoxy-1,3-oxathiolane

Coupling of **3** with silylated pyrimidines (uracil, thymine and cytosine) under reflux with TMSOTf (2-3 days) produced a mixture of the desired <u>cis</u> and <u>trans</u> nucleoside analogues in 1:1 ratio and low yields. Similarly, the purine analogues were obtained by coupling of **3** with the corresponding silylated 6-chloropurine and 2-amino-6-chloropurine derivatives[20]. In order to circumvent the difficulties encountered with the ethoxy leaving group, 2-benzoyloxymethyl-5-acetoxy-1,3-oxathiolane **4** was developed. Indeed, coupling of this reagent with silylated bases under mild conditions substantially improved the yields of the nucleoside products[21]. In a related sense, Abbott scientists[22] and Chu et al.[23] reported that trimethylsilyl triflate achieved good yield but low stereoselectivity in their synthesis of ± dioxolane-T and related analogues.

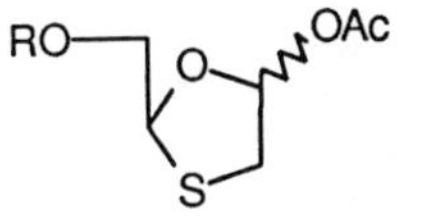

4; R = PhCO
5; R = TBDPS

An interesting approach to the synthesis of oxathiolane pyrimidine nucleosides in high cis diastereoselectivity was recently reported by Liotta et al. *In situ* complexation of stannic chloride with silylated pyrimidine bases directs the stereochemistry of N-glycosylation with **5** to the exclusive formation of the cis adduct due to the participation of the Lewis acid in the reaction transition state[24]. In this manner the racemic cytosine analogue **BCH-189** was prepared in good yields.

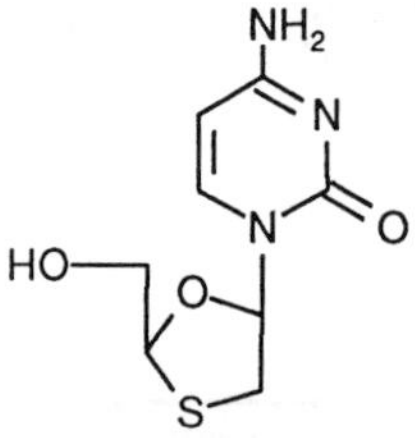

BCH-189

It is important to note that in subsequent work both Chu et al.[27] and Jones et al.[28] reported that complete racemization occurred upon the coupling of silylated cytosine with optically pure oxathiolane sugar precursors. Therefore, the advantage of controlling the relative stereochemistry at C5 by stannic chloride[24] is limited to racemic oxthiolane precursors.

Utilizing approaches based upon carbohydrate chirons, Chu et al., achieved the synthesis of the (+) enantiomer of BCH-189, (+) 2S-hydroxymethyl-5R-(cytosin-1'-yl)-1,3-oxathiolane, from D-mannose[25] in 20 steps and in 18 steps from D-galactose[26] via the intermediacy of 1,6-thioanhydro sugars (Scheme 2).

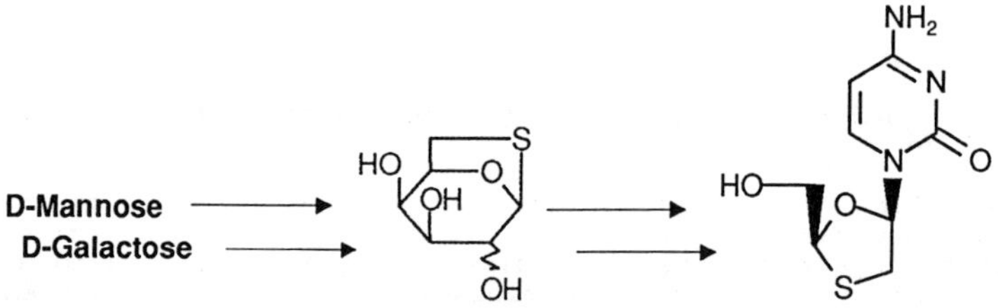

Scheme 2. Synthesis of (+) 2S-hydroxymethyl-5R-(cytosin-1'-yl)-1,3-oxathiolane from D-mannose or D-galactose.

The biologically significant (-) enantiomer **3TC** was recently obtained in 16 steps from L-gulose following a similar approach (Scheme 3)[27].

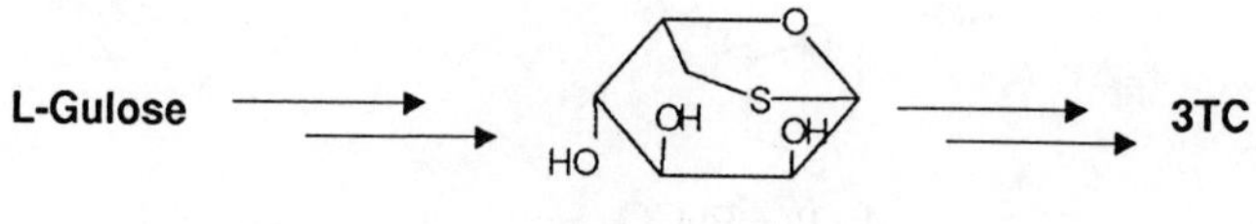

Scheme 3. Synthesis of (-) 2R-hydroxymethyl-5S-(cytosin-1'-yl)-1,3-oxathiolane (3TC) from L-gulose.

Neither of these approaches are practical for scale-up purposes. An attractive and efficient synthesis of **3TC** from (-) thiolactic acid was completed in 5 steps in collaboration with Jones et al. (Scheme 4)[28]. The coupling was achieved mediated by silyl Lewis Acid to furnish optically pure products.

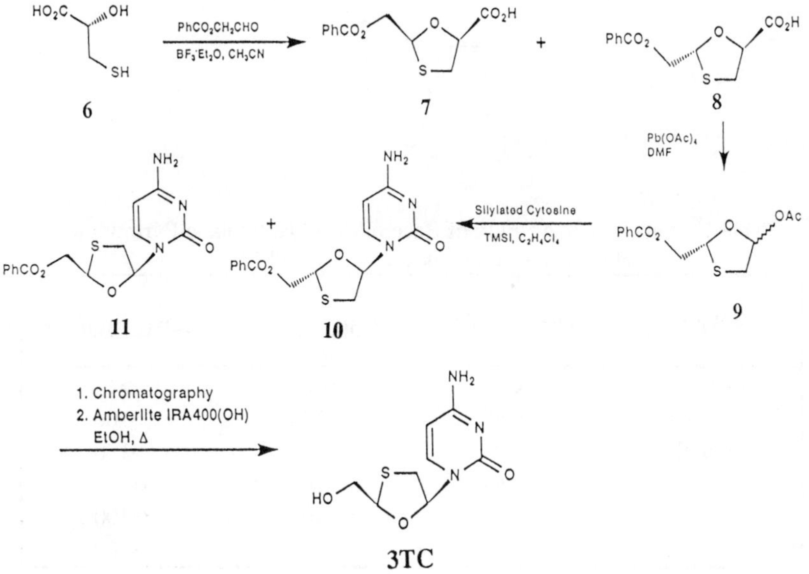

Scheme 4. Synthesis of 3TC from (-) thiolactic acid.

Anti-HIV Activity

The anti HIV-1 activity and cytotoxicities in the primary MT-4 formazan assay of racemic 1,3-oxathiolane pyrimidine and purine nucleoside analogues are summarized in Table 1 and 2. The anti-HIV activity is recorded as IC_{50} (50% inhibitory concentration) and the cytotoxicity in MT-4 cells is recorded as ID_{50} (50% inhibitory dose). The cis cytosine analogue **BCH-189** (Table 1) displayed potent anti-HIV activity without cytotoxicity. Two purine analogues (Table 2), the cis guanine and the trans-2,6-diaminopurine derivatives exhibited weak anti-HIV activity.

With the emergence of BCH-189 as a possible anti HIV-agent, its two enantiomers were sought for further biological evaluation.

Thus, the HPLC separated enantiomers of **BCH-189** were tested against HIV-1 and HIV-2 and found to be equipotent in a selection of T-lymphocyte cells lines[3]. However, the (-) enantiomer was 20- to 100-fold less cytotoxic than the (+) enantiomer in a wide range of cells lines (CEM,H9,JM,U937 and C8166). Comparative studies of the (-) enantiomer with **AZT, ddC** and **ddI** in cells infected with HIV-1 or HIV-2 showed it to be a potent and selective inhibitor of HIV-1 and HIV-2 replication in vitro.[29] This (-) enantiomer of **BCH-189** is referred to as **3TC** (2',3'-dideoxy-3'-thiacytidine) and possesses the L-sugar configuration with respect to oxygen (as determined by X-ray, enzymatic and synthetic routes). **3TC** is currently undergoing clinical trials for the potential treatment of AIDS, HIV infections and hepatitis B following demonstration of potent activity against human hepatitis B *in vitro* and against human HBV in a primate model.[30] This is of significant interest since HBV infections are common in patients who are infected with HIV. The anti-HIV[31] and anti-HBV[32] activity of **3TC** *in vitro* has been independently reported by the Georgia group.[31]

Table 1. Anti-HIV-1 Activity of Racemic 1,3-Oxathiolane Pyrimidine Nucleoside Analogues in MT-4 Cell Lines

Base	Stereochemistry	IC_{50}, µg/mL	ID_{50}, µg/ml
cytosine	cis	0.13	>1
cytosine	trans	>100	>100
uracil	cis	>100	>100
uracil	trans	>100	>100
thymine	cis	>100	>100
thymine	trans	>100	>100

Table 2. Anti-HIV-1 Activity of Racemic 1,3-Oxathiolane Purine Nucleoside Analogues in MT-4 Cells Lines

Base	Stereochemistry	IC_{50}, µg/mL	ID_{50}, µg/ml
adenine	cis	>100	>100
adenine	trans	>100	>100
2,6-diaminopurine	cis	62	>100
2,6-diaminopurine	trans	6	>100
hypoxanthine	cis	>100	>100
hypoxanthine	trans	>100	>100
guanine	cis	17	>100
guanine	trans	>100	>100

5   1,3-Dioxolane Nucleoside Analogues
Synthesis

Since the anti-HIV activity of racemic dioxolane nucleoside analogues include both the <u>trans</u> and <u>cis</u> anomers[34] (Figure 5), we sought to access all the stereoisomers in this new series of **ddN** analogues for antiviral evaluation.

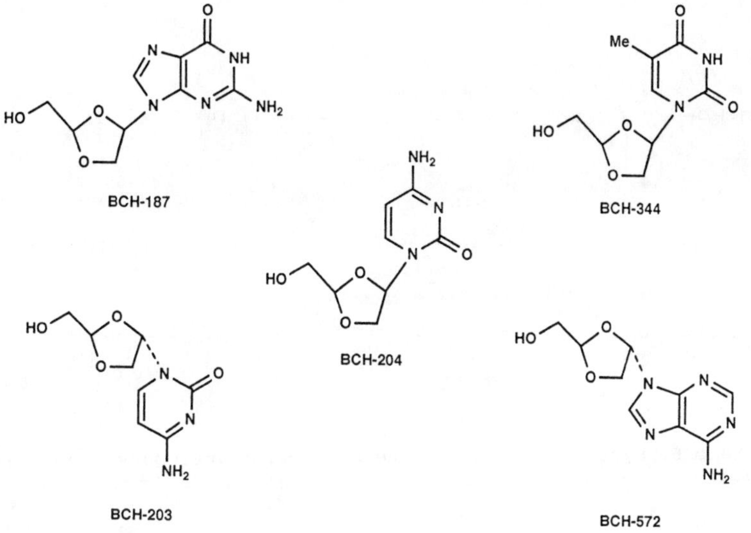

BCH-187

BCH-344

BCH-204

BCH-203

BCH-572

Figure 5. HIV active racemic 1,3-dioxolane nucleoside analogues.

Thus, upon treatment of L-ascorbic acid **12** with benzyoxyacetaldehyde dimethylacetal **13** two dioxolane diastereomers **14** and **15** were formed. The <u>trans</u>-diastereomer **14** (configuration established by single crystal X-ray diffraction study) was converted to **16** by successive oxidative degradation. In a similar fashion, degradation of a mixture enriched with **15** afforded **17** (Scheme 5). The use of silyl Lewis acids (TMSOTf,TMSI) in the coupling of **16** and **17** with silylated N-acetylcytosine was advantageous owing to the substantial racemization in the isolated nucleoside products with SnCl₄ and TiCl₂(O-iPr)₂.[35] This methodology was further extended to the synthesis of thymine, 5'-fluorocytosine, adenine and guanine analogues.

In a separate work, starting from 1,6-anhydro-D-mannose, Chu et al. demonstrated the synthesis of pyrimidine dioxolane nucleosides corresponding to 2R hydroxymethyl configuration (Scheme 6).[36]

Figure 6 illustrates an HPLC chromatogram containing the mixture of two racemic cytosine analogues **BCH-203** and **BCH-204** and the individual stereoisomers provided by the synthetic route. By this method the optical and diastereomeric purity of the final product can be assured.[37]

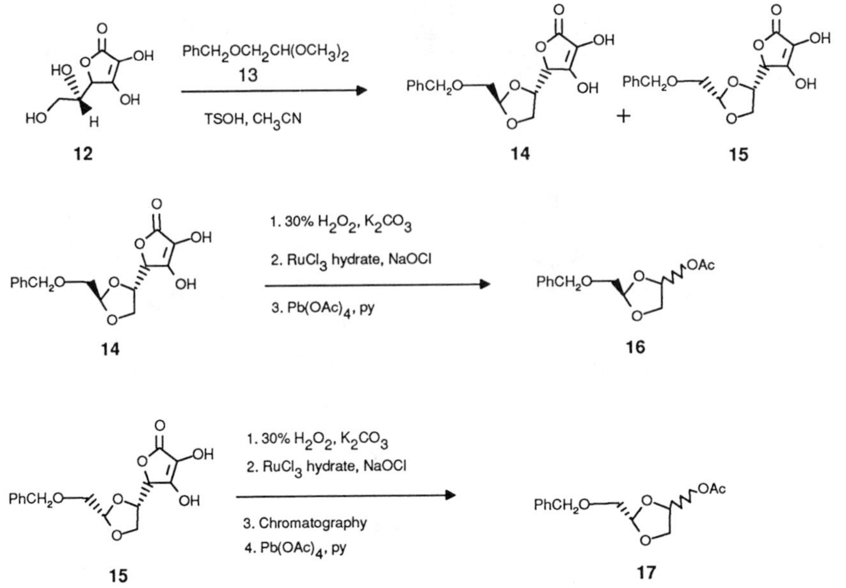

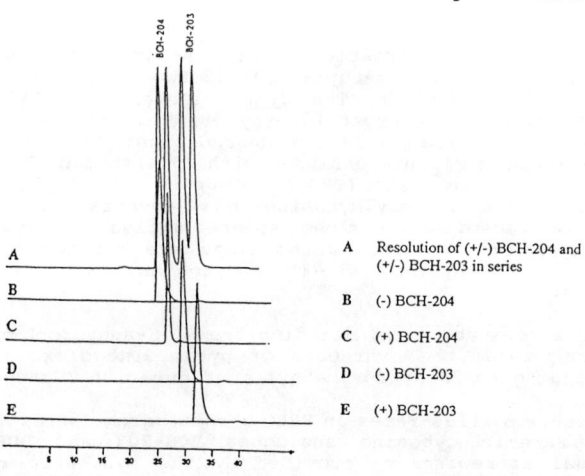

Scheme 5. Synthesis of dioxolane nucleoside analogues from L-ascorbic acid.

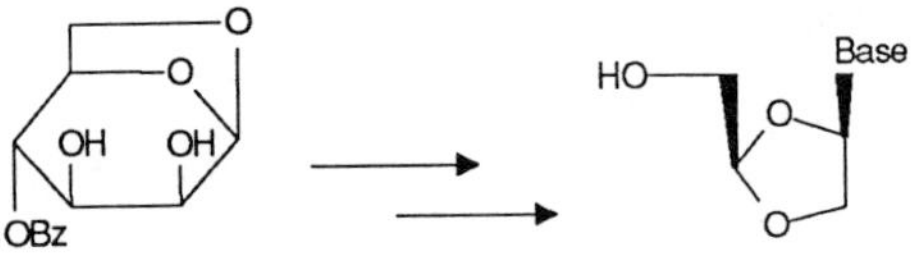

Scheme 6. Synthesis of dioxolane nucleoside analogues from D-mannose.

A	Resolution of (+/-) BCH-204 and (+/-) BCH-203 in series
B	(-) BCH-204
C	(+) BCH-204
D	(-) BCH-203
E	(+) BCH-203

Figure 6. HPLC chromatogram of **BCH-203** and **BCH-204** stereoisomers.

Enzymatic resolution by adenosine deaminase (Ada) of the racemic *cis* adenine analogue indicated that one enantiomer is deaminated selectively. Deamination of the *cis* 2,6-diaminopurine analogue by Ada furnished the (-) guanine derivative selectively. The (+) guanine derivative was obtained by prolonged exposure to Ada at 37°C. Correlation of the stereochemical outcome from the enzymatic and synthetic routes is in progress.

Anti-HIV Activity

The anti-HIV activities of pyrimidine nucleosides are displayed in Tables 3 and 4. Contrary to the 1,3-oxathiolane cytosine analogues, the anti-HIV activity of dioxolane cytosine derivatives resides in one enantiomer possessing the (2R,4R) absolute configuration (D- sugar). However, the HIV active cytosine analogue is cytotoxic. The thymine analogue (2R,4R) is weakly active in MT-4 cells. This compound, however, is reported to be very active in PBM cells ($IC_{50}=0.39\mu M$; ID_{50} > $100\mu M$).[36]

In the purine series the (-) *cis* adenine compound is active ($IC_{50}=.03\mu g\backslash ml$) and cytotoxic at $10\mu g/ml$. The (-) *cis* guanine analogue has good activity ($IC_{50}=3.2\mu g\backslash ml$) with no detectable cytotoxicity ($100\mu g/ml$).

Table 3. Anti HIV Activity of Cis-Dioxolane Pyrimidine Analogues in MT-4 Cells

Base	IC_{50}, μg/ mL	ID_{50}, μg/ ml	IC_{50}, μg/ mL	ID_{50}, μg/ ml
thymine	28	>100	>100	>100
cytosine	.04	10	>0.1	0.1

Table 4. Anti-HIV Activity of Trans-Dioxolane Pyrimidine Analogues in MT-4 Cells

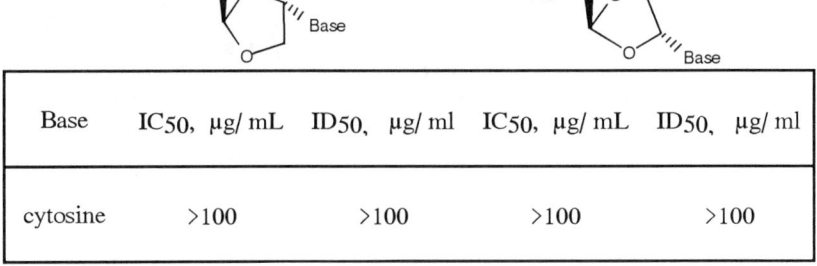

Base	IC_{50}, μg/ mL	ID_{50}, μg/ ml	IC_{50}, μg/ mL	ID_{50}, μg/ ml
cytosine	>100	>100	>100	>100

In order to correlate the structure-activity relationships of 1,3-dioxolane and 1,3-oxathiolane nucleoside analogues, the corresponding 2,5-tetrahydrofuranyl cytosine analogues were synthesized in enantiomerically pure form and their anti-HIV activities and cytotoxicities were determined. The dioxolane cytosine analogue having the same relative configuration as **3TC** showed substantial cytotoxicity. The corresponding tetrahydrofuranyl analogue is weakly active. In contrast, analogues possessing the "natural sugar configuration" are cytotoxic although all are potent anti-HIV agents.

These findings serve the need to exercise caution in attributing the favourable antiviral profile of **3TC** to its unique configuration. The greater specificity of **3TC** as a selective, non-toxic antiviral agent is remarkable. However, the exact reasons for this favourable selectivity remain to be determined.

In conclusion, efficient synthetic methodologies to access oxathiolane and dioxolane nucleoside analogues have been described. A selective and potent nucleoside analogue **3TC** emerged as a clinical candidate for the treatment of AIDS and HBV infections. The mere replacement of a methylene group by the isosteres sulfur or oxygen moieties had remarkable effects on the biological activity.

Acknowledgements

The authors wish to thank the Virology Department of Glaxo Group Research Ltd for antiviral testing, Dr. J.W. Gillard for his useful comments and discussions, Drs. R. Storer, C.M. Beels, P. Ravenscroft of GGR and G. Dionne (BioChem Pharma Inc.) for frequent discussions. We also thank Mr. P. Bouça and J. Dugas for technical assistance.

References

1.   H. Mitsuya, K.J. Weinhold, P.A. Furman, M.H. St-Clair, S.N. Lehrman, R.C. Gallo, D. Bolognesi, D.W. Barry and S. Broder, Proc. Natl. Acad. Sci. U.S.A. 1985, 82, 7096.

2.   H. Mitsuya and S. Broder, Proc. Natl. Acad. Sci. U.S.A. 1986, 83, 1911.

3.   3TC/J.A.V. Coates, N. Cammack, H.J. Jenkinson, I.M. Mutton, B.A. Pearson, R.Storer, J.M. Cameron and C.R. Penn, Antimicrob. Agents Chemother. 1992, 36, 202.

4.   FLT/P. Herdewijn, J. Balzarini, E. DeClercq, R. Pauwels, M. Baba, S. Broder and H. Vanderhaeghe, J. Med. Chem., 1987, 30, 1270.
      D₄T/M.J.M. Hitchcock, Antiviral Chem. Chemother. 1991, 2, 125.

5.   J. Saunders and R. Storer, Drug News Perspect. 1992, 5, 153, K.L. Dueholm and E.B. Pedersen, Synthesis 1992, 1.

6.   K.J. Connolly and S.M. Hammer, Antimicrob. Agents Chemother. 1992, 36, 245.

7.   C.K. Tan, R. Cival, A.M. Mian, A.G. So and K.M. Downey, Biochemistry, 1991,30, 4831.

8.   J.C. Martin, M.J.M. Hitchcock, A. Fridland, I. Ghazzouli, S. Kaul, L.M. Dunkle, R.Z. Sterzycki and M.M. Mansuri, Ann NY Acad. Sci. 1990, 616, 22; V.E. Marquez, C.K.H. Tseng, H. Mitsuya, S. Aoki, J.A. Kelley, H. Ford, J.S. Roth, S. Broder, J.G. Johns and J.S. Driscoll, J. Med. Chem., 1990, 33, 978.

9.   A. Matsuda, M. Satoh, T. Ueda, H. Machida and T. Sasaki, Nucl. Nucl., 1990, 9, 587.

10.   K. Hirota, H. Hosono, Y. Kitade, Y. Maki, C.K. Chu, R.F. Schinazi, H. Nakane and K. Ono, Chem. Pharm. Bull., 1990, 38, 2597.

11.   D.M. Huryn, B.C. Sluboski, S.Y. Tam, L.J. Todara and M. Wiegele, Tetrahedron Lett., 1989, 30 6259; K.B. Frank, E.V. Connell, M.J. Holman, D.M. Huryn, B.C. Sluboski, S.Y. Tam, L.J. Todara, M. Wiegele, D.D. Richman, H. Mitsuya, S. Broder and I.S. Sim, Ann NY Acad. Sci. 1990, 616, 408.

12.   M.F. Jones, S.A. Noble, C.A. Robertson and R. Storer, Tetrahedron Lett., 1991, 32, 247.

13.   K.E. Ng. and L.E. Orgel, J. Med. Chem., 1989, 32, 1754; M.L.

Peterson and R. Vince, <u>J. Med. Chem.</u>, 1991 <u>34</u>, 2787.

14. M.J. Bamford, D.C. Humber and R. Storer, <u>Tetrahedron Lett.</u>, 1991, <u>32</u>, 271.
15. Y. Terao, M. Akamatsu and K. Achiwa, <u>Chem. Pharm. Bull.</u>, 1991, <u>39</u>, 823.
16. M.R. Harnden and R.L. Jarvest, <u>Tetrahedron Lett.</u>, 1991, <u>32</u>, 3863; T.S. Mansour and H. Jin, <u>BioMed. Chem. Lett.</u>, 1991, <u>1</u>, 757.
17. Y.H. Lee, H.K. Kim, I.K. Youn and Y.B. Chae, <u>BioMed. Chem. Lett.</u>, 1991, <u>1</u>, 287
18. B. Belleau, D. Dixit, N. Nguyen-Ba and J.L. Kraus, <u>5th Int. AIDS Conf.</u>, 1989, 515; M.A. Wainberg, M. Stern, R. Martel, B. Belleau and H. Soudeyns, <u>5th Int. AIDS Conf.</u> 1989, 552.
19. J.L. Kraus and G. Attardo, <u>Synthesis</u>, 1991, 1046.
20. B. Belleau, N. Nguyen-Ba, B. Zacharie and J.L. Kraus to be submittted.
21. B. Belleau and T.S. Mansour to be published.
22. D.W. Norbeck, S. Spanton, S. Broder and H. Mitsuya, <u>Tetrahedron Lett.</u>, 1989, <u>30</u>, 6263.
23. C.K. Chu, S.K. Ahn, H.O. Kim, J.W. Beach, A.J. Alves, L.S. Jeong, Q. Islam, P. Van Roey and R.F. Schinazi, <u>Tetrahedron Lett.</u>, 1991, <u>32</u>, 3791.
24. W-B Choi, L.J. Wilson, S. Yeola, D.C. Liotta and R.F. Schinazi, <u>J. Am. Chem. Soc.</u>, 1991, <u>113</u>, 9377.
25. C.K. Chu, J.W. Beach, L.S. Jeong, B.G. Choi, F.I. Comer, A.J. Alves and R.F. Schinazi, <u>J. Org. Chem.</u>, 1991, <u>56</u>, 6503.
26. L.S. Jeong, A.J. Alves, S.W. Carrignan, H.O. Kim, J.W. Beach and C.K. Chu, <u>Tetrahedron Lett.</u>, 1992, <u>33</u>, 595.
27. J.W. Beach, L.S. Jeong, A.J. Alves, D. Pohl, H.O. Kim, C.N. Chang, S.L. Doong, R.F. Schinazi, Y.C. Cheng and C.K. Chu, <u>J. Org. Chem.</u>, 1992, <u>57</u>, 2217.
28. D.C. Humber, M.F. Jones, J.J. Payne, M.V.J. Ramsay, B. Zacharie, H. Jin, A. Siddiqui, C.A. Evans, H.L.A. Tse and T.S. Mansour, <u>Tetrahedron Lett.</u>, 1992, <u>33</u>, 0000.
29. J.A.V. Coates, N. Cammack, H.J. Jenkinson, A.J. Jowett, M.I. Jowett, B.A. Pearson, C.R. Penn, P.L. Rouse, K.C. Viner and J.M. Cameron, <u>Antimicrob. Agents Chemother.</u>, 1992, <u>36</u>, 733.
30. D.L. Tyrrell, Fifth Int. Conf. Antiviral Res., Vancouver, B.C. 1992
31. R.F. Schinazi, C.K. Chu, A. Peck, A. McMillan, R. Mathis, D. Cannon, L.S. Jeong, J.W. Beach, W.B. Choi, S. Yeola and D.C. Liotta, <u>Antimicrob. Agents Chemother.</u>, 1992, <u>36</u>, 672.
32. S.L. Doong, C.H. Tsai, R.F. Schinazi, D.C. Liotta and Y.C. Cheng, <u>Proc. Natl. Acad. Sci. U.S.A.</u>, 1991, <u>88</u>, 8495.
33. PCT. WO 92/08717 May 29, 1992 (BioChem Pharma).
34. B.Belleau, D.M. Dixit, N. Nguyen-Ba, W.L. Brown and U. Gulini, to be submitted.
35. B. Belleau, C.A. Evans, H.L.A. Tse, H. Jin, D.M. Dixit and T.S. Mansour, to be submitted.
36. H.O. Kim, S.K. Ahn, A.J. Alves, J.W. Beach, L.S. Jeong, B.G. Choi, P. Van Roey, R.F. Schinazi and C.K. Chu, <u>J. Med. Chem.</u>, 1992, <u>35</u>, 1987.
37. M.P. DiMarco, D.M. Dixit, C.A. Evans, H. Jin, H.L.A. Tse and T.S. Mansour, to be submitted.
38. M. M. Mansuri, V. Farina, J. E. Starrett, D. A. Bengini, V. Brankovan and J. C. Martin, <u>BioMed. Chem. Lett.</u>, 1991, <u>1</u>, 65.

24
Crystallographic Studies of Influenza Virus Neuraminidase and Implications for New Anti-influenza Drugs

P. M. Colman

BIOMOLECULAR RESEARCH INSTITUTE, 343 ROYAL PARADE, PARKVILLE, VICTORIA 3052, AUSTRALIA

The discovery of enzymatic activity on the surface of influenza virus is attributable to Hirst[1], who observed that it was not possible to reagglutinate red blood cells (RBC) after they had once been agglutinated by the virus. The modification to the cells resulting from the first contact with virus is now known to be the specific removal of sialic acid (N-acetylneuraminic acid) from glycoconjugates on the RBC surface by the viral neuraminidase. The enzyme action of the neuraminidase destroys the receptors for the viral hemagglutinin. Bacterial neuraminidases, as found for example in culture filtrates of *Vibrio cholerae*, also render RBC nonagglutinable by influenza virus.[2]

Neuraminidase catalyses the removal of terminal sialic acid residues from oligosaccharide chains. N-acetylneuraminic acid is but one member of a family of neuraminic acid derivatives known collectively as sialic acids. The enzyme from influenza and other ortho- and para-myxoviruses preferably cleaves the α2-3 linkage between terminal N-acetylneuraminic acid and the penultimate sugar.[3-5]

The biological consequences of this activity are not well understood. Cells of the upper respiratory tract are a common target for infection by influenza virus. These cells are bathed in mucosal secretions rich in sialic acid-containing macromolecules.[6-8] Entrapment of the virus in these secretions by the action of the hemagglutinin is likely[9,10] in the absence of neuraminidase activity. The virus can also be immobilized at the surface of infected cells. Newly synthesized virus particles bud from the plasma membrane of infected cells. The hemagglutinin and neuraminidase molecules are both glycosylated by host-cell processes; in particular, terminal sialic acid is added to these oligosaccharide chains.[11] Cleavage of these sugars is a prerequisite for elution of virus from infected cells.[12-14] Release of virus from cells is also controlled by cell-surface sialic acid.[15]

Neuraminidase inhibitors have been shown to be active in preventing multi-cycle replication of the virus in tissue culture and electron microscope evidence suggests that inhibition of the enzyme leads to aggregation of progeny virus particles at the surface of infected cells.[16,17]

Influenza virus neuraminidase is also an important viral antigen. Anti-neuraminidase antibodies appear incapable of neutralising viral infectivity, except at very high concentrations.[18] They do, however, modify the disease in favour of the host by reducing both the levels of virus in lungs and the extent of lung lesions.[19] Antigenic variation of the neuraminidase polypeptides is as extensive as that of the hemagglutinin, and neuraminidase variants can be selected by monoclonal antibodies.[20] It is therefore likely that antibodies to neuraminidase have an important role in the epidemiology of influenza.

The neuraminidase is a tetramer of 50KDa glycosylated polypeptides, with a β sheet propeller fold (fig.1), which displays C_4 symmetry.[21]

Sialic acid is the product of the reaction catalysed by the enzyme and its binding site has previously been characterised as a cavity on the upper surface of the subunit, directly above the six, parallel, central strands of the propeller structure[22] (fig.1). The site is unusual for the concentration of charged amino acids which project into it, in contrast to the sialic acid binding site of the viral haemagglutinin which has one histidine and one glutamic acid residue.[23]

The active site has been conserved throughout viral evolution. Host antibodies appear incapable of binding exclusively to active site amino acids, since such antibodies would protect against infection by all known strains. Studies of the three dimensional structure of antibody-neuraminidase complexes[24,25] show that antibodies bind to the antigen over a surface area of more than $800Å^2$, i.e. larger than the area of the active site. Since a single mutation on an antigen within an antibody binding site can abolish binding, enzymatically viable variants to which antibodies do not bind can emerge.[26]

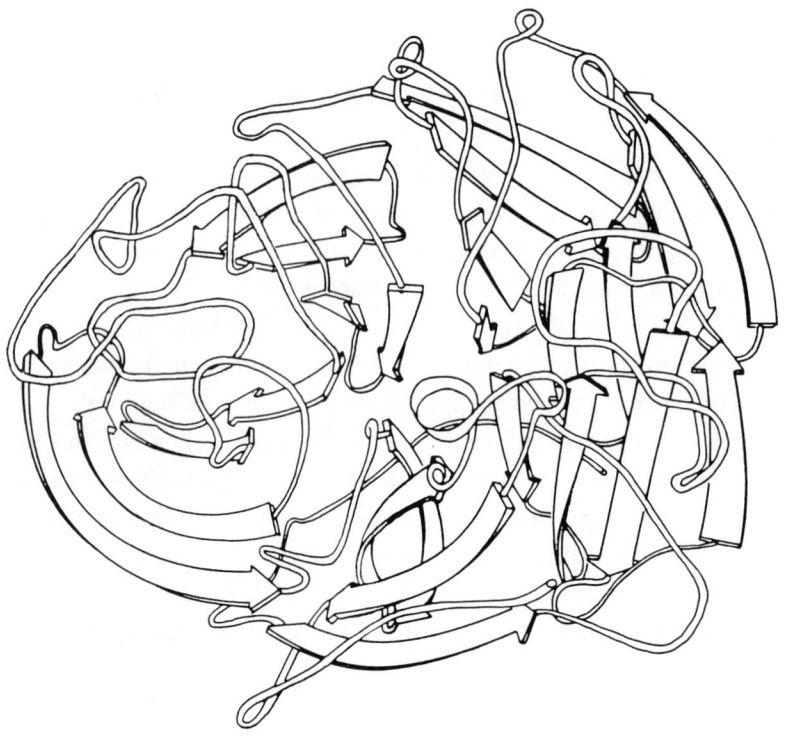

Figure 1
Chain folding diagram of one subunit of neuraminidase. The active site is on
the upper surface over the six central parallel β strands.

Crystallographic studies of neuraminidase complexed with sialic acid
and unsaturated derivatives of sialic acid (2-deoxy-2,3-dehydro-N-acetyl
neuraminic acid) lead to the conclusion that sialic acid binds to
neuraminidase in a distorted configuration with the carboxylate in the
equatorial position.[27]

The final atomic model for the sialic acid complex is shown
schematically in figure 2. Sialic acid binds to the enzyme through the same
face as is used in the interaction with haemagglutin.[28] The NH group of the
acetamido side chain is directed towards the floor of the active site cavity
where it makes a hydrogen bond to a bound water molecule. The
carboxylate group, axial towards the floor in the undistorted structure of the
sugar, is held equatorial, possibly by reason of interactions it makes with the
three guanidinium groups of arginine residues at sequence positions 118, 292

and 371. Refinement of the bound sialic acid has resulted in a boat configuration for the pyranose ring. The oxygen of the acetamido group is hydrogen bonded to Arg152 and the methyl group lies in a hydrophobic pocket near Ile222 and Trp178, but not over the six-membered ring of the tryptophan residue as observed in the haemagglutinin complex.[28] Hydroxyl groups of the glycerol side chain are hydrogen bonded to Glu276, and the 4-hydroxyl is directed towards Glu119. Features of the saccharide-protein interaction which are generally observed in other systems[29] include the participation of many planar, polar amino acids and the underlying network of hydrogen bonds in the protein which hold those groups in position to receive the sugar.

As a consequence of the distortion at C2 raising the carboxylate to the equatorial position, the orientation of the glycosidic oxygen is axial and directed vertically out of the binding pocket. In this position its closest neighbour on the enzyme is Asp151, OD2 (fig.2). This protein atom is remarkable because of its failure to participate in the complex hydrogen

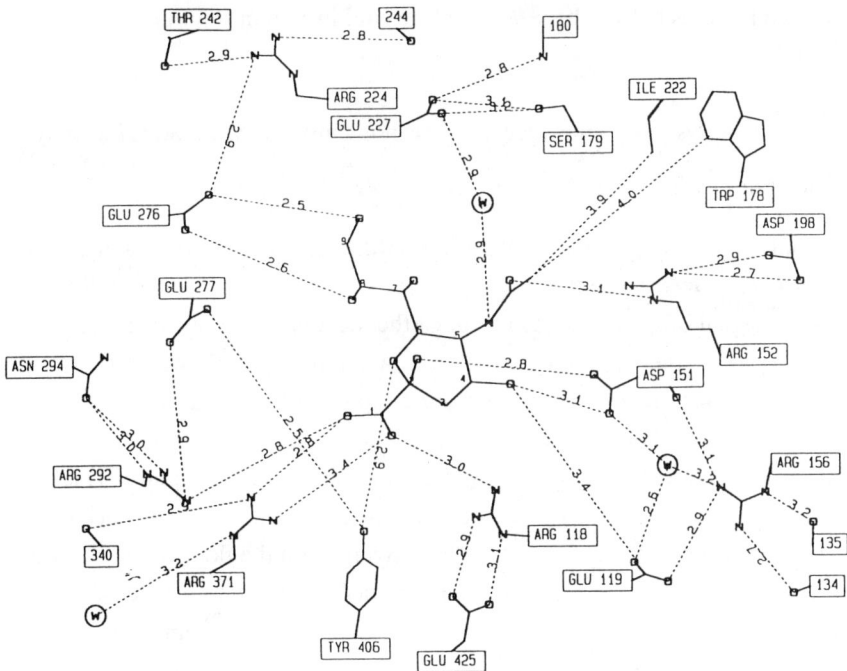

Figure 2

Details of interaction between α-sialic acid and neuraminidase. Coordinate errors are 0.3Å.

bonding network lining the active site. A parallel with hen egg white lysozyme is suggested whereby Asp151 acts as a general acid in the catalysis by protonating the glycosidic oxygen.

Glycosidic bond cleavage by lysozyme[30] is facilitated by a pair of carboxylate groups on the enzyme. One of them, Glu35, is believed to be the general acid and the other, Asp52, is thought to stabilise the developing positive charge on a carbonium ion intermediate. The bacteriophage T4 lysozyme contains an analogue of Asp52 but the proton donor is uncertain and probably acts indirectly.[31] There is no carboxylate analogue of Asp52 in the neuraminidase structure. The closest approach of neuraminidase to the hetero-atom of the pyranose ring is through the phenolic oxygen of Tyr406 which may direct one of its lone pairs at the ring oxygen. The sialic acid carboxylate itself may also assist in the charge stabilisation of the intermediate. The reaction could be completed through the activation of a water molecule by the deprotonated Asp151, and its attack on the carbonium ion, resulting in the formation of the α anomer of sialic acid. This is the form in which the product is released from the active site of bacterial neuraminidases before the α/β anomeric equilibrium in solution is established.[32]

This 'lysozyme-like' mechanism is not consistent with the pH activity profile for the enzyme[33], with critical groups titrating at around pH 4.5 and 9.0, or with data from site directed mutants[34] implicating His274 in modulating the high pH activity profile. Although there has been some controversy over the geometrical distortion of lysozyme substrates[35], there is a clear signal of such distortion through the location of the sialic acid carboxylate in the studies reported here. A combination of hydrogen bonding and steric interference is believed to underlie the strained conformation of N-acetylmuramic acid in lysozyme.[35] In neuraminidase, the electrostatic effect of the tri-arginyl cluster (residues 119, 292 and 371) is one influence on the geometry of the bound sialic acid. Unfavourable packing and electrostatic interactions of the carboxylate with the floor of the site are other factors. The contribution of this strained geometry to the cleavage reaction remains to be determined. Molecular mechanics[36] and dynamics[37] calculations both suggest that the role of mechanical strain in the lysozyme hydrolytic mechanism is minimal.

The position of the glycosidic oxygen atom in the bound substrate suggests that there is limited potential for binding interactions between the enzyme and the aglycon. If this is indeed the case, then most of the binding energy must derive from interactions between the enzyme and the product, α-sialic acid. Nevertheless, release of product from the active site is favoured by the preferred β conformation for the sugar in solution; β anomers of sialic acid are neither substrates nor inhibitors of the enzyme.[38]

A number of approaches to designing enzyme inhibitors based on target site structure have been described. These include methods which focus on issues of chemical complementarity,[39] spatial complementarity[40] and some combination of these aspects of the problem.[41] These various methods are being applied to the active site structure described here with the aim of producing tightly binding inhibitors of neuraminidase for testing as antiviral agents.

1. G.K. Hirst, Exp.Med., 1942, 76, 195.
2. F.M. Burnet and J.D. Stone, Aust.J.Exp.Biol.Med.Sci., 1947, 26, 381.
3. A.P. Corfield, M. Wember, R. Schauer and R. Rott, Eur.J.Biochem., 1982, 1124, 521.
4. A.P. Corfield, H. Higa, J.C. Paulson, and R. Schauer, Biochim.Biophys.Acta, 1983, 744, 121.
5. J.C. Paulson, J. Weinstein, L. Dorland, H. Van Halbeek, and J.F.G. Vliegenthart, J.Biol.Chem., 1982, 257, 12734.
6. A. Gottschalk, Adv.Enzymol., 1958, 20, 135.
7. A. Gottschalk, 'Glycoproteins, Their Composition, Structure and Function', Ed. A. Gottschalk, Elsevier, Amsterdam, 1972, p.2.
8. A. Allen, J.Gen.Virol., 1983, 37, 625.
9. F.M. Burnet, J.F. McCrea and S.G. Anderson, Nature, London, 1947, 160, 404.
10. F.M. Burnet, Aust.J.Exp.Biol.Med.Sci., 1948, 26, 381.
11. S. Basak, M. Tomana and R.W. Compans, Virus Res., 1985, 2, 61.
12. P. Palese, K. Tobita, M. Ueda and R.W. Compans, Virology, 1974, 61, 397.
13. P. Palese and R.W. Compans, J.Gen.Virol., 1976, 33, 159.
14. J.A. Griffin and R.W. Compans, J.Exp.Med., 1979, 150, 379.
15. J.A. Griffin, S. Basak and R.W. Compans, Virology, 1983, 125, 324.

16. P.Meindl, G. Bodo, P. Palese, J. Schulman and H. Tuppy, <u>Virology</u>, 1974, <u>58</u>, 457.

17. P. Palese, J. Schulman, G. Bodo and P. Meindl, <u>Virology</u>, 1974, <u>59</u>, 490.

18. E.D. Kilbourne, W.G. Laver, J.L. Schulman and R.G. Webster, <u>J.Virol.</u>, 1968, <u>2</u>, 281.

19. J.L. Schulman, 'The Influenza Viruses and Influenza', Ed. E.D. Kilbourne, Academic, New York, 1975, p.373.

20. R.G. Webster, V.S. Hinshaw and W.G. Laver, <u>Virology</u>, 1982, <u>117</u>, 93.

21. J.N. Varghese, W.G. Laver and P.M. Colman, <u>Nature</u>, London, 1983, <u>303</u>, 35.

22. P.M. Colman, J.N. Varghese, W.G. Laver, <u>Nature</u>, London, 1983, <u>303</u>, 41.

23. W. Weis, J.H. Brown, S. Cusack, J.C. Paulson, J.J. Skehel and D.C. Wiley, <u>Nature</u>, London, 1988, <u>333</u>, 426.

24. P.M. Colman, W.G. Laver, J.N. Varghese, A.T. Baker, P.A. Tulloch, G.M. Air, R.G. Webster, <u>Nature</u>, 1987, <u>326</u>, 358.

25. W.R. Tulip, J.N. Varghese, W.G. Laver, R.G. Webster and P.M. Colman, <u>J.Mol.Biol.</u>, 1992. In press.

26. P.M. Colman, 'The Influenza Viruses', Ed. R.M. Klug, Plenum, 1989, p.175.

27. J.N. Varghese, J. McKimm-Breschkin, J.B. Caldwell, A.A. Kortt and P.M. Colman, <u>Proteins: Structure Function and Genetics</u>, 1992. In press.

28. W. Weis, J.H. Brown, S. Cusack, J.C. Paulson, J.J. Skehel and D.C. Wiley, <u>Nature</u>, London, 1988, <u>333</u>, 426.

29. F. Quiocho, <u>Annu.Rev.Biochem.</u> , 1986, <u>55</u>, 287.

30. D.C. Phillips, <u>Sci.Am.</u> , 1966, <u>215</u>, 78.

31. W.F. Anderson, M.G. Grutter, S.G. Remington, L.H. Weaver and B.W. Matthews, <u>J.Mol.Biol.</u>, 1981. <u>147</u>, 523.

32. H. Friebolin, R. Brossmer, G. Keilich, D Ziegler, M. Supp, <u>Hoppe-Seyler's Z. Physiol.Chem.</u>, 1980, <u>361</u>, 697.

33. A.K.J. Chong, M.S. Pegg, M. von Itzstein, <u>Biochem.Int.</u>, 1991, <u>24</u>, 165.

34. M.R. Lentz, R.G. Webster, G.M. Air, <u>Biochemistry</u>, 1987, <u>26</u>, 5351.

35. N.C.J. Strynadka and M.N.G. James, <u>J.Mol.Biol.</u>, 1991, <u>220</u>, 401.

36. A. Warshel and M. Levitt, <u>J.Mol.Biol.</u>, 1976, <u>103</u>, 227.

37. C.B. Post and M. Karplus, <u>J.Am.Chem.Soc.</u> , 1986, <u>108</u>, 1317.

38. P. Meidl and H. Tuppy, <u>Mitt.Mh Chem.</u>, 1965, <u>96</u>, 816.

39.   P.J. Goodford, <u>J.Med.Chem</u>., 1985, <u>28</u>, 849.

40.   I.D. Kuntz, J.M. Blaney, S.S. Oakley, R. Langridge and T.E. Ferrin, <u>J.Mol.Biol</u>., 1982, <u>161</u>, 269.

41.   M.C. Lawrence and P.C. Davis, <u>Proteins: Structure Function and Genetics</u>, 1992, <u>12</u>, 31.

Subject Index